PLASTICS: SURFACE AND FINISH
Second Edition

Frontispiece *Celluloid (cellulose nitrate) and casein were particularly decorative: an 'ivory' relief-moulded photo frame, lidded box, hand-mirror, and 'amber' brush grooved and decorated with gilt, 1880s–1920s. Celluloid 'ivory' hair combs, one with ornate hand-cut design, another painted and inlaid with diamante, 1890s. Celluloid 'tortoise-shell' tray, 1880s–1920s. Travel combs of casein decorated with diamante and metal, and one in pearlized hand-painted celluloid, 1920s–1930s. Amber and jade effect cast phenolic bracelets, 1930s*
(Photo Stephen Brayne, Collection Sylvia Katz)

Plastics: Surface and Finish

Second Edition

Edited by

W. Gordon Simpson, MA, MPhil, AIIMR, MIMgt, FIDiagE, FIM

A catalogue record for this book is available from the British Library.

ISBN: 0-85186-209-8

Published by The Royal Society of Chemistry,
Thomas Graham House, Science Park, Cambridge CB4 4WF

Typeset by Computape (Pickering) Ltd, Pickering, North Yorkshire
Printed by Bookcraft (Bath) Ltd.

Contents

Preface

ISO 472 ('Plastics—Vocabulary') provides the following definition of the first word in the title of this book:

plastic (noun): A material which contains as an essential ingredient a high polymer and which at some stage in its processing into finished products can be shaped by flow.

—and goes on to note, in part:

Elastomeric materials, which also are shaped by flow, are not considered as plastics.

The Standard of course also defines 'polymer', giving as reference in this instance terms recommended by the IUPAC Macromolecular Nomenclature Commission; in most instance it makes a distinction between the polymer and the plastic (so seeming perhaps unduly long-winded) but lists and describes about forty different groups of materials, each of which is of some significance in commerce and industry. Understandably enough, the Standard does not suggest in order to justify the name 'plastic' what proportion of 'a material' should be a high polymer (should it be more than 50%, 30%, or—say—2%?: as gradually monomers become more costly, there may be a temptation if possible, for some uses at least, to reduce by small amounts the polymer contents of compositions). In this connection too it must be expected that in future there will be increasing incorporation of re-worked materials recovered from factory scrap, domestic waste, motor vehicle breakers' yards, and other miscellaneous sources: such re-worked material in a composition may be akin to the original polymer but is unlikely to be precisely the same.

ISO 472 defines 'finishing' thus:

The process of developing desired surface characteristics on plastic products by appropriate operations such as tumbling, grinding, sanding, polishing, coating and electroplating.

This definition again invites further comment and elaboration. The implication may be that 'finishing' is begun after a product is 'shaped by flow', but this is not always the case and the surfaces of many products are finished in part while compositions are hot or warm still. It mentions as examples six techniques for finishing but the Standard itself includes many more than this and it is worth remembering too that intermediates such as film, sheet, and mouldings (which require further fabrication before they become 'plastic products') also must be finished satisfactorily; the properties of the surface of an intermediate can be essential in obtaining the results required in the ultimate product.

Many plastics are available in a variety of forms (most commonly as film, sheet, or mouldings—but also as fibres, filaments, extruded profiles, additives, coatings, and dispersions of various kinds, among others) and not all of them can be prepared and fabricated satisfactorily by the same or similar techniques. The principal advantage of the plastics as a group—that materials can be provided in a range of forms and with characteristics to satisfy differing specifications—means also that the method of finishing has to be suitable in each instance. It must be applicable to a plastic, and be employed in a manner appropriate to the material selected (which may be hard or soft, brittle or resilient, conductive or non-conductive, and so forth).

The basic methods of finishing used for plastics and for other substances like paper, textiles, ferrous and non-ferrous metals were developed originally from each other, so necessarily there are similarities between them—but the evolution of a technology specific for plastics was essential and therefore inevitable.

The original edition of 'Plastics: Surface and Finish' was an attempt to review progress towards this objective—a book of some 242 pages (not including the Index) published in London by Butterworths in 1971 and prepared jointly by Dr S. H. Pinner and the present editor (editorial credit was settled by placing the surnames in alphabetical order). It contained contributed chapters as follows:

- 'Embossing of thermoplastics' (Dr G. O'Donnell and K. D. Reid)
- 'Surface treatments for plastics films and containers' (J. Gray)
- 'A general review of printing processes for plastics' (L. Leese)
- 'Gravure printing on plastics films' (W. A. S. Fry)
- 'Silk-screen printing on plastics films' (E. S. Snyder)
- 'The formulation of printing inks for plastics' (Dr D. G. Hare and C. H. Smith)
- 'Adhesives for plastics fabrication' (Dr P. J. C. Counsell)
- 'The sealing and welding of thermoplastic materials' (D. F. Neale)
- 'The vacuum metallizing of plastics' (D. W. Barker and B. J. Williams)
- 'Developments in the electroplating of plastics' (Dr R. R. Smith)
- 'Decorative laminates' (W. M. Hunter)
- 'Aspects of surface chemistry and morphology' (Dr B. W. Cherry).

Almost all the contributors were employed at the time by manufacturers in the fields concerned. The contents were based on papers given at a symposium arranged by Dr Pinner, one of a series he ran at the former Borough Polytechnic, London (some on other themes, including 'Modern Packaging Films' and 'The Weathering and Degradation of Plastics', also led to publication in book form; looked at again today, some parts at least of those books still seem far-sighted and useful): for 'Plastics: Surface and Finish', contributions from the symposium were supplemented by chapters invited from a few additional authors. Reviewers of the first edition were kindly, and we have been encouraged while preparing the new one by some favourable comments about the usefulness of its predecessor. However, in due course parts at least of the original text became out-of-date, and the publisher allowed the book to go out of print in 1983.

So long a period of years brings many changes and for several reasons the present edition is rather different from the first. The editors were conscious (as they wrote in the Foreword) that no attempt was made to cover every aspect of the decorating and finishing of plastics. This time, not so restricted to topics of current interest as one is by the nature of a symposium, we hoped at least to try to be more systematic and complete—to take methods of finishing through from basic fabrication to the more recent technical ingenuities that seem sometimes to change radically the key characteristics of the materials (not only giving them the appearance of other substances, but making soft surfaces hard, electrical insulators into electrical conductors, changing brittle to resilient, and so forth). The main difficulty would be finding the expert contributors who would be prepared to extend their time and knowledge to such a project.

As a first step, letters were sent to the original authors at the most recent addresses known to us, with copies of their chapters and invitations to bring the subject covered up-to-date for the new edition. A few of the authors still were in the same or similar appointments and responded immediately agreeing to take part. However, many had retired and did not feel sufficiently in touch with more recent technology to contribute effectively. Some of these former authors suggested other individuals or enterprises that might be approached to replace them. In a third and the most difficult group the original contributors could not be found—not only must it be assumed that they had moved elsewhere but in some cases the former employers also had disappeared. It became thereafter a question of extending invitations to consultants, to manufacturing firms, research organizations, and individuals known to be active in particular aspects of finishing, explaining our aspirations, and compiling the present edited work as the product of these efforts, the generosity and goodwill of those concerned. A complete list of the contributors this time appears on page xv. It is particularly pleasing that (as in the first edition) most of them have close practical experience of manufacture. We are grateful to all who took part, and to their secretaries, both for their willingness to contribute and their patience in dealing with our queries. Special acknowledgments are due also to Mr P. Reboul of the Plastics Historical Society, for help with photographs, and to Mrs S. Katz for items lent from her collection for the frontispiece and illustrations elsewhere.

Briefly stated, an outline of the contents is as follows:

'Surface recognition' (K. Nakajima and Y. Sato)
'Basic finishing techniques' (W. G. Simpson)
'Calendered thermoplastics' (R. A. Fairbairn)
'The sealing and welding of thermoplastic materials' (C. Hughes)
'Adhesives for plastics fabrication' (Dr P. J. C. Counsell)
'Decorative laminates' (P. Allen and M. F. Kemp)
'Mouldings—their surface and finish' (A. Whelan)
'Extruded surfaces' (A. Whelan)
'Electroplating and electroless plating on plastics' (A. C. Hart)
'Vacuum metallizing' (R. R. Read)
'Painting' (T. A. Wilde)
'Surface treatments for plastic films and containers' (P. B. Sherman and M. P. Garrard)
'A General Review of Printing Processes for Plastics' (J. W. Davison)
'Wall Coverings' (W. G. Niven).

All the information is included in good faith but it should be made clear at this stage that no guarantees or endorsements are given (nor should they be taken to be implied) by the Royal Society of Chemistry, by the editor, or by any individual contributor. In more specific fields—such as health, safety, complaints from and liability to the public—it remains the responsibility of the employer and the operator to ensure that all relevant legislation is obeyed and that appropriate precautions are taken at all times. The mere fact of inclusion in this book of information such as a description of a technique or its application does not imply that the information is reliable, that the technique or application will be appropriate in all or particular circumstances, or that it will be safe and effective in use. (In short, no responsibility is accepted.)

The original edition would not have come about but for the energy and initiative of Dr Pinner. The editor met him for the first time in the early 'sixties at the former Lawford Place Research Station in North-East Essex, to which Pinner recently had returned after being engaged elsewhere in a joint research project for the cross-linking of plastics by radiation. He was both (as those organizations then were called) a Fellow of the Royal Institute of Chemistry and of the Plastics Institute, and continued as a visiting teacher both at Borough and elsewhere, in addition to his work at Lawford. It was a period still of considerable optimism in Britain—a move away from war-time controls over manufacturing at last was beginning to take place and competition from overseas suppliers of polymers, films, and other materials was not yet serious. The post-war official sorties against business and industry that led to high costs and the devastation we see now were only in their early stages. The plastics industry as such already was a century old (resinous compositions and articles made from bone or horn of course went back thousands of years), but new technologies and products were being brought forward. Raw materials derived from oil remained comparatively inexpensive (some were regarded by the industry virtually as waste for which it was desirable to find and to develop new markets). It was possible for an oil refinery like that at Grangemouth in

Scotland to be made the centre of an inter-linked group of jointly-owned companies and plants, taking monomers or other precursors and producing from them an ever-widening range of polymers, copolymers, and additives of various kinds. To a degree, there was an atmosphere of ease and friendly co-operation (the old-established company which at that time shared ownership of Lawford Place became eventually part of the group linked with Grangemouth).

As often was the case, the Research Station originally was an attractive small country house; it was used during World War II as a convalescence home and in parts of the grounds afterwards a series of laboratories and workshops was built. The staff (perhaps 80 to 100 people in total) included gardeners, and there were lawns (with miniature golf), roses, tennis, and an ornamental pond. There were also a plastics development department (with pilot plant), a physics department, analytical and testing laboratories, technical services (with processing machinery such as mills, presses, and equipment for moulding and thermoforming), a library, and various specialized groups working on electronics, new products, technical literature, and so forth. The philosophy behind so considerable a financial commitment was that a company wishing to expand needed to have its own facilities of this kind. (There were various 'research associations', in which manufacturers in different fields co-operated, but they were not appropriate for developing new proprietary products.)

For some years Dr Pinner played an active part in the technical work, particularly the modification by various means of films and of resins for making them. Later he was given a new job in the headquarters of the company at London, becoming one of the many travellers each day from Colchester to Liverpool Street railway station; he was engaged in trying to envisage the plastics industry twenty years ahead and in preparing reports which were intended to assist senior management with strategic planning. (If they could be found, those reports, written around 1970, might be interesting to see now.) In time more immediate problems for the company swept strategic planning aside and until his death a few years later he was a consultant.

There must be many in the industry who remember Dr Pinner still—as a teacher, a development scientist, or a manager—and who could contribute good stories about his activities in those days. He was an enthusiast for plastics. If one wrote a paper or a chapter for a book of his and was inclined to use a detached, scholarly style one was likely to find the page proofs speckled with words he had added, among which 'phenomenal' and 'unique' would be characteristic. We recall him as one of a number of practical British scientists who tried hard to move all-too-sceptical directors in ways they did not want to go.

(In time also Lawford Place was closed, the staff scattered, and the land sold for building. To clear space for more homes, the laboratories and workshops were demolished (in one instance, presumably by accident, burned) and at the date of writing only the shell of the original house remains—stripped of lead and other fittings—in its small way an allegory.)

Just now a young person could not be blamed for thinking hard before entering a career in manufacturing industry. From standpoints such as income and status, far more attractive possibilities exist—but it is to be hoped that a few at least will be interested enough in the various ways of making useful and decorative objects to wish to become proficient and to contribute by this route. We hope that this second edition of 'Plastics: Surface and Finish' will be of help to them, at least so far as plastics are concerned—perhaps especially in the early stages of their lives in preparing for the time when they are able and sufficiently experienced to follow their own ideas about how goods should be presented and made.

W. G. S.

Contributors

P. Allen
Formica Limited

P. J. C. Counsell, BSc, PhD
former Senior Scientist, Evode Group PLC

J. W. Davison, PhD, LRSC, MIOP
Managing Director, Davison Chemographics Limited

R. A. Fairbairn

M. P. Gerrard, BSc, PhD
General Manager and a Director, Sherman Treaters Limited

A. C. Hart, BSc, CChem MRSC, FIMF
Hart Coating Technology; Consultant, Nickel Development Institute

C. Hughes
Technical Manager, Radyne Limited

M. F. Kemp, MIQA
Formica Limited

K. Nakajima, BE
Sekisui Chemical Company Limited

W. G. Niven, MBE, BSc
Technical Director, Forbo-Kingfisher Limited

R. R. Read, LRSC, FIMF, FICorrST
Consultant in Surface Finishing

Y. Sato, ME
Sekisui Chemical Company Limited

P. B. Sherman
Founder and Managing Director, Sherman Treaters Limited

A. Whelan, MSc
Consultant

The Editor wishes to thank also Dr D. Simpson, Analysis For Industry, for her help in preparing several of the Figures.

Extracts from British Standards are reproduced here with the permission of BSI Standards; complete copies of the documents concerned may be obtained from: BSI Sales, Linford Wood, Milton Keynes, MK14 6LE, England.

The extracts from International Standards are reproduced by permission of ISO Central Secretariat, Case Postale 56, CH-1211, Geneva 20, Switzerland.

CHAPTER 1

Introduction

W. G. SIMPSON

'Natural' and Other Finishes

The finish of an uncoloured plastic film or moulding is said in the industry to be 'natural'; such finishes can be serviceable and perfectly adequate in some uses (as in pulley wheels or other unseen parts in mechanical assemblies).

However, coloured or uncoloured materials may be finished in a great variety of ways. For certain purposes (even for 'natural'), quenching or annealing immediately after processing are necessary in order to obtain the molecular arrangements and therefore the mechanical or other properties required. In other words, the finishing is an important factor in deciding the suitability of the object for its purpose, and whether or not it will meet a specification.

The world would be more uniform, less colourful and stimulating, if it were not for the human habit of decorating and finishing objects of all kinds—both useful and ceremonial—often almost regardless of the types of substance from which they were made (like gilded or lacquered timber). Brooches and other items of jewellery, buttons, fasteners, seals, and more mundane things have been made for centuries by working bone or horn, and from malleable compositions like clays or filled resins, which were set hard just by drying under ambient conditions or were fired. A striking example of benefit from the habit was the skills developed in the nineteenth century in finishing a common material like cast iron (for a balustrade in the Royal Pavilion at Brighton, Sussex, for instance, this heavy dull metal was transformed by clever casting and painting into an unusually tough and long-lasting bamboo). Besides being converted by such means into a variety of 'timbers' iron was made to represent different types of stonework and decorative plaster (it was the 'versatile' building material of its day).

When 'natural', many plastics are transparent or translucent—which means, in effect, that they can be produced as transparent goods as well as in translucent or opaque colours. With materials that essentially are transparent, the range of colours possible is very wide. Other plastics are opaque, or dark in original colour—so that the opportunities for coloration by blending and mixing are more restricted.

With some plastics, the 'natural' appearance of the surface (especially in the case of opaque materials) is matt and not especially attractive; on the other hand, a polymer such as polystyrene in 'natural' form has a glossy surface which is described sometimes as 'glass-like'. Paradoxically enough, it may be the object of the technologist by finishing a material in a special way to make it more attractive and arresting in appearance than otherwise it would be—or, on the other hand, to make a glossy and rather brittle material less so. Success in finishing may mean the consistent production of a 'rough' finish as well as a smooth one.

It always is difficult to generalize, but apart from the desire to change and to improve the appearance of a substrate, perhaps to disguise it, there can be a variety of technical reasons for selecting methods of finishing (at times with objectives which may seem confusing or even contradictory)—such as to change the chemical, electrical, or physical properties, to match the appearance of some adjacent components, or merely to convey information by colour-coding or printing. Some possible approaches to and applications for finishing technology are summarized in the form of a list below. (It is not proposed in this volume to attempt to encompass the use of plastics and synthetic resins in finishing other substances—in forms such as paints, coatings, or coloured films applied (say) to protect timber boards, and sheets in ferrous or non-ferrous metal.)

(i) To provide a skin on the surface of a product and so to impart altered or improved properties (as examples, to make the surface conductive electrically, sensitive to light, more resistant to weathering—including degradation by ultra-violet light, and harder or stronger in a mechanical sense)

(ii) to provide a distinctive surface, either generally, or in a particular part (such as to abrade other substances, or to resist wear; to break or to tear easily in pre-arranged circumstances)

(iii) to assist in fabrication by marking out and cutting lines or shapes for sealing or welding later

(iv) to apply to the surface of the plastic a layer of another material—perhaps another plastic or something of different nature entirely (like organic fibres or silicates)—so that the resulting composite combines the properties of added materials with those of the original (typically, in packaging a film of another plastic might be applied in order to enhance resistance by the base material to permeation by gases, or to improve adhesion)

(v) to make for some specific purpose a new composite material (like a laminate in different colours–for engraving or otherwise machining for fancy effects, a flexible film to make signs that will reflect light, or a rigid but light 'sandwich' for structural purposes in aircraft and other applications where weight is an important consideration)

(vi) so that a surface will be more suitable for further processing (in preparation for electroplating, lamination, metallizing, printing, and so forth)

(vii) to mimic, deceive, or mislead, by giving the plastic the appearance of other materials (like brass, ivory, marble, leather, or wood)

(viii) to make an object similar in appearance (and therefore a match to or a contrast with) adjacent objects in other materials

(ix) to extend a material (and thus to reduce its cost)—perhaps physically by stretching it in some way, or (through the introduction of 'blowing agents' or gaseous expansion) by separating the particles or portions of the feedstock by means of closed or open voids

(x) to make decorative items such as prints for furnishings in domestic premises, hotels, and other public buildings, accessories for clothing, and so forth

(xi) for advertising and to convey information by printing logotypes, trade marks, instructions for use, details of composition, computer codes, *etc.*

(xii) as the reclamation and re-cycling of waste material become more important it will be an international requirement that all plastics are identified visibly by means of a standard code system—which, it is envisaged, will be printed or embossed on the surface.

Considerable ingenuity and ranges of equipment have been developed at various times for purposes such as these. For the continuous production of articles of a required quality perhaps two overall requirements are applicable in most if not all cases—the maintenance of consistency in the plastic to be finished, and standardization of the conditions of treatment. To comment briefly on both of these in turn:

Consistency in the Plastic

Techniques and plants for polymerization have become more precise and specific but there is a possibility still that similar grades of the same material made in different units may differ in practice (in features such as the distribution of molecular weights, and colour). It will be appreciated too that many polymers and copolymers are used in combination with other substances—stabilizers, fillers, and miscellaneous additives—all of which (and especially those occurring naturally, like China clay and some types of plasticizer) may themselves differ appreciably from batch to batch.

Differences between the ingredients in a formulation may be essentially of a chemical nature, but also may be physical (such as the ranges of particle sizes)—and quite often both chemical and physical variations are found in practice. In order to match a particular material consistently it is not sufficient merely to know even the precise formulation in terms of polymer, plasticizer, heat stabilizer, filler, 'anti-static' agent, colour, and so forth; each of the components in the formulation must be supplied consistently to a tight specification, and (furthermore) the blending and mixing must be carried out with similar equipment in the same manner each time.

When making special effects (as examples, 'marble' or 'wood-grain' sheets) the principal components may be quite different in appearance but still must be compatible and capable of being combined.

Lastly, besides inconsistency in the polymer and additives, there may be fortuitous differences between mixes of what supposedly is the same composition—such as the water content changing between summer and winter (or by night and day under some climatic conditions). Thus, provision must be made for drying and the removal of other volatiles, and (especially if the processing is by batches and components are held in store at intermediate stages) drying may be required at more than one point in the manufacture.

Standardization of Conditions

For the reasons indicated above, consistent and successful production of plastics articles at times in the past was regarded as more of an art than a science (with the production manager rather more of a master chef than a scientist or engineer).

The first hurdle on which success in production depended (assuming consistent raw materials, stored satisfactorily before use) was the blending and mixing of ingredients with some assurance that homogeneity was achieved and that obvious features (like the shade of colour) in the finished batch would be the same (or approximate closely to) its predecessors and those following. Since very few polymers were used alone, effective blending and mixing (with the exclusion of all forms of contamination) soon were understood to be essential.

Typical contaminants include general dust and dirt, other foreign matter (pieces of sticking plaster, gloves, nuts and bolts, pins, insects, fragments of paper, feathers, and fibres)—almost anything. Ideally, all processing should be in automatic units with full air-conditioning but in practice this is not possible so great care must be taken not only by securing long hair and all clothing and equipment, but to locate suitable detectors before nips and other critical stages in the process. Such detectors always should be as close as possible to the inward ('feed') end, with automatic cut-outs.

Static electrical charges in the machinery and stock can be important in attracting contamination, and facilities for electrical discharge at appropriate points also will be needed.

In all types of thermal processing (calendering, extrusion, or whatever it might be) the control of temperatures is critical, but precise control of equipment all the time is difficult. There were engineering considerations—notably the heating and cooling of large volumes of metal and alloys both quickly and effectively, and economic ones; usually, greater precision in control called for more expensive facilities. A third factor of importance was the effectiveness of operators in keeping to the standards required—and, sometimes, their preparedness to do so at the expense of increased rates of output.

In some types of work there has been an increase, following the example of the pharmaceutical industry, in the taking of samples at intermediate stages in the production and testing them with a view to ensuring control of quality: many convenient instruments and techniques are available for purposes such as this.

Subject to satisfactory solution to the questions of engineering and related economies, the manufacturing operations in general today (not only thermal processing) may be controlled by means such as:

(i) mechanical devices (like spring-loaded cut-outs, trips, balances, levers, punched cards or paper rolls, pneumatic switches, hydraulics)
(ii) electrical and electronic devices (including sensors and computer programmes)
(iii) robots
(iv) human operators.

The first approach might be characterized as rather inflexible and suited to more simple routine work, although punched cards and rolls open the opportunity for variations in some fixed procedures: the second and third possibilities are more versatile but still somewhat less flexible than human operators. The approaches (i) to (iii), though capable no doubt of much further extension, are most appropriate for production in large numbers and for repetitive work: also, while they may well be far more precise in long runs than human operators, computers and robots can produce rejects in larger numbers more quickly. For all such methods of control the ability to unlock and to vary at times is necessary still—provided a variation is made deliberately and for a sound technical reason.

CHAPTER 2

Surface Recognition

K. NAKAJIMA and Y. SATO

Introduction

Plastics, like many industrial products, are manufactured with a view to being useful in society—and with the approach of a new century the words 'useful in society' have widening implications. In other chapters of this book emphasis is placed upon progress in science and technology leading to new processes and products, but as part of this advance the influence of technology on environments should be considered too. Still more, the product is required to satisfy not only physical and functional needs laid down but should respond to the sensitivity of those in contact with it. The implication is a technology in harmony with environments, and also with human feelings.

It is suggested therefore that the topic here—'surface recognition'—can be seen as more than an approach to inspecting for surface defects but rather as a contribution towards goods offering genuine appeal for highly sensitive functions of human beings (like eyesight and touch). The range of materials and uses is wide and the present review can hardly cover all the details, but publications elsewhere (such as in the technical press) should help to meet further interests in this field.

In the first stages of assessing surfaces of plastics (and aside from the analysis of data) reliance is placed largely on human senses. Hence, in any transition to a mechanical system of recognition, the tendency is to regard the human inspector as the 'automatic' system and to use him or her as a standard for comparison. Similarly in this chapter we introduce first the method of recognition, with visual sensing and accepted software, then the technology intended to systematize and to integrate all aspects, including economy.

However, systems have been evolved with performance different from and in some ways superior to those associated with recognition by skilled inspectors.

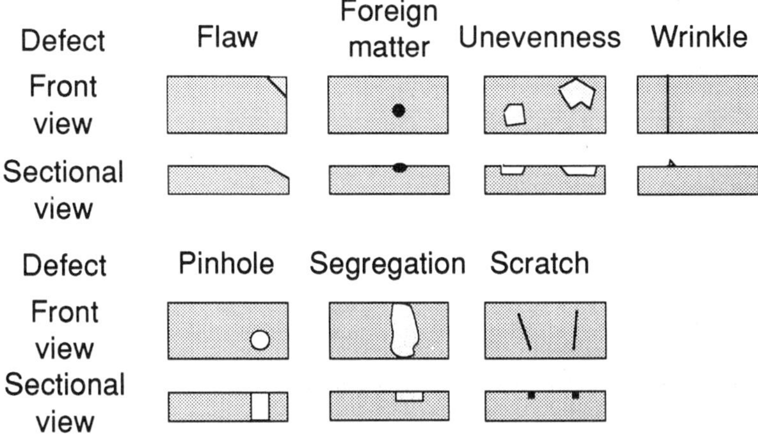

Figure 1 *Graphic models: front and sectional views of typical surface defects*

Specific schemes such as these have been brought into use, and in general the technology is being advanced rapidly. With a view to taking the pace of development into account it was considered appropriate here to pay particular attention to the principles that underly the design of satisfactory systems.

Types of Surface Defects

Defects are mentioned in other chapters in connection with plastics in a variety of forms, manufactured in various ways (extrusions, films, mouldings, sheet, and so forth). As demands for quality and uniformity become more stringent it is important to limit and to localize all defects, and as a first step it is necessary to identify and to classify them. Figure 1 is an attempt to do so for seven types of fault, as follows:

- (i) 'flaw', occurring in many sizes, mainly in sheets, mouldings, and pipes
- (ii) 'foreign matter', which can occur with any product, consequential upon contamination of the raw materials used
- (iii) 'pinholes', which occur mainly in film and sheet, often from bubbles of trapped air
- (iv) 'scratches'—marks made on the surfaces by tool edges or other hard objects
- (v) 'segregation'—defects of transparency or colour most apparent in films
- (vi) 'unevenness'—dents or protruberances at the surface
- (vii) 'wrinkles'—faults in flat-lying, again most apparent with films.

Types of Methods for Recognition

Many methods are available for classifying and appraising defects, and selection depends largely on the purpose and the nature of the system to be employed. Signals are obtained from the surface and from a defect by an

TABLE I *Comparison of methods of recognition*

Requirement	*Method and assessment*			
	Photo-sensor		*Laser beam scanning*	*Ultrasonic sensor*
	One-dimensional	*Two-dimensional*		
Resolution	medium	medium to small	large	small*
Flexibility	medium	large	small	small
Size of system	small	small	large	large
Speed	medium	low	fast	low
Application to samples	medium	wide	narrow	narrow

* Minimum: 1 mm.

appropriate method of 'sensing'—by the radiation, say, of electromagnetic or ultrasonic waves and then recording the transmission or reflection of such waves so that comparison is made with data for satisfactory surfaces and the size, shape, and position of any defect noted.

For the ultrasonic approach, receiving apparatus is arranged round the material under observation. The waves are generated and aimed at the surface, their reflection and transmission then being recorded. Defects are identified by differences between normal and abnormal wave shapes.

In methods with light, the surface to be examined may be irradiated uniformly, a photo-sensor being used to measure its distribution; alternatively, a laser may scan the surface in regular patterns, a sensor noting any differences in the amounts of light collected.

Table I provides a schematic comparison between photo-sensor, laser, and ultrasonic methods, the different sensor systems being considered in more detail below.

Photo-Sensors

Figures 2 and 3 show, respectively, one-dimensional and two-dimensional photo-sensing systems, with representations of the images obtained in both cases.

Essentially, the one-dimensional system comprises a photo-electrical transformer in which reflected and transmitted light are collected by lenses and converted into electrical signals. A surface defect results in a light level different from that of the normal signal and as evaluation can be rapid such methods are suitable for assessing faults arising, for example, on fast-moving production lines for film.

In the second example, the defect is observed by means of a video camera as a two-dimensional signal—that is, as an image. Recognition takes place, as with the one-dimensional approach, when data returned are different from those of a normal surface—but since the information embraces more than a single

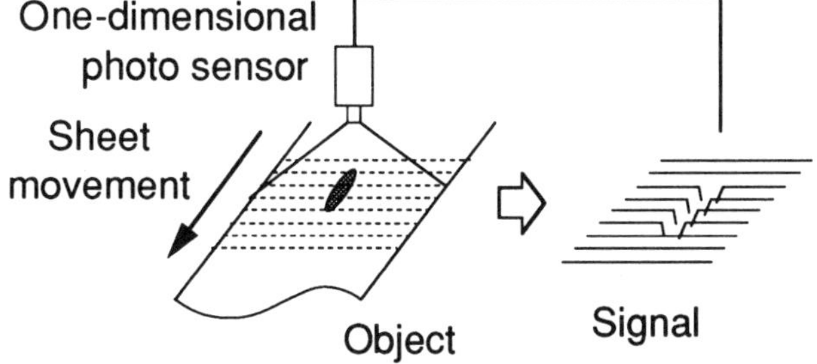

Figure 2 *Using a one-dimensional photo-sensor*

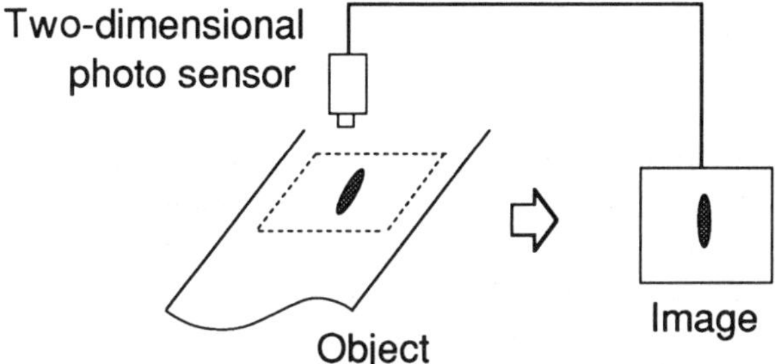

Figure 3 *Using a two-dimensional photo-sensor*

dimension the notation and recognition of the form of a defect can be more complete. On the other hand, as the information returned becomes more complex longer periods of time are necessary for its analysis and for taking suitable corrective action; because of this, a two-dimensional technique may not be suitable for fast-moving lines.

Laser Sensing

Figure 4 illustrates a system of the laser-scanning type. The laser scans the surface and detection occurs because in a defect the intensity of light trans-

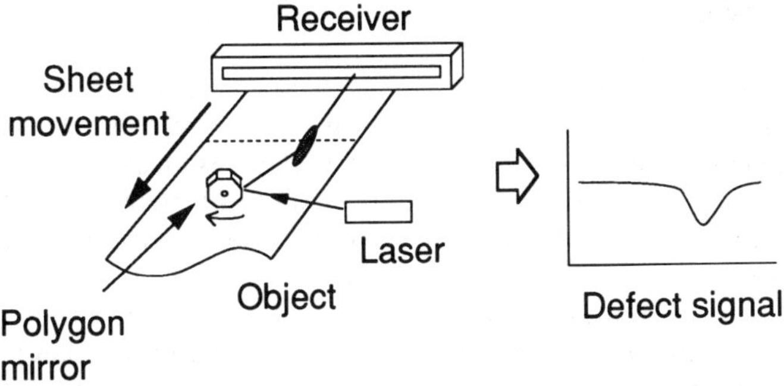

Figure 4 *Sensing with laser scanning (reflection system)*

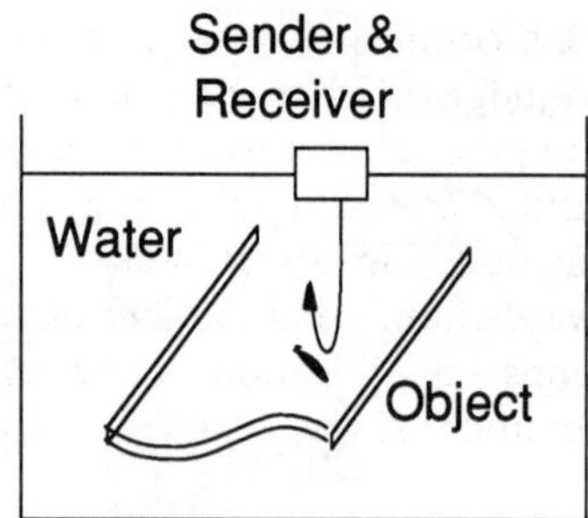

Figure 5 *Using an ultrasonic sensor (reflection system)*

mitted or reflected differs from a normal region. Variations in reflection can indicate unevenness of surface and, in a transmission system, differences in the transmission of light may indicate pinholes or foreign matter in films.

Such a system will detect any variation from the normal intensity of the light, the equipment being effective across the whole width scanned. An optical receiver is employed and this collects all light entering from any position and converts it into an electrical signal. The laser is concentrated on a single small spot and by rotation of the polygon mirror fast scanning in the transverse direction can be achieved. Defects are recognized as the spot moves and variations in signal are detected.

Ultrasonic Sensing

Figure 5 illustrates an ultrasonic system. The waves are transmitted and impinge on the material under examination, returning to a receiver; variations in the shape of waves received indicate the defects.

Transmission of ultrasonic waves in air is difficult (the intensities received are low) and to overcome this the material to be examined must be immersed in water. A further limitation follows from the frequencies that may be used with plastics, and in practice application is mainly to hard and to uniformly dense products in which the defects are not more than one millimetre in diameter.

Technology for Image Sensing

Among the techniques for recognition mentioned earlier we refer now more specifically to sensing by two-dimensional photo systems.

In a general system of this type, information from the object under study is converted into electrical signals, the resulting image is recognized and judged in a data processing unit, and signals assessing the type and quality of the object are then passed to the control procedure or actuator.

For the most part such systems are used to automate visual inspection, so helping to save labour and to free employees from the necessary periods of intense concentration, which at times could lead even to clinical difficulties like headache or eyestrain. Unfortunately, though, they hardly have reached the levels of recognition and judgment that are achievable by human eyesight and brains. To some extent, in practice, this weakness can be met by improvements in the sensing: the qualities of the lighting and image sensing are of such importance that they may determine the standard of performance of the entire system. Approaches towards improving the sensing aspect of recognition may be classified in two groups—modification of the sensor as such, and of the illumination of the object under study.

(i) The Sensor

It is more than half a century since the development of the image tube in which scanning was by electron beam, while study of the solid-state sensor was started in the 'sixties—a variety of configurations of elements and structures having been proposed—such as (in the first instance) Scanistor.[1] In view of the complexity of the solid-state image sensor (which integrates several hundred thousand features in a single semi-conductor substrate) many technical problems required solution before it could be employed in the factory. Some help in this came from innovation and progress with memory devices, to the extent that the solid-state sensor now has many industrial applications (though not in cameras for television, where very high clarity is essential). Some of the key qualities of the device are presented in Table II. Referring briefly to the Table, the sensor offers obvious advantages and is capable of excellent performance. This is true especially with regard to the question of distortion and with each unit arranged in regular order geometrically no significant distortion should occur—a factor of the greatest importance when recognition and measurement with geometrical precision are required.

In the solid-state image sensor, transducers with photo-electric converting functions are arrayed, the electrical signal generated at each element being switched regularly in space and time by pulse signals from the scanning circuit and then read as signals presented in time sequence from a single output wire. In the early stages of development the current flowing in a transducer was

TABLE II *Summary of features of solid-state image sensor*

Performance
Very little distortion of image
The standard of resolution of peripheral image is uniform throughout the picture
Hardly any residual image
Low noise, high sensitivity, wide dynamic range
Specification
Activation at low voltage and low power consumption
Small size, light weight
Reliable; resistant to impact and vibration
Affected little by a magnetic field
Production possible on a large scale, with low cost per unit

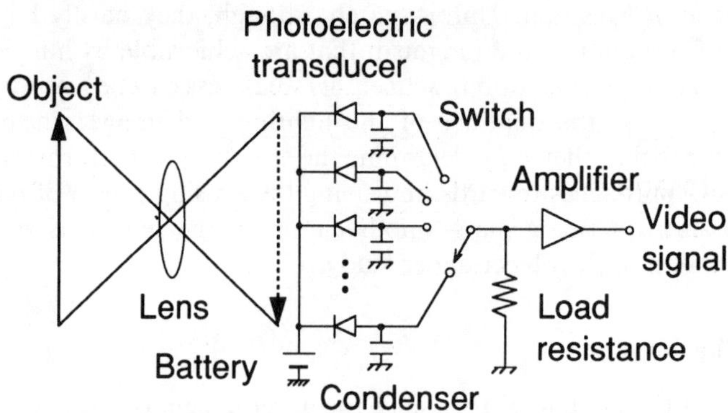

Figure 6 *Imaging by electric charge integration*

detected only when it had access to the scanning circuit and sensitivity therefore was low. However, as illustrated in Figure 6, a charge integrating mode[2] was conceived in which each transducer was provided with a condenser connected in parallel; the current generated in the transducer during the cycle of pulse received from the scanning unit was accumulated as an electrical charge and each time information was transferred the integral value of the stored charge was detected. As operation was integral, in this mode the image for a fast-moving object was not very clear—but the approach nevertheless is highly sensitive and most image sensors in practical use are of this type.

Solid-state image sensors consist of photo-electric transducers, switches, and scanning circuit, and there are many types with various configurations. In general terms, the metal oxide semi-conductor ('MOS') type has been under study a long time while the charge-coupled device ('CCD') type was developed in 1970. Both have both advantages and disadvantages, so it is hard to decide in absolute terms which is superior: products of the MOS type were the first to be offered commercially, with the CCD type coming later.

Figure 7 illustrates the operating principle of a CCD system—essentially, the electrical charge or signal is passed sequentially in a certain direction, just as in a relay of fire buckets. An insulating oxide film (SiO_2) covers an earthed ('grounded') p-type silicon substrate and when positive voltage is applied holes in the silicon allow penetration to the substrate and the formation there of deep hollows of electrical potential ('potential wells') in which free electrons arising from irradiation by light will gather. The imposition of a three-phase pulse takes the electrons to the depth of the well and carries them in a single direction as indicated in the Figure. The efficiency of transfer of the electrons in each step is 99.999%—that is, there is almost no loss in the electric signal.

Table III is a summary presentation of the CCD image sensor products of various companies in Japan, while Figure 8 shows the change in numbers of elements for photo-electric transducers. In the background of this change there was a great improvement in the clarity of pictures in home video systems (the photo-sensor for surface recognition of course was not developed exclusively for these measurements but was adapted from the general type).

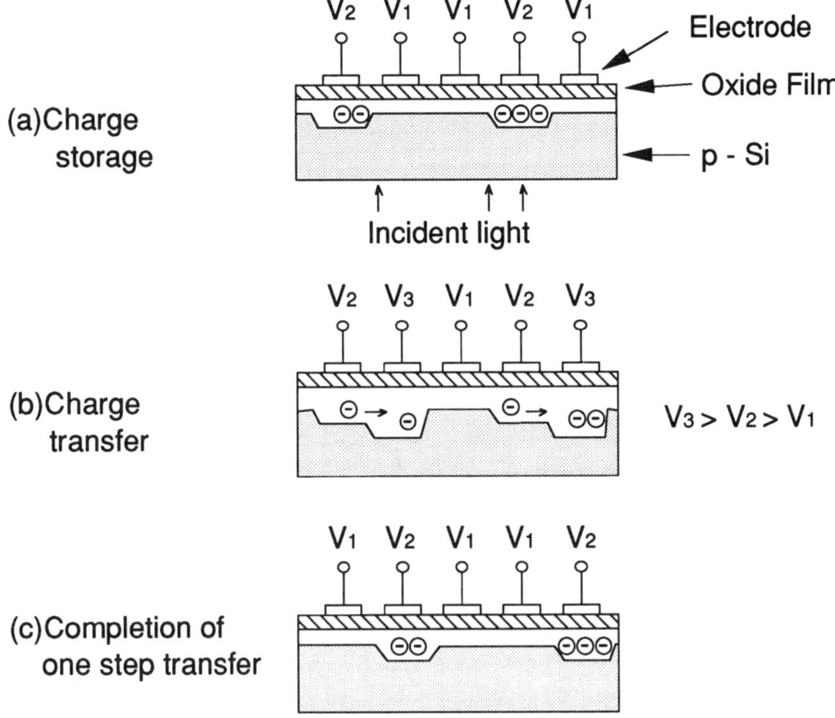

Figure 7 *Principle of the transference of charge in a charge-coupled device*

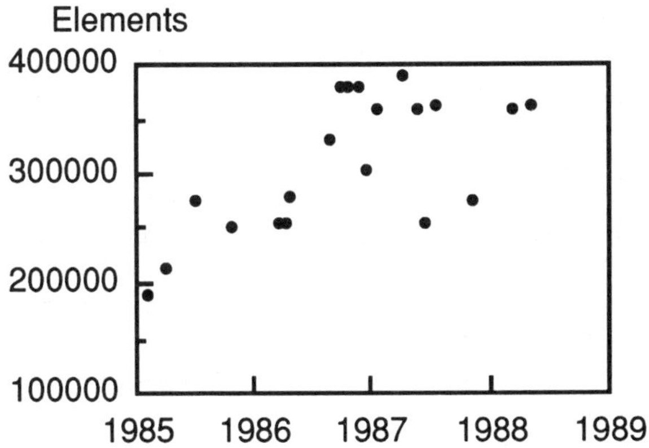

Figure 8 *Change in number of elements for photo-electric transducers*

Information gathered by the two-dimensional photo-sensor is essentially that of the distribution of visible light (the image). However, such images may embrace a variety of types, including:

(*a*) monochrome still image (like a black-and-white photograph)

(*b*) stereo image (that is, extended in a third direction)

TABLE III *Comparison of CCD image sensor products*

Feature	*Name of product*					
	Matsushita	Mitsubishi	NEC	Sanyo	Sony	Tokyo Electronic
Optical size of element (inch)	$\frac{1}{2}$	$\frac{1}{2}$	$\frac{2}{3}$	$\frac{2}{3}$	$\frac{2}{3}$	$\frac{2}{3}$
Number of elements						
horizontal	570	768	728	800	768	768
vertical	485	494	493	490	493	492
Chip size (micrometres)			12.3 by 13.5	11.5 by 13.5	11 by 13	
Lowest sensible intensity of object (1 ×)	0.5	5	2		5	2
Resolution (number of television lines)						
horizontal	420	470	540	560	570	570
vertical	350		480		485	350
dB	48	48	50		56	50

(*c*) image measured against a time axis
(*d*) image expressed as a spectrum, or as more than one spectrum
(*e*) animated or dynamic image
(*f*) a colour, stereo image, with animation (that is, changing over time)—said to be 'five-dimensional'

In the last instance especially a very large amount of information can be assessed—reaching, in a single television picture, 2×10^6 bits. In general, the greater the amount of information the higher the standard of recognition but associated with greater complexity come the possibilities of increased technical difficulty and higher costs. In practice therefore, image sensors will be selected on the basis of providing the data that really are needed.

(ii) Illumination of Image and its Significance

If the surface to be studied is darker than the minimum illumination required by the photo-sensor (that is, above the noise level) then the details of the surface will be hard to distinguish. Hence, an appropriate source of light should be employed for illumination—an incandescent electric lamp, fluorescent lamp or xenon lamp. The selection should be appropriate for the purpose and with the wavelength of light emitted in mind. Arrangement of the lighting also is important—whether it be diffused, transmitted, or directional—and again the designer must select the method most appropriate for the surface being studied and the types of defect involved. Some indications of means by which the most significant characteristics may be assessed are given as follows:

(*a*) use of wavelength—especially when an object has an absorptive band for a specific wavelength, or when the colour information is important
(*b*) use of shadow—such as when the roughness or otherwise of a surface is to be assessed—directional lighting being employed to create shadows: the angle of incidence of the directional light should be linked with the degree of roughness anticipated
(*c*) regular reflection—after allowing for a certain standard or regular reflection by the object, roughness can be detected with remarkably clear contrast (this assumes, however, that it is possible with the object under study to establish 'regular' reflection—which is not always the case)
(*d*) polarized light; when an object gives full or near-total reflection, polarized light filters fitted to both illuminating and image-sensing equipment will indicate defects
(*e*) shadow-less lighting; when an object gives strong and complicated regular reflection it is helpful to employ 'shadow-less' illumination (that is, highly-diffused lighting).

Figure 9 illustrates an example in which the object has a uniform diffusion surface of area S, reflectance R, the focal length of the lens f, the distance from the surface of the object to the principal point on the front of the lens is a, and the distance from the principal point on the reverse of the lens to the sensor plane is b—giving the well-known formula:

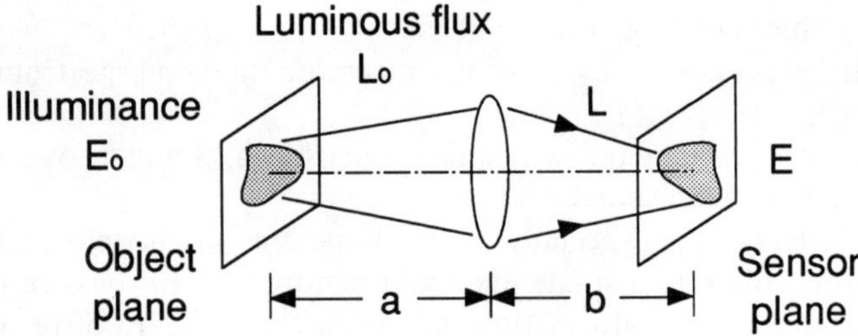

Figure 9 *Optical image system*

$$\frac{1}{a}+\frac{1}{b}=\frac{1}{f} \tag{1}$$

then, the scaling-up ratio for the image, M, will be:

$$M=\frac{b}{a} \tag{2}$$

and the area on the sensor $M^2 S$. Luminous flux L (lumen) passing the lens when light L_0 is applied is:

$$L=\frac{L_0 RT}{4F^2}\cdot\frac{M^2}{(1+M^2)} \tag{3}$$

Where T = transmission by the lens, F = the ratio of focal length to effective diameter of the lens, and E_0 = illumination of the object plane (1 ×) and that of the image plane E, the formula would be:

$$E=\frac{E_0 RT}{4F^2(1+M^2)} \tag{4}$$

In the ordinary recognition system a is greater than b, so the result is:

$$E=\frac{E_0 RT}{4F^2} \tag{5}$$

As an example, if $R = 0.7$, $T = 0.3$, $F = 2.8$, and 20 (1 ×) is the illumination required on the image plane, then 3000 (1 ×) are necessary on the object plane.

It should be remembered that the diameter and thickness of the lens are limited and peripheral light from the object will be at an angle to the optical axis of the lens, less light being transmitted as the sectional area passing the lens becomes smaller. Therefore, if the lens is too close to the object there may be shading at the edges—that is to say, the periphery of the image will be dark.

Recognition of Image

Usually, when a person looks at a surface he or she can tell very easily whether an abnormality is present. The human brain notes a difference from a pattern seen previously, or from information perceived about surrounding areas. The technology for processing images is a substantiation of these activities, the

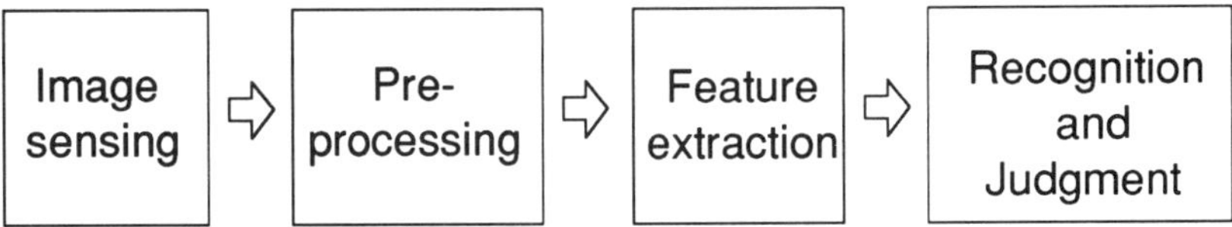

Figure 10 *Flow chart for processing the image*

computer working with information gathered from a sensor such as a CCD camera and the flow of data being as shown in Figure 10, namely:

(*a*) sensing of image—basically the distribution of light in varying intensities on a surface, but for use within a computer expressed in numerical values and converted to digital signals

(*b*) pre-processing—in which adjustments necessary are made for interference in the input ('noise'), for geometrical distortion or distortion in greyness by the lens—the original image generally being restored

(*c*) extraction of features—the selection in particular of data in respect of greyness (or of colour in the case of a coloured image), the shapes, and other characteristics of a surface or defects

(*d*) recognition and judgment—on the basis of features extracted and compared with standard data, the object thus being delineated and classified.

An image is expressed in digital terms in the form of a large number of discrete dots known as picture elements or 'pixels', each dot being recorded in terms of brightness. Figure 11 indicates how the picture obtained from sampling becomes the digital image.

The sampling converts the image in space into a set of dispersing dots: initially it comprises information distributed on a two-dimensional plane and the first step is conversion by scanning into 'one-dimensional' signals. A typical method of scanning would be horizontally across the two-dimensional plane at specific intervals in regular order from the top, with recording of values for brightness; since the intervals between signals are known the line of dispersing dots can be created.

Factors that determine the intervals at which sampling is carried out include the extent of the minute changes of brightness in the image, the extent to which these changes can be accepted, and the area involved. Strictly speaking, for the

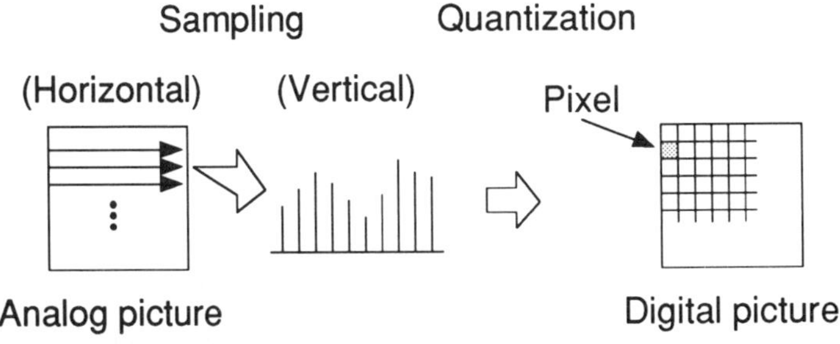

Figure 11 *Composition of digital image*

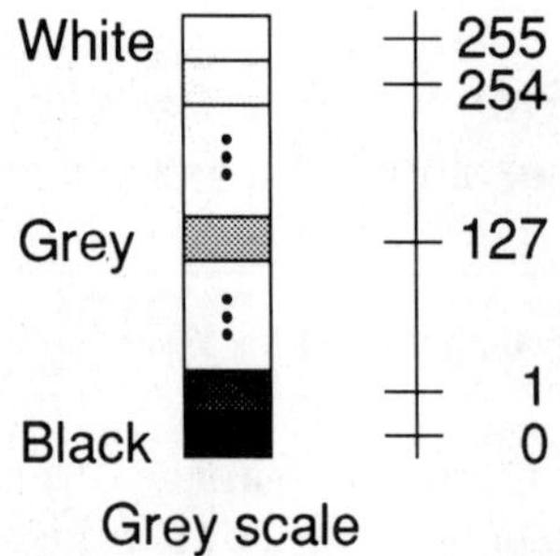

Figure 12 *Quantization of grey image*

best results, sampling theory should be applied, but in practice the divisions mainly are either 256 by 256, or 512 by 512.

While the image is resolved into picture elements conversion of the brightness values into numerical values ('quantization') also is required: Figure 12 shows how the continuous density values black, through grey, to white are quantized on a scale from 0 to 255; since the values in each different range are converted on the same basis there may be some variation from the true basis; this is known as the 'quantization error', and the calculation reported is the one that limits this error to a minimum.[3]

Processing of the Grey Image

A number of approaches may be employed to process the grey image into digital signals in order to obtain a more useful result. Pre-processing may include:

(*a*) transformation in which the distribution of grey level in the image becomes uniform
(*b*) shift of the image to a standard position
(*c*) filtering to eliminate noise and emphasize the defects.

Figure 13 shows the effect of filtering to emphasize the edges of a defect in a film.

The task of filtering is viewed mathematically. Edge in the image means at that part a great change in the grey level. The image is two-dimensional, so isotropic property—that of being constant against rotation—is applicable to edges in many directions. In a Laplacian formula:

$$\nabla^2 f \equiv \frac{\partial^2 f}{\partial x^2} + \frac{\partial^2 f}{\partial y^2} \tag{6}$$

the approximate dispersion value in a digital image:

$$\nabla^2 f(i,j) \equiv \Delta_x^2 f(i,j) + \Delta_y^2 f(i,j) \tag{7}$$

with $f(i,j)$ as the grey level when x co-ordinates are at position i, and y co-ordinates at j:

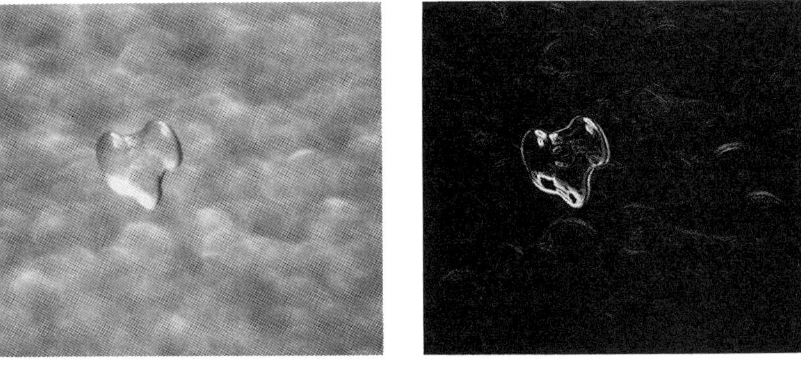

(a)Original (b)Filtering result

Figure 13 *Emphasizing the edges of the image by spatial filtering*

$$\Delta_x f(i,j) \equiv f(i,j) - f(i-1,j) \tag{8}$$

$$\Delta_y f(i,j) \equiv f(i,j) - f(i,j-1) \tag{9}$$

the formula (7) above becomes eventually:

$$\nabla^2 f(i,j) \equiv f(i+1,j) + f(i-1,j) + f(i,j+1) + f(i,j-1) - 4f(i,j) \tag{10}$$

Figure 14 shows the formula (10) in the form of an operator: filtering is the most useful technique in image processing and a mere change of value on the operator will bring a variety of effects, so many kinds of operators are proposed and applied.

Figure 15 illustrates the value of the histogram of grey level, the horizontal axis being that level and the vertical axis the frequency or probability of occurrence. It may be used to calculate the number of pixels in the image for each different grey level and indicates the values of features in the grey image. Average values, variance, and (by Fourier transformation of the image) distribution of the element of spatial frequency also may be employed.

0	1	0
1	-4	1
0	1	0

Figure 14 *Laplacian operator*

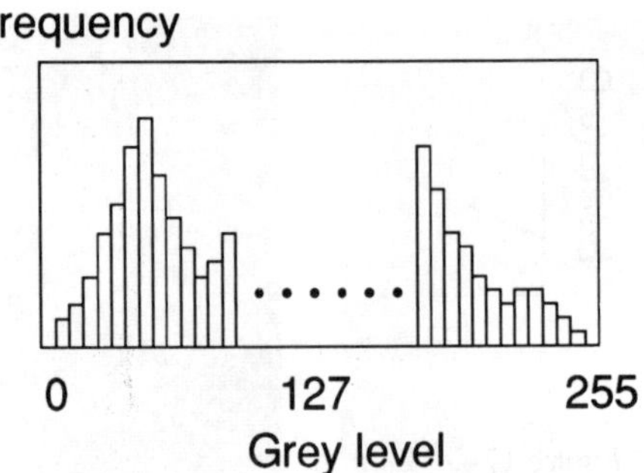

Figure 15 *Histogram of grey level*

Binary Image Processing

In a binary approach the image is processed in white and black colours—which may be called 'two levels quantization'. It provides information in the most compact form, and so is used widely in practical systems where high speed and low costs are requirements.

The binary is derived from the grey image by 'thresholding'; in the formula, threshold t is set for the grey image:

$$f_t(i,j) = \begin{Bmatrix} 1; & \text{for } f(i,j) \geqq t \\ 0; & \text{for } f(i,j) < t \end{Bmatrix} \tag{11}$$

and Figure 16 expresses the formula in pictorial form; in this manner an object such as a surface defect may be identified and distinguished from the background.

In this system, interconnected pixels form groups and are expressed in geometrical form—hence, the values for features can represent parameters such as area, length at periphery, degree of circularity, centre of gravity, and width.

The values for features in both grey and binary images are related to whole values and to other parameters; in recognition, a key factor is the effectiveness or otherwise of the features and values selected—as these decide its precision.

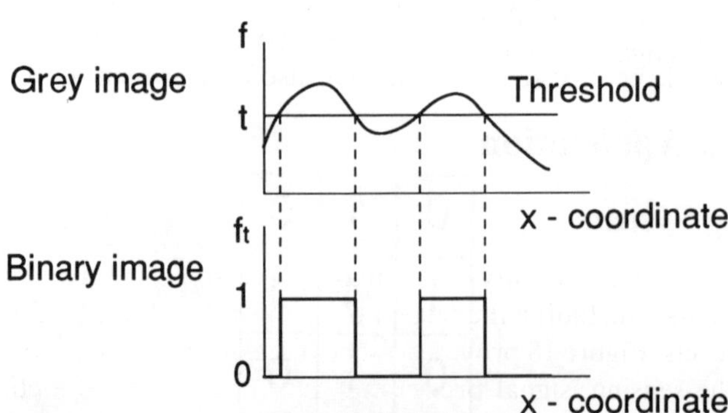

Figure 16 *'Thresholding'*

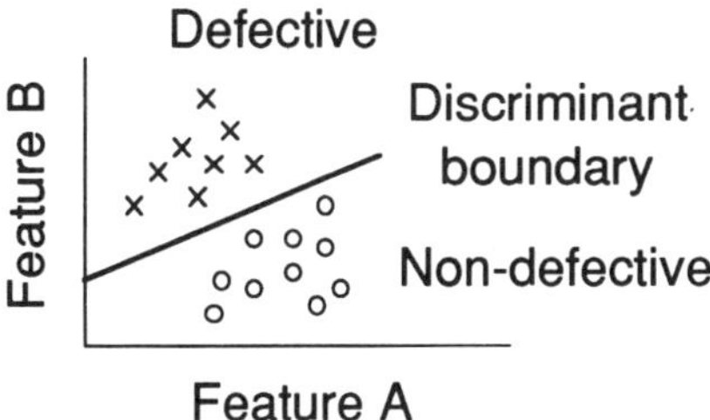

Figure 17 *Differentiation of feature space*

Recognition and Judgment

Appropriate statistical techniques are applied in order to detect, classify, and assess defects at the surface; the procedures employed are well understood and are applied to features of images as for other purposes.

In assessment, the values extracted from the image are collated in feature spaces; in other words, a number m of feature values extracted $(y_1, y_2, \ldots y_m)$ are converted into an m-dimensional feature vector of:

$$Y = [y_1, y_2, \ldots y_m]^t \tag{12}$$

and, as the values recorded are gathered in the feature space, a 'cluster' emerges and indicates the importance or otherwise of the classification. Figure 17 illustrates the division of results into two categories—'defective' and 'non-defective', depending in each case upon the location of the values for the features recorded. When applying this in a practical sense it is important that the features and values selected for use as criteria are separate and distinct—not related closely to each other.

In recent studies emphasis has been placed upon binary image processing, in which filtering is not appropriate and the extraction of defects depends mainly upon the method of sensing. The introduction of systems offering the advantages of high performance and low cost, together with recognition at high speed, made possible practical use of both grey and coloured image processing, and also the recognition of images not possible previously.

Industrial Application

Processing System

As outlined above, the totality of the image processing system includes a variety of requirements—including the recognition, control and transmission of signals at various levels. Figure 18 provides a schematic outline of the complete system, including the sensing, signal processing, and input/output functions. In particular, signal processing comprises an 'interface section' in which signals received from the camera are sensed and converted into digital form, a 'memory

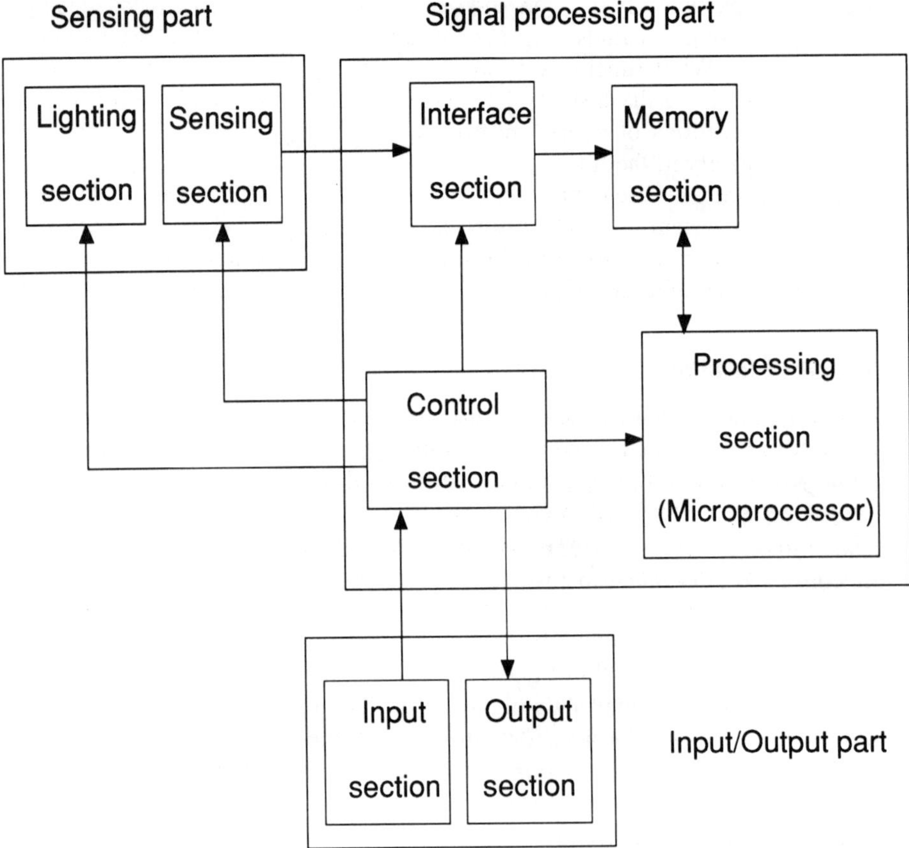

Figure 18 *Systematic diagram of image processing system*

section' in which inputs of data, results of processing, and other information at various levels are accumulated, a 'processing section' in which information may be handled, features extracted and recognized, and a 'control section' supervising input and output both from and to the separate items of equipment that constitute the system as a whole.

The microprocessor forms an integral part of the processing section and in it are stored the programmes developed for each type of work. A high-performance 32-bit device is used, and LSI ('large scale integration') developed specifically for image processing also is installed; together these facilitate fast operation. The Figure does not show the transmitting section but when data and signals are to be sent over long distances to the signal processing part the techniques employed for transmission must be of high standard.

In essence, a satisfactory system should be suitable for the purpose and as compact as possible. In order to achieve this, the requirements must be stated clearly at the outset, and the design made consistent with them. The equipment ('hardware' and 'software') must be appropriate for the purpose and for the design—and particularly there should be consideration of the needs of sensing and the processing of signals, together with the relationship and correct balance between them. What functions of signal processing are to be allotted to the electronic circuit, to the LSI, the hardware (the computer), and the software? Other considerations of course include costs, and possible requirements for increased capacity in the future.

Further comments are made below with regard to the design and implementation of such systems—which may be used also to give robots a capacity for 'visual' recognition, and so control production. Several systems are used quite widely in the manufacture of plastics, including new materials in this field.

Design of System

From the point of view of the recognition system, the many different forms of plastics can be classified conveniently as in Figure 19.

For practical design, many different aspects of the work must be taken into account, as are exemplified in Figure 20.

In addition, novel concepts such as 'computer-integrated manufacturing' (CIM) may require not only integration with the recognition system of robots but also generation and processing of management data. These aspects of design, and how they might be considered in relationship to a system for recognition, are illustrated in Figure 21.

Concepts such as these should be considered and the design developed according to which aspects will be required and how best these requirements can be met.

Some Essential Factors

The central factor in realizing the concept of design is attaining *highly reliable* and *steady recognition of the subject*. For this, suitable investigation is necessary of both image sensing and lighting (which might be regarded as the most important questions in the technology for systematization). It provides the foundation for the Development Steps (1) shown in Figure 20 and general examples related to the forms of the materials are provided in Table IV.

Referring to the Table, the fluorescent light indicated is not as in domestic lighting but (to avoid unsatisfactory environmental effects) employing a transmitter of high-frequency type (that is, approximately 20 kHz): normally the

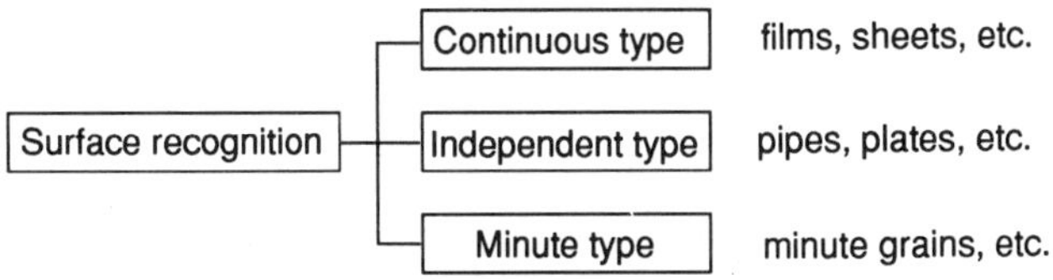

Figure 19 *Classification of products to be observed*

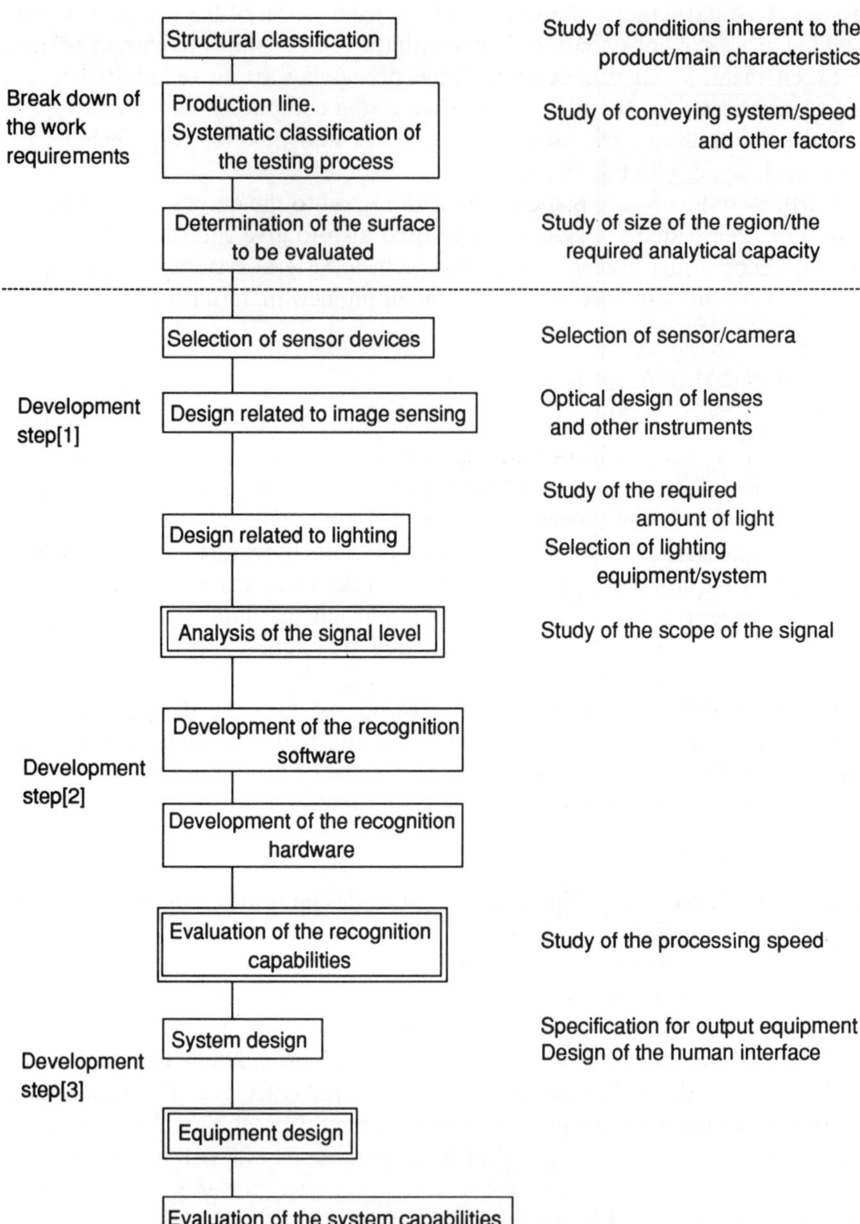

Figure 20 *Systematic diagram of stages in design*

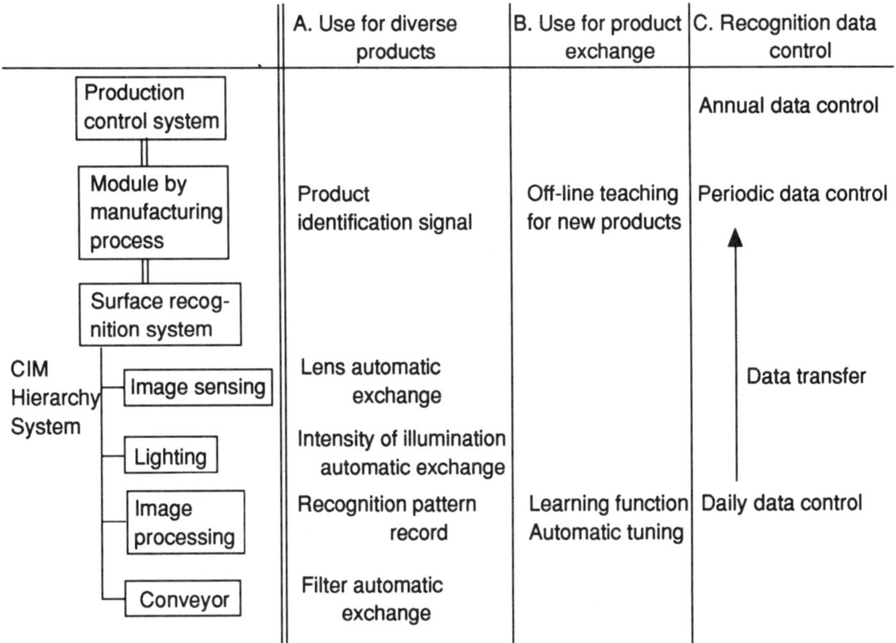

Figure 21 *Exemplification of computer-integrated manufacturing*

TABLE IV *Types of sensor, image sensing system and lighting*

Form of product	*Sensor*	*Image sensing system*	*Lighting system*
Continuous	Line sensor	Lens, filter	Special fluorescent
Independent	CCD camera	Lens, filter	Special fluorescent
Minute	CCD camera	Microscope	Halogen

inner surface of the lighting cylinder is covered with a rare earth element. In applications with continuous production its configuration would be a straight tube, but for independent production the most common would be the circular or ring type.

In the manufacture of plastics in forms such as minute particles, special illumination is needed—typically, the combination of a halogen lamp with optical fibres. Recently, equipment of high illuminating capacity using a short arc has been marketed for work of this nature.

For the image sensing unit an optical technique using macrolenses is presented. In most such systems it is important that information from the subject be projected to the CCD terminal without distortion. Figure 22 illustrates equipment for this purpose based on macrolenses. Equation (13) indicates the basis for calculating optical magnification.

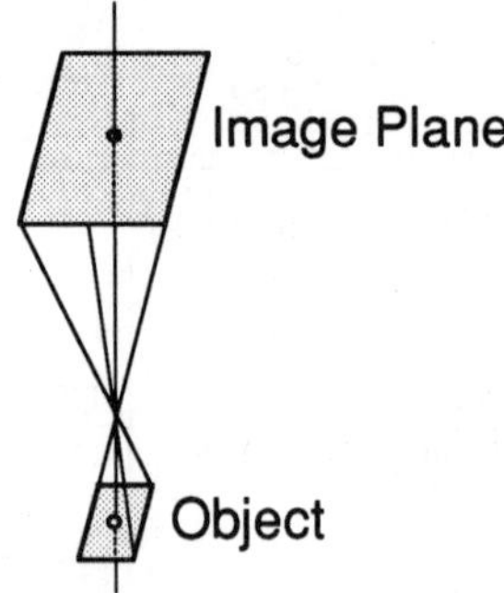

Figure 22 *The image-sensing system*

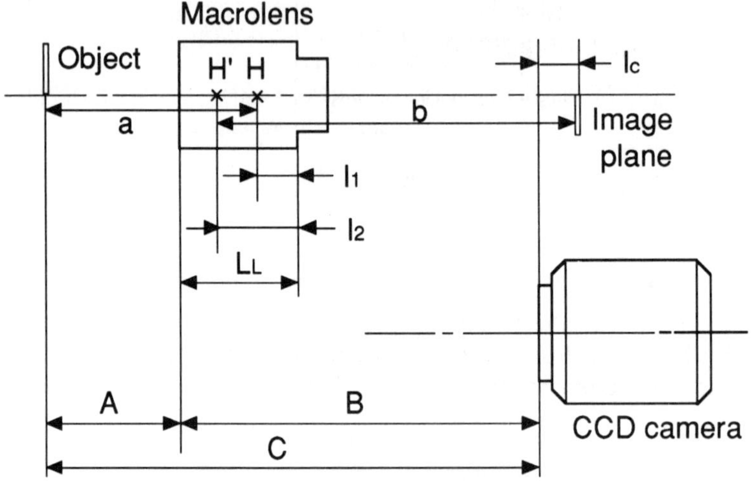

Figure 23 *Optical system with CCD camera*

$$\text{Optical Magnification } (M) = \frac{\text{(Minimum Perimeter of the Projected Image)}}{\text{(Minimum Perimeter of the Field of Vision)}} \tag{13}$$

Figure 23 is a representation of the use of composite lenses for such work.

The special distance values L_L, l_1, and l_2 relate to the focal points in composite lenses H and H'; the focal distance is f_0. The basic optical parameters in the image sensing systems A, B, and C are calculated as follows:

$$a = f_0(1 + M)/M \tag{14}$$

$$L = a(1 + M) \tag{15}$$

$$b = L - a \tag{16}$$

$$A = a - (L_L - l_1) \tag{17}$$

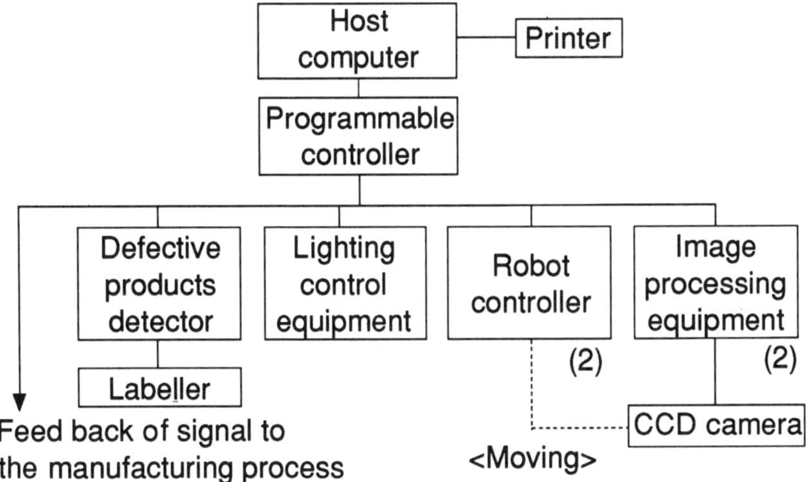

Figure 24 *Diagrammatic representation of surface inspection system*

$$B = (L_L - l_2) \times b - l_C \tag{18}$$

$$C = A + B \tag{19}$$

where l_C represents the distance from the lens surface to the CCD terminal.

Further technical advances may be anticipated in the scope and ease of operation of equipment for recognition, but these are not likely to diminish the importance for good results of satisfactory design of the sensing and lighting functions.

Recognition System

For plastics at least examining systems are required most often for goods that are in continuous production (like film or sheet), and an example intended for work of this kind will be described. It evaluates the surface roughness of a product used widely in vehicle fittings; the material undergoes an expansion process so the surface tends to be uneven—but a high degree of uniformity in fact is required.

Figure 24 provides an outline of the structure of the system concerned. By adjustment of the specimen width (1.3 m) to the speed of the conveyor, a uniaxial robot can move the camera and lighting unit in parallel, making possible the examination of the entire surface of the product. The camera and lighting are depicted in Figure 25.

The unit shown has the following features:

(i) the amount of undesirable movement in the specimen on the conveyor is kept to a minimum

(ii) the image sensing and lighting are on the most suitable optical axes

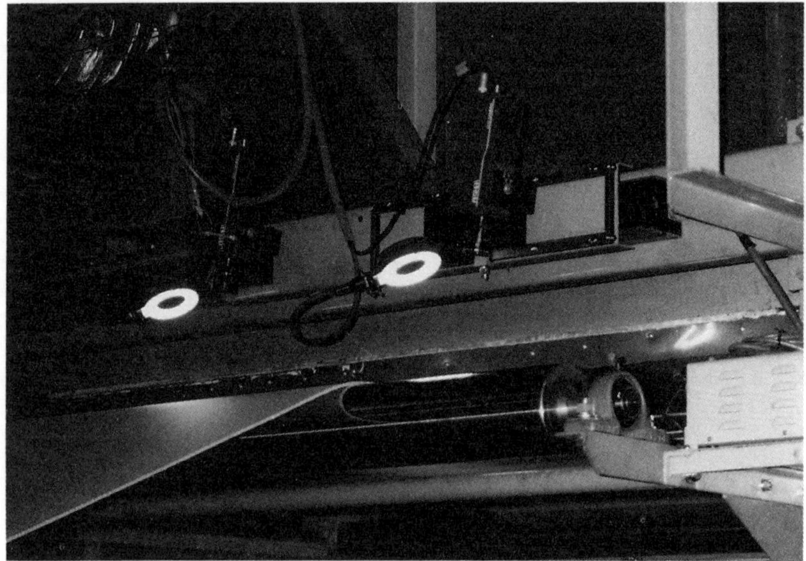

Figure 25 *Camera and lighting unit*

(iii) all equipment and movements are arranged so as to minimize the risk of foreign material falling on the product.

In some instances spark-free and explosion-proof equipment may be necessary.

Figure 26 indicates how in future robot machines may be incorporated in these systems. It is anticipated that further attention will be devoted to methods of knowledge management and recent research has been directed towards achieving automatic tuning of the level of examination, by acting upon recognition information from complex input sensors incorporated in a neural network. The assessment of surface patterns is extremely difficult but as the development of microcomputers goes on it is anticipated that applications in this field also will increase in number.

In Summary

Automatic systems for surface recognition have been demonstrated but technical problems and limitations remain, including:

(i) the performance of the microcomputer (limitation of processing speed)
(ii) performance of the image sensor (limitation of resolution)
(iii) method of manufacture (limitation of conditions of input).

Taken together, such factors mean that at present the most practical systems are those designed exclusively for particular types of work, and that equipment for general use still is some way off. This applies equally for plastics, where the differences in the engineering of methods of production and in the forms in which material is made give rise to differing needs. A further point is that

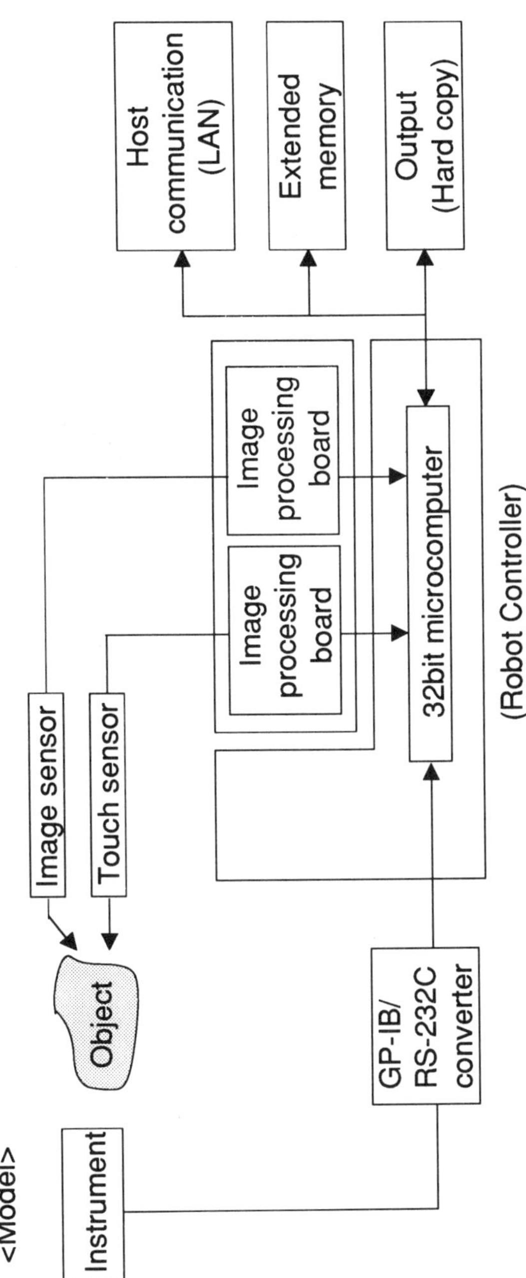

Figure 26 *A recognition system envisaged*

hitherto the emphasis has been on the inspection of products at the final stages of production, when identification of defects may well lead only to the rejection of a batch of material. It is desirable that in future the recognition system be integrated with control of quality during the process, and indeed that perhaps it might supervise the key factors in the manufacturing or finishing.

The improvements necessary for achieving these objectives have been indicated already but, briefly stated they are:

(i) superior hardware (image sensor and computer)

(ii) more sophisticated software for the recognition and processing of data

(iii) the ability to derive from surface characteristics information other than that of image alone

(iv) more rigorous comparison of derived information with recorded parameters and specifications

(v) the means to apply such comparisons in practice (that is, in the course of operations).

The technology in the field is evolving still, and success in attaining objectives such as those noted here will depend more than a little on the flexibility of the minds of research workers involved, and whether they have courage and confidence enough to try new approaches whenever innovations make them feasible.

References

1. J. W. Horton, R. v. Mazza and H. Dym, *Proc. IEEE*, 1964, **52**, 1513.
2. G. Weckler, *IEEE J. Solid-State Cirts.*, 1967, **SC-2**, 65.
3. J. Max, *IRE Trans. Informat. Theory*, 1960, **IT-6**, 7.

CHAPTER 3

Basic Finishing Techniques

W. G. SIMPSON

Pre-History

It may be deduced from numerous useful and decorative articles to be seen in centres like the British Museum or the National Archeological Museum at Athens that considerable practical knowledge of the properties and applications of materials (not just of stone, iron, and bronze) pre-dated written economic and social history. The dramatists, philosophers, and sociologists appeared it seems some time after the craftsmen and technicians: philosophy and certain other studies somewhat later than materials science.

The specialized tradespeople of those days knew the pre-requisites for successful design and fabrication in a wide range of materials, including the selection and the use of stone, the mining, smelting, and working of common and noble metals, alloying, moulding clay, painting, firing and glazing, weaving and dyeing, and the cutting and preparation of bone, timber, and a range of other natural substances like papyrus and jute.

They made in a variety of materials, applying their knowledge and skill, not only considerable works of art but also the precursors of objects familiar today—tools, weapons, clothing, furniture, household goods, games, toys, transport vehicles, impedimenta, articles of jewellery, and public monuments.

Traditional Attitudes

The surviving trade goods, the impressive ruins of engineering achievements large and small (ranging from statutary to circuses), and such records as have been found and deciphered, confirm the availability of an array of specialists—the people, slaves or free, who kept the accounts, built the aqueducts and ships, and made useful and attractive objects of all kinds.

No doubt, when such practitioners were being chosen, there would be a preference for honest professionals who knew how to draw a straight line, to calculate forces, or simply to add up. Occasionally, mistakes would be made, and a man or woman engaged in good faith, even with the best of references, would prove to be incompetent, over-ambitious, or a mere charlatan. The good servant, on the other hand, would be a master of his craft and at the same time fully aware of his limitations. He might be open-minded, or he might be inclined to think that no better way of producing his 'lines' could be envisaged (believing perhaps that beating by hand was the only way to obtain fine gold leaf; that glass bottles could be made only by human breath blowing the molten stock; that there was no substitute for woad, and so forth).

Such attitudes of mind are not hard to understand, and even to justify—business is running smoothly, complaints are few, and new ideas only create problems. Unfortunately, however, some people always are inclined to be curious, even eager to see what happens when a material or condition is changed; such spirits upset the apple cart—aside from the commercial advantage to be gained by making serviceable objects more attractive or cheaper than hitherto there is an intrinsic appeal in applying ingenuity and seeking innovation. Probably the wisest course for the business person would be to be always sceptical but open-minded. With innovators always somewhere about the traditionalist man or woman who sought no changes was likely to find that they took place in any event.

Familiarity Breeds Acceptance

In more recent times, through parallel developments in chemical science and engineering, natural materials that once formed the basis for colouring, dyeing, and flavouring, for foodstuffs, perfumery, pharmaceuticals, and textiles, gradually have been superseded by substances similar or even the same from the point of view of chemical nature, but which have little direct connection with the *flora* and *fauna* of the planet. (The substitutes are facsimiles, not extracts.)

Oddly enough, there has been willing, even enthusiastic, acceptance on the part of the public all over the world of synthetic pharmaceuticals and foodstuffs, such as substitutes for herbal remedies, butter, cream, milk, *etc*. Some synthetics are advertised vaguely as if somehow they were linked with Nature (with pictures of flowers, farm gates perhaps, or cows), but many also are stated quite honestly to be of limited nutritional value—'low calorie', *etc*.) Soap still is used in a large way, but synthetic washing products and toiletries also have found widespread approval.

This easy acceptance is rather different from popular attitudes towards synthetic plastics, notwithstanding the historic use of horn and ivory, their evident versatility and utility. Many plastics are not very different in 'natural' appearance from bone or hides (and of course they are 'greener' than materials of animal origin) but a part of the trouble perhaps may be that unlike drugs or food some plastics at least are a little too unfamiliar and distinct from the substances they replace—like metals and wood. They may be seen still as complicated and unknown, sometimes even as threatening to the environment and to life.

The 'alien' aspect may be illustrated by polystyrene, which is glossy and transparent like glass (though a little stronger) and makes when dropped on a hard surface a sound like a bit of tinplate. Nothing else is quite like it.

One response was to try to make plastics more like familiar materials but unless this was done well it too could have a negative effect. (There have been occasions when the differences were all too obvious.) At times the industry itself might have been a little unwise; some years ago (before the Organization of Petroleum-Exporting Countries became effective and the prices of oil and its derivatives were increased) there was to be found a certain aggressive attitude, a desire to develop new uses and in so doing, if need be, to displace other substances and methods. The plastics engineers and technicians were the innovators; positions sometimes were taken that could be thought 'technological arrogance'. Sometimes the consequence would be success, but on other occasions disappointment—and the latter in particular was remembered.

The industry still is concerned to achieve growth, but its former advantage of somewhat lower basic costs has been eroded, and it has had the chance to learn from experience that greater volume is not necessarily more profitable: the materials and techniques available continue to offer great opportunities, and it is in the general interest both now and for the future that lessons be learned and use be made of them in ways that are likely to earn more general approval.

Early Days

I Cellulosics

The industrial change and expansion of the nineteenth century had many strands and among them attention was given to man-made replacements for resinous compositions and horn. Alexander Parkes, a prolific inventor and manufacturer, was involved closely with the search for commercial materials; he showed articles of 'Parkesine' (a cellulosic) at the Universal Exhibition in London in 1862. Further investigations and development led eventually in Britain, Germany, the USA, and elsewhere to the industry based on a cellulose nitrate plasticized with camphor and (somewhat later) to cellulose acetate and to other cellulose plastics (cellulose acetate butyrate, ethyl cellulose, *etc.*).

An early important commercial use for the cellulosics was to replace ivory in making billiard balls, and while today this certainly would win approval by 'green' enthusiasts it is not far-fetched to imagine at the time some habitues of the tables grumbling that the plastic balls 'just were not the same'. When perfected eventually in a commercial sense cellulose nitrate plastics were more consistent in appearance and quality than tusks, and duly replaced them too for uses such as piano keys and handles for table cutlery: in products like these they were an economical and practical substitute but in stiff collars and cuffs—another important early application (eventually, millions were made) to help keep clerks, nannies, and others looking smart throughout the working week—the rigid white shiny plastic definitely was not the same as starched linen.

Displays in the Victorian period of articles made from cellulose plastics included products much as might be expected today in more novel substances—combs, trinket boxes, buttons, manicure sets, and a variety of fancy goods. Since they were attractive and comparatively inexpensive such items were popular and many ladies' dressing tables were embellished with coloured or patterned trays, boxes, brush- and mirror-handles cut and made up from cellulose nitrate film or sheet.* (The effects available included translucent and pearlescent colours, and 'tortoise-shell': a tortoise-shell tray and mirror are included among the items shown in the Frontispiece to this book.) A representation of a grand piano (a few centimetres high), complete with liftable lid, was another familiar ornament; it was to be seen on numerous mantle-shelves (a photograph including one was published in *The Link*, **2** 1 (Spring 1992, page 2).

A typical small trinket box from that period, machined and formed from tortoise-shell sheet, has a chamfered and hinged lid and four machined legs (each about 3 cm in length) all with representations of claw feet. The interior is lined with velveteen and one interesting feature is that the tops of the sides and front were machined so that the lid closed on five pegs (each less than a millimetre in diameter) which were raised only very slightly above the rest of the surface; thus a flat closure and ease of opening were assured.

Another example was octagonal in shape, made from a double-layer of film to give an impression of greater solidity without the actual cost of thicker sheet (which in any case would have been heavier and more clumsy). This box had a comparatively short hinge (on one only of the octagonal sides) but more substantial machined feet. The problem of easy opening was solved by forming a narrow raised ridge of material about 1 mm in width along the whole length of the inside edge of the lid.

To judge from a gate-mark, a matching circular tray 1 cm in diameter was made by moulding, while a lidded box for face powder—6 cm deep and 7 cm in diameter, with matching concave lid—apparently was drawn from film with a circular rim inserted. Such examples of tray and powder box were embossed on the base with the words 'MADE IN ENGLAND' in small capital letters.

Cellulose film in translucent and opaque colours was cut and folded to make colourful vanes for countless toy windmills to amuse children at the seaside—the vanes being fixed on a central wire spindle by means of a metal eyelet embedded in them (the spindle in turn being fitted to a handle consisting usually of a short piece of dowel).

Later, the uses for transparent sheet included rules and set-squares for geometrical drawing, while pearlescent materials were highly effective for show purposes—as in drum kits for dance bands, accordions, and other musical instruments. Some pearl finishes were produced much as veneers, for embellishing small items like the handles of pen-knives.

In large part, purposes such as these were served by cutting or machining

* Planar material of thickness less than 0.010 inch, or 'ten thou.' (0.25 mm), was classified as film.

shapes required from sheets or blocks of the material, followed by assembly, cleaning and polishing. Cellulosic sheets, particularly when in darker colours, had something of the appearance and other characteristics of hides—but of course were rectangular and more uniform in composition, dimensions, and thickness.

Cellulose nitrate and cellulose acetate plastics could be made transparent, in translucent or opaque colours, and the film, sheet, or blocks given gloss or matt finishes. Rectangular blocks were produced by subjecting the material to heat and pressure, and sheets then were cut from them by planing (by 'skiving', to use the word from the leather trade).* The sheets were surfaced in 'multi-daylight' presses, between metal plates which usually had either a mirror finish (to give a gloss effect) or the negative of the matt surface required. In the pressing, the surface of the metal was replicated on the plastic.

Besides its importance in providing the surface finish, pressing was essential for reasonable flat-lying qualities in the skived sheet. It could be employed also for lamination, making thicker sheets and special effects such as two or more colours interleaved. (Many signs were made from sheets that sandwiched two or three contrasting colours—the upper layer or layers being cut away by engraving to form lettering and designs in the different colours underneath.)

Table XI, page 144, gives indications of temperatures of processing for a number of materials. With cellulose nitrate plastics in particular there was a risk of explosion if over-heating took place, so besides all normal safety precautions for press-work the operating temperature of presses was subject to strict control.

The production of cellulose nitrate lacquers used to be (and in some forms remains) an important business; the resulting films were strong, resistant to moisture, flexible, and attractive (normally glossy) in appearance. Similarly, cellulosic films were made by casting from solution, the nitrates being displaced as time went on by the less-hazardous acetates (and, for certain uses, triacetates). Surfaces of high quality often were required—as examples, for film for decorative purposes or photography—and were obtained by continuous casting on moving bands of polished metal. Non-photographic uses for such superior films included the more costly forms of packaging (as transparent or translucent boxes for fancy confectionery) and to protect the surfaces of paper or board products like maps or the sleeves of long-playing records, the film often being laminated over notes or art-work printed on the paper or board. For uses like this it was especially important not only that the film itself was uniform and free from imperfections but also that the surface of the substrate was smooth, with no dust particles and contaminants of any kind, and compatible with the laminate.

Cellulosics could be moulded and extruded, and were employed in forms such as transparent or translucent tubes for packaging, and in multi-coloured effects for objects such as the barrels of fountain pens. Transparent tubes of high quality still are produced for use in packaging today.

* More details of the manufacture of film and sheet are given in V. E. Yarsley *et al.*, 'Cellulosic Plastics' (see Bibliography).

The sheets were sold for many different purposes, an important one in terms of tonnage and value being the manufacture of spectacle frames. In this application the requirement might a plain colour, a laminate of transparent with translucent or coloured material, or special effects such as tortoise-shell.

Colours or straightforward laminates would be made by hot-pressing as in the finishing of sheet, but the production of tortoise-shells was rather more elaborate. In these, the effects resembled the apparently random patterns of dark and light which camouflage turtles of all sorts in their natural environments. As a first stage in their preparation, sheets were made in different shades and thicknesses of the predominant colour required (typically, browns or greens). They then were cut and interleaved by hand in jigs, using a variety of complex patterns, which were written down (it is interesting to consider how such methods could be modernized today, with the aid of computer programmes controlling automatic sorting and collation): the interleaved sheets were pressed again into blocks, from which the 'tortoise-shells' were skived for final surfacing.

The maker of spectacle frames would receive his requirements for colours, laminates, and effects in sheet form and would work these accordingly. If the stock proved too thick to be cut satisfactorily in one operation by a press the rectangular blanks which ultimately would be eye-pieces were made on a guillotine or by sawing. Much of the rest of the material then would be cut away with tools adapted from carpentry; where appropriate (as for fancy effects) the surface layers were removed by knives to reveal the contrasting colour or colours underneath. Since the material was thermoplastic, heat could be applied through patterns in small hand-presses to help soften it and make the bridges, or for bending ear-pieces. Hand-operated or power drills were used to prepare for the attachment of hinges, and (if required) to introduce wire re-inforcements.

Many of the workshops in the trade were small and often the employees would make components in batches, which were carried in small skips from bench to bench, rather than in continuous operations on moving assembly belts or tables.

Machined parts were given matt or polished surfaces by tumbling in appropriate media. Trade-marks and stock numbers were applied by 'hot-blocking' in small hand-operated presses, using heat and pressure to emboss and to imprint contrasting colour from film or foil.

Today, almost universally, workshop techniques such as these are regarded as being too slow and requiring too much labour—so spectacle frames are produced largely by moulding from other materials rather than in this way—but it is worth recalling the former approach since there could be occasions still for machining from sheet when producing simple items in small numbers. In all such instances, some attention should be given to the extent to which the work can be continuous in sequence (it is desirable to adopt such an arrangement if the volume and nature of the requirements allow). Machining and assembly may be arranged linearly, with operators seated at stages on each side of a moving belt, or at circular tables—the entire table-top being driven so as to rotate and so pass the pieces to the next stage. Some operators are believed to prefer working together in groups at such tables.

Even yet experts say that the best combs are saw-cut from flat uniform pieces of bone (and such combs remain on sale in more-expensive shops). Combs can be cut in the same way from plastic sheet, achieving comparable quality, but the stock must be prepared suitably and the work as a whole is more costly than the commercial method of making plastic combs in large numbers—injection moulding using materials such as nylon or polypropylene; from the moulding process, after removal of the sprues, fully finished, saleable products (teeth included) can be obtained in a few seconds.

The cellulose nitrate plastics were adaptable, easy to work, and quite durable, but had the disadvantages of being plasticized with camphor and flammability (they burnt, in fact, quite violently). At one time the smell of camphor was thought pleasant (even therapeutic) but instances have been cited of harm to children attributed to inhaling this vapour. So, while the acetates were not quite so attractive in terms of ease of fabrication it was inevitable eventually that the so-called 'safety' plastic would supersede its forerunner. Among other uses it was stitched into motor car tonneau covers as flexible glazing, and thicker transparent sheet was cut and formed into cockpit canopies for aircraft [though by that time the material preferred for this purpose was poly(methyl methacrylate)]. Later on, in their turn, sheets made from cellulose acetate butyrate and propionate took over some of the uses of earlier commercial materials.

II Phenol–Formaldehyde, Urea–Formaldehyde, and Melamine

Moulded phenolic resins were first used commercially in 1910: it is convenient here to take these resins as a group with urea–formaldehyde and melamine, since together they were complementary to the earlier thermoplastic materials and because their particular qualities helped to open new markets.

The thermosetting plastics were used in substantial quantities as counters and other pieces (like dice-boxes) for toys and games but the trade was developed far beyond the ranges of decorative and fancy goods typical of the cellulosics, particularly into the comparatively new and expanding electrical and motor businesses—as mouldings for telephones, switch covers, and other electrical fittings, distributor caps, instrument panels, and so forth. The range of colours available was more restricted (from the nature of the resins, black was quite usual—hence the once-ubiquitous black telephone), and even the best finishes obtainable could not match the high gloss of the cellulosics, but they were suitable for moulding techniques that made it possible to produce complex components rapidly and cheaply (since machining either was not necessary at all, or was needed to a much more limited extent). In compression moulding the pieces were shaped and cured under heat and pressure; transfer and (later) methods akin to injection moulding also were used.

A selection of items fabricated from thermosetting materials is shown in Figures 27 and 28.

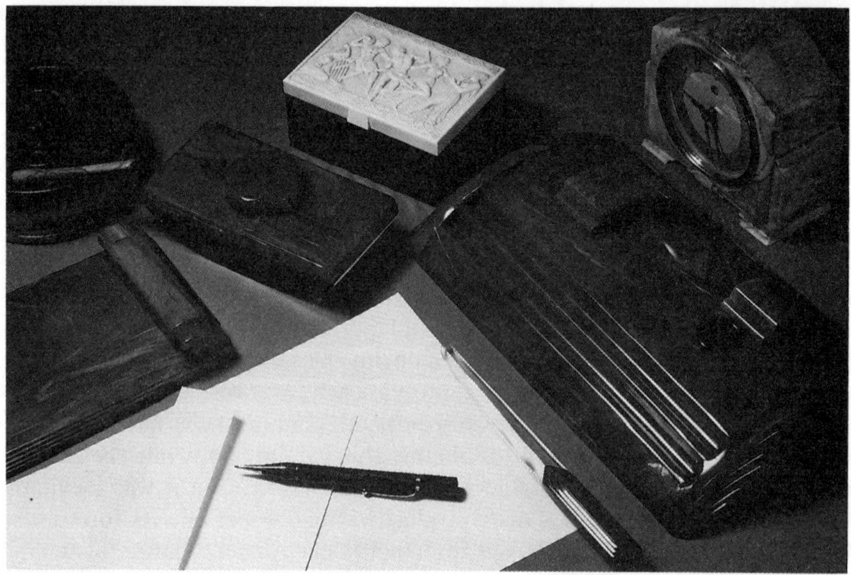

Figure 27 *Synthetic plastics imitated successfully exclusive natural materials: a Bakelite ash tray with built-in match holder and striker in 'marble'* Roanoid, *with marbleized, brass-tipped cigarette holder, 1930s.* Carvacraft *desk set in 'amber' cast phenolic, 1948–51; double inkwell, notepad holder, blotter and paper knife with clear acrylic blade.* Smiths Sectric *alarm clock with marbleized urea–formaldehyde case, 1932.* Ardath Tobacco *cigarette box in ivory and black urea–formaldehyde with lid moulded to simulate a classical relief, 1935. Propelling pencil with black and orange mottle typical of vulcanite, 1930s*
(Photo Stephen Brayne, Collection Sylvia Katz)

Figure 28 *Right: richly-mottled Bakelite wool holder, 1940s. Front: toast rack and* Bandalasta *picnic ware and salt and pepper set in urea–thiourea–formaldehyde, 1927–32. The powdered form of thiourea made it possible to create softer mottled effects than with granular urea–formaldehyde*
(Photo Stephen Brayne, Collection Sylvia Katz)

Typical moulding materials would include various additives and often were filled with cotton, wood flour, or other economical and unreactive substances; usually, the type and amounts of filler were important factors in deciding the properties of the mouldings, and these took priority over mere appearance. Since the filler affected the finish, this in turn frequently was rather a dull matt. When finish was considered unimportant it might be left essentially as that imparted by the inner surfaces of the mould (after removal of flash)—or mouldings might be polished if required by buffing or by tumbling in appropriate media.

Melamine resins found a wider range of uses, including tableware (cups, saucers, plates), and in decorative and industrial laminates. One might think that melamine tableware never was likely to rival bone china in delicacy and appeal but it was attractive enough for the canteen, kitchen, or nursery, and less fragile even than delft or unglazed earthenware. The mouldings were available in a variety of opaque and translucent colours, and for a time enjoyed considerable popularity.

Melamine decorative laminates are made by curing and pressing, most usually as flat sheets which are then shaped for fitting to benches, table-tops, fascias, and so forth. They are still, of course, an important industry. Colours, patterns, photographs, and designs carried on sheets of paper can be incorporated in the laminates, giving opportunities for both fancy and practical uses ranging from bright disco decor to institutional furnishing. 'Wood-grain' effects can be obtained by the inclusion of veneers or even of colour photographs of teak, ebony, or other valuable timbers, while further ingenious special patterns have been created.

III Casein Plastics

During the period between the first and second world wars a plastics business was developed based upon rennet casein from cow's milk, from soya beans or other pulses; the reaction with formaldehyde gave a rigid plastic material which could be produced in light colours, mottle and pearlescent effects, and was capable of being machined and polished to a high gloss.

The reaction of casein with aluminium salts in water resulted in a thermoplastic which could be extruded as rod or tube.

Quantities of casein plastics were cut from rod stock into button blanks, and finished by drilling (to form holes for the thread) and by polishing. For a time the plastic knitting needle (for home knitting) was another important use for the materials, but such needles were broken easily and were unsuitable for delicate work.

However, although the casein plastics extended the range of colours and effects available from the contemporary materials they were not distinctive enough to survive in competition with plastics based on oil, and have no commercial significance today.

Modern Times

As noted in the Preface, some forty different groups of plastic compositions now are of commercial and industrial significance; the modern era might be said to commence with the discovery in the early 'thirties of polyethylene, and later evolution of the markets for polyolefins, for vinyl compositions, polystyrene, and its derivatives. These materials (which sometimes were called 'commodity plastics') offered wide ranges of properties and possibilities for exploitation—and as time went on more specialized and exacting requirements were met by newer materials, including some proprietary products.

As a general observation, plastics which offered (say) improved resistance to heat while in service were also more difficult to fabricate successfully in thermal processes (additional heating was needed to bring them to the processing temperatures, and consequently longer periods of time in which to cool; if they were not handled appropriately, there could be distortion of the products). In thermal processing, even of similar materials, there could be on the one hand a wide range of temperature at which satisfactory results were possible, or, in some instances, a rather limited range of temperature—depending upon the distribution of molecular weight in the polymer (and particularly the proportion of material of comparatively low molecular weight) and also upon the nature and amounts of additives in the composition as a whole.

On mechanical rather than thermal properties, plastics which were harder were also more difficult to cut and to fabricate satisfactorily. Chemically, plastics which resisted organic solvents were also more difficult to print safely and with good results, and so forth. Often, in practice, intractable properties were modified to the required degree by the inclusion of small amounts of other monomers (vinyl acetate in vinyl chloride; α-olefins in polypropylene, *etc.*)—just as an unwanted property like brittleness in polystyrene was ameliorated by adding to it styrene–butadiene.

The user and, more particularly, the fabricator of plastics is offered an array of materials that in fact can be bewildering, even without the competing claims of manufacturers offering different proprietary grades of essentially similar things. Another factor is the increasing pressure upon fabricators to take up 're-processed' plastics (often re-worked scrap, or waste), which in reality may be blends with properties rather different from those of the originals. There can be some difficulty in giving advice about satisfactory methods of fabrication when the substances to be worked can vary so widely. Fortunately, most manufacturers are conscious of the dangers and there is a well-established tradition of workshop practice and trials before attempting operations on a full scale.

For the reasons indicated it is not possible here to cover each material in detail and before going on to make some general observations about basic techniques for finishing one should emphasize that trials in advance with the actual substances to be worked usually are necessary. The testing should cover every aspect relevant to what is required of a component, and obviously it is helpful in this regard if reference can be made to a written Standard or specification. Tests may be of chemical behaviour, electrical, mechanical, physical, and thermal properties, or of some other aspect like performance when exposed to fire or to strong sunlight. The tests should be on samples that are representative of the production, should be appropriate for the parameters required, and be carried out under proper supervision. Reference samples and written records should be kept. The testing normally would be carried out at 'room temperature' (23 °C) under ambient conditions and for many parameters different results would be obtained in different circumstances; because of this, for some products, testing under a range of relevant conditions is essential.

Safety is important in testing as well as manufacture, and in physical trials especially every care should be taken to ensure that equipment is guarded (and that pieces shattering will be captured and held); the wearing of goggles or other protective clothing may be appropriate; dust, swarf, fumes, and decomposition products should be extracted effectively; and static electricity should be discharged without danger of fire or explosion. A trial should be carried out under clean conditions, with the materials used free from contaminants of all kinds. (Contaminants may include water, general dust, hair, fibre, scraps of paper, metals, and other substances; in testing just as in full-scale production effective means must be found of ensuring their absence always.)

There is still within the industry the appealing feature of technology in progress and willingness when new products are under development to consider novel approaches. In routine production so far as possible, simplicity is to be preferred and while on the one hand there can be a danger that complex requirements will bring about a similar response the fewer the materials and stages that are represented in a finished object the easier it will be to maintain production at high standards and, when the time comes, to reclaim materials from scrap. In other words, there is scope for the engineer or technician to think ahead and to design in advance the system for fabrication that gives results required in the most convenient way.

Many of the terms for finishing techniques used by the industry are listed and defined in international Standards; they are familiar and there is no need to give definitions again here. However, as an indication it does seem worthwhile to note the following groups of examples of terminology, which are selected from the current edition if ISO 472:

(i) *Film and Sheet*
calendering; casting; coating (process); cold and deep drawing; embossing; film blowing, casting and extrusion; laminate (noun); plate mark; plug-assist vacuum and pressure thermoforming; roll coating; shrink packaging, shrink wrapping; slip thermoforming; slitting; stretch thermoforming; thermoforming and vacuum thermoforming

(ii) *Extrusion*
extrusion coating; pultrusion

(iii) *Moulding*
centrifugal casting; cold, compression, and contact moulding; ejection; embedding; injection blow and injection moulding; insert; insert pin; mould mark; organosol and plastisol fusion; reaction injection moulding; rotary moulding; rotational casting; sink or shrink mark; slush casting or moulding; transfer moulding; weld line, knit line or weld mark

(iv) *Fabrication and Machining*
blast finishing; deflashing; die cutting; dip coating; drawing; encapsulation; filament winding; flame spray coating; fluidized bed coating;

forming; friction or spin welding; heat sealing; high-frequency, dielectric or RF welding; hot-gas welding; hot stamping; impregnation; impulse sealing or thermal impulse sealing; metallized plastic; pressure welding; seam welding; slip; solvent bonding or welding; solvent polishing; spot welding; tumble or barrel polishing; ultrasonic welding.

Many of these expressions are descriptive and self-explanatory. In addition to terminology developed for and applied specifically to plastics, numerous more general expressions in relation to fabrication and machining also are used throughout the industry.

Reference has been made already to the working of plastics by means of hand tools derived from carpentry; suitable materials can be cut by knives and saws, drilled, shaped on lathes, cut and cropped by guillotine, and so on. Tables of properties summarizing the mechanical characteristics of resins and intermediate products (like blocks, sheet, or tube) are available from manufacturers and are published in standard references. It is sufficient to write here that the energy available should be appropriate to the task (in relation to the type of material, the thickness, and the type of cut required), the tools should be suitable, and the speeds of cutting must be adjusted to allow dissipation of heat generated (if this is not done there will be a tendency for tools to bind and for workpieces to be deformed). Dust and swarf should be removed at frequent intervals, and preferably by some continuous means such as a vacuum system or washing (the removal of waste is important from the point of view of obtaining satisfactory finish and also to safeguard health; in Europe the limits for presence in atmosphere of dust, swarf, mist, and vapour from decomposition products or coolants are all prescribed by law). If possible, systems for removal of waste should be designed so that the scrap can be re-claimed and re-cycled.

The feature of thermoplastics—that they become harder and more brittle when cooled and soften and more flexible when heated—should be kept in mind. Such changes may be particularly noticeable when rigid materials are being machined: the behaviour of apparently similar materials may be quite different on a cold day in winter than in hot weather in July. Usually it is undesirable to bring material straight from store and to attempt machining immediately (it should be allowed to warm to room temperature throughout) but on occasion the alteration of behaviour with temperature can be useful—with flexible products being machined more uniformly and successfully when cold, and rigid perhaps giving better results if first heated slightly.

Unlike a fibrous material such as wood, most plastics respond more satisfactorily to cutting by blade rather than by saw teeth; if it is necessary to use a saw (for example, when cutting thick sheet or pipe) the best results are obtained by avoiding the extremes of fine-tooth systems (as in hack-saws) and coarse (as in a rip-saw); the former will tend to bind and split the material and coarse teeth give jagged edges.

All knife systems, including drilling and engraving, should be set up in such a way as to slice rather than to make an abrupt cut in a single action. Again, the speed at which the work is done should be adjusted according to the type of material—especially to its thermal properties and ductility, or lack of it. Suitable equipment and systems can be run at high speed.

The manufacturers of drills and other cutting tools have developed designs intended for use with plastics, and details of these, with recommendations about speeds, are available from the suppliers.

When bags or other components are being made from flexible films or expanded materials (rigid like polystyrene, or flexible) it often is convenient to make cuts by means of a simple hot-wire device. Proprietary machines based on a reciprocating action are available for cutting flexible poly(vinyl chloride) films in quantity around patterns or templates.

In transparent workpieces the patterns of strain can be checked visually with the help of a viewer using polarized light. Systems are available for identifying and depicting strains in opaque articles but they can be expensive and are not so convenient to use. Another difficulty is that strain may not be distributed in the same way in different articles in the same series—so its prediction may not be possible. If articles are worked in which there is the possibility of strain at random (and, hence, of local brittleness and unpredictability of behaviour) it is of especial importance to ensure that equipment is guarded, preferably enclosed, so that pieces or fragments breaking off will be retained. At times, particularly with rigid materials, quite violent forces can be released, and if there is any chance of some freak escape of material all the operators should be equipped with, and wear, suitable safety goggles, helmets, and other adequate shielding.

Sometimes, when cutting or machining, water may be used as a coolant, but this should not be done with plastics that have a propensity to absorb moisture (and, as a consequence, might distort to a greater or lesser degree), or if there is risk of damage through the corrosion of any parts, with the formation and deposition of rust.

In recent years there has been considerable interest in the development of cutting systems using lasers, which can be applied for drilling, making blanks or patterns, reaming, and so forth. The equipment can be controlled precisely and employed for a variety of types of work; its effect is to vaporize the plastic—so conventional systems for cooling and removing swarf are not necessary. However, the difficulty has been noted that in thicker material it is hard to make cuts or holes of uniform cross-section (they taper inwards from the top surface).

Presses, Press Brakes, and Guillotines

Engineers have developed for plastics an impressive range of special processing machinery—for extrusion, forming, injection moulding, and so forth—but perhaps it is just worth remembering that for certain types of work (especially when various runs of comparatively simple shapes are required) press techniques akin to those for sheet metal can be appropriate. Examples include:

bending — the cold-forming of film or sheet

blanking—cutting patterns from film or sheet by means of hardened-steel knives (as examples, to make gaskets or washers)

clipping (also 'shaving' and 'trimming')—cutting partly finished components to remove surplus
curling—shaping rounded corners or ends
cutting or 'shearing'—make from flat stock rectangles, slats, or other straight angular shapes
drawing—to make shallow boxes or trays
flattening—to encourage flat-lying (and perhaps to impart surface effects)
indenting—making a rectangular impression in flat stock
joining—to form simple folded joins
piercing—cutting circular shapes (in a similar technique, eyelets may be located in the workpieces, for joining, suspending, or fixing).
Figures 29 to 36 illustrate some of these.

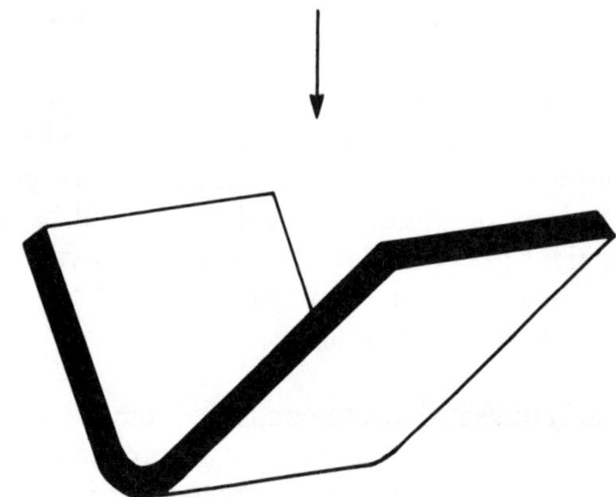

Figure 29 *Bending*

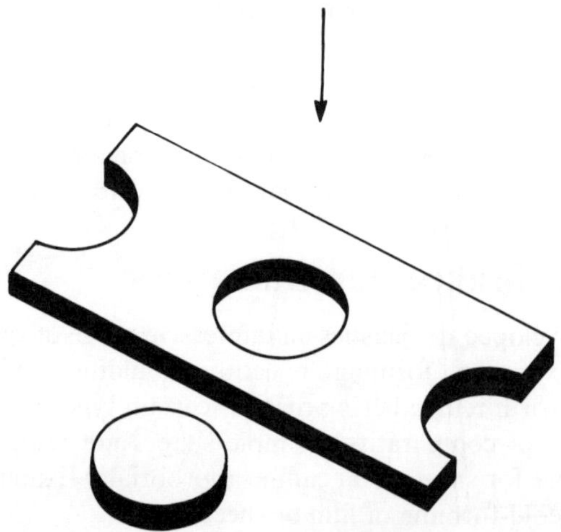

Figure 30 *Blanking*

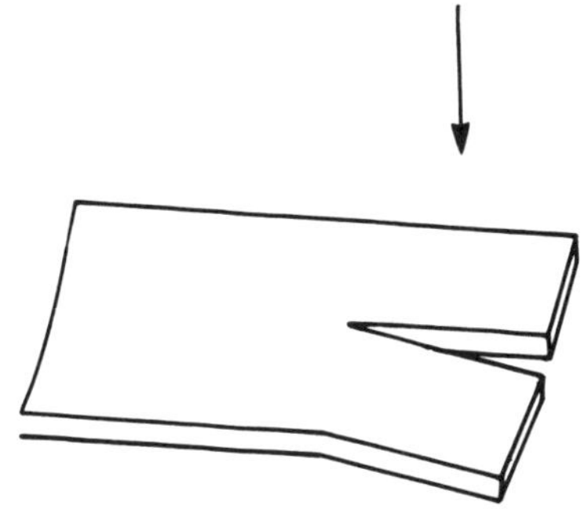

Figure 31 *Clipping*

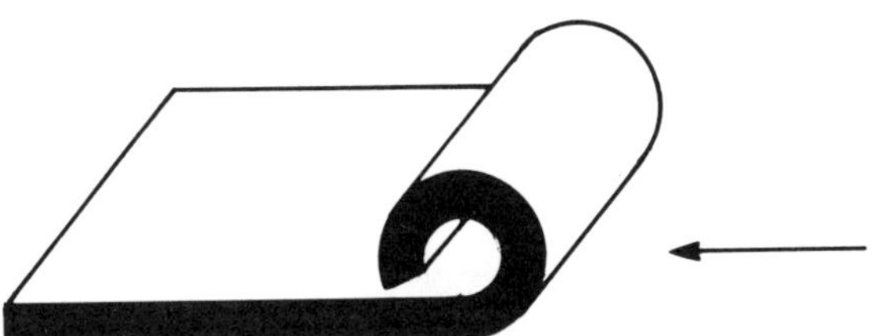

Figure 32 *Curling*

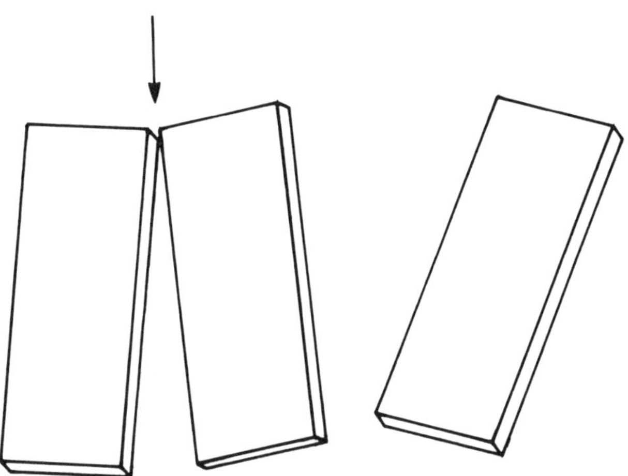

Figure 33 *Cutting or shearing*

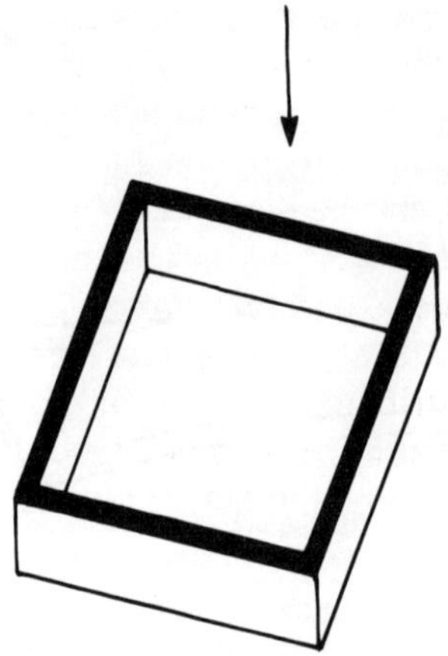

Figure 34 *Drawing*

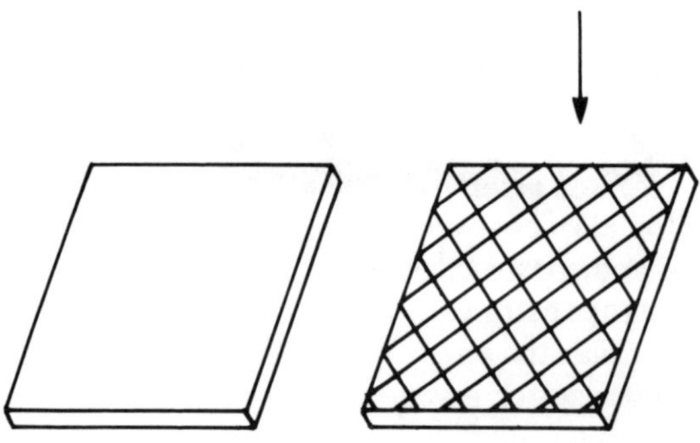

Figure 35 *Flattening*

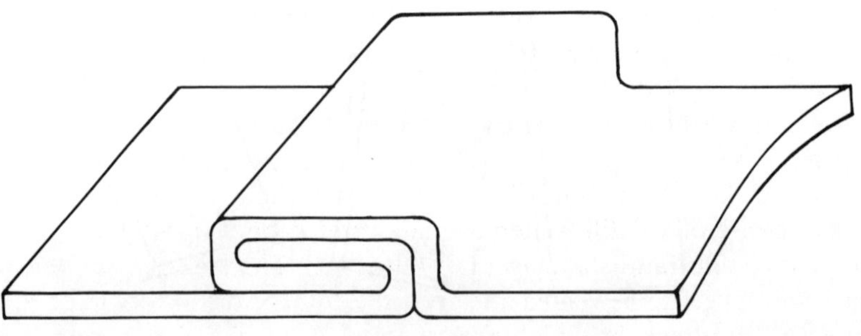

Figure 36 *Joining*

Of course in general the mechanical properties of plastics film and sheet will be rather different from those of metal foil and strip, whether ferrous or non-ferrous, and equipment and conditions may have to be adapted to allow for this. Sometimes cold techniques are modified to take advantage of the thermal properties of plastics but in their simplest forms they are limited to the compounds with which cold working is possible—that is, those of suitable hardness, ductility, and finish—and that can be fed in the forms of sheet, roll, or strip. So as to avoid frequent interruptions of the cycle the stock must be uniform and any strain distributed in a regular manner. Fast processes requiring the application of pressure are not suitable for hard, brittle materials, nor for any produced in such a way as to give concentrations of strain, perhaps in areas that differ with circumstances.

Subject to this, the main advantage is that the heating of the materials is not needed and costs associated with a thermal cycle do not arise; without the time taken by heating and cooling it may be that components can be finished more quickly. Such advantages should be kept in mind and items to be made simplified at an early stage whenever possible so that the approach can be employed. Continuous operations can be designed in which (as an instance) film is slit to width then blanked in the next stage to make gaskets or sleeves. Subject to appropriate safeguards, processes like this may be controlled conveniently by computer programmes and robots.

Dies and Tools

Dies and tools for plastics—whether for pressing, moulding, or extrusion—must be designed and constructed for the purpose. Simple tools not involving application of heat (like cutting blanks or patterns) may be made from ordinary strip steel of appropriate thickness. When tools are to cut shapes from flat stock it is important to design for minimum amounts of trimmings and waste. For short runs of uncomplicated items, applying little more than atmospheric pressure and low temperatures (as in the thermoforming of film or sheet), tools can be made from wood and soft metals requiring little more than basic good workshop practice. However, for high-pressure work and long runs substantial tooling is essential.

Usually it is necessary to know the precise type of material to be employed and its characteristic behaviour under the conditions in the machine concerned. The sheet or compound must be free to flow within suitable limits in response to the forces involved; the rates of flow and cycle times at the relevant temperatures and pressures must be calculated in advance and accommodated in the design.

Even without heating it is likely that in the course of each pass there will be some release of volatile material—like water vapour, plasticizer, extender, or processing aids (anti-static agent or lubricant). There must be opportunity for such substances to be vented and carried away; the die or tool must be capable of resisting attack by volatiles (suitably plated if they are corrosive), and designed to prevent an accumulation of released materials. Special hardened steels may be necessary, and several processes are offered commercially in order to give protective coatings to the metal. When in use, tools should be inspected regularly to ensure that they are not being affected unduly by the work. Damaged tools, and those beyond an acceptable degree of wear, must be repaired or replaced.

Not only should the designer allow for expansion of the material at the start of the cycle, the tool must be of the appropriate size to allow for the considerable degree of shrinkage that can take place on cooling. It should be as fool-proof as possible: there should be no need for constant adjustment as the work proceeds. Precise pieces or volumes of material should be admitted for each pass, and stops or jig movements should be automatic and effective throughout the life of the set-up.

An important advantage of the plastic is that in thermal processing the finish on the surface of the tool—whether it be matt, leather grain, jazzy pattern, or what you will—can be transferred directly to the workpieces. One proviso when a design is applied in this way is that the tool should not wear quickly (which will result in loss of definition of the decoration) or, worse, that it be scratched or damaged so that the unwanted mark is carried over to the products.

Static electric charges and the comparative lightness in weight of small items can combine to make plastics less easy then other materials to separate from the die or tool when working is complete. Many plastics have in their formulation 'anti-static', 'anti-blocking', slip agents and lubricants but it may be necessary also to use mould-release films or sprayable release agents; for regular fast cycles, mechanical release devices such as a ram, suction, and air blast should be fitted.

With fast working it is important to have provision for swift (preferably automatic) shut-down when tools bind or for any reason workpieces come out of alignment. Once again, safety shields and grilles should be present, and designed to stop the escape even of small pieces (not just a component in entirety).

Removing Flash, Buffing, and Polishing

Often, equipment for joining, moulding, or welding plastics is designed so that, at the end of each cycle, the feed system or separate pieces are cut away and an excess of material does not present a problem. In other processes, surplus may be removed and returned automatically for re-cycling. Otherwise, the removal of flash and finishing of items might be carried out in a simple hand operation or by mechanical stages of increasing complexity as are needed.

Flash and swarf from machining are attracted by static electricity to the extent that some parts can appear hairy through extraneous material adhering. Detritus can be taken off by wiping but action of this nature will add to the static charge and hence to the drawing power of the surface; in consequence, unless the excess material is transported away from the vicinity it will be attracted back to the component. Difficulties of this nature can arise in all finishing involving abrasion with coarse or fine glass, sand or emery cloths, with filing, trimming lathes, and so on. It is necessary to remove dust and swarf for the sake of health and safety and also to ensure cleanliness in the items being produced. Systems for humidfying an area or for moistening surfaces with water (so that static charges are dissipated) may be of help at times but for obvious reasons are of limited application.

With filled, re-inforced, and composite products the separation of unwanted material can present special problems, particularly if fibres or fibrous substances are incorporated: there may be a tendency for the fibres to hold unwanted portions, while attempts at cutting or smoothing with abrasives can take away resin, filler, and re-inforcement selectively. Because of this it is difficult with such products to obtain finish of high quality unless a suitable film or some other coating is applied later.

Individual items may be buffed and polished by means of appropriate cloths or mops fitted to rotating heads; polishing in bulk usually is by tumbling in barrels with agent or agents selected from the range of proprietary compounds that is available for the work. Tumbling necessitates a subsequent stage in which the pieces are separated from the unwanted flash and polishing medium, and sometimes this can be time-consuming.

Thermoforming

The thermoforming of film and sheet has been covered extensively in journals and other publications and it is not necessary here to go into great detail. The process has numerous variations. In essence, a thermoplastic film or sheet first is clamped in such a way that it can be heated evenly, and when softened it is by various means shaped to fit a mould or moulds. The heated material may be blown to form a bubble, then drawn downwards by vacuum, or it may be shaped by plugs before being brought—by air or gas pressure, or by suction—into contact with the mould proper. Film may be extruded and formed into packs which then are filled and sealed in a continuous operation. In such instances, the film used may be printed before forming and the pattern of printing is designed to allow for this and so that the final result will not appear distorted. Other products made in similar manner range from disposable boxes and jars to large rigid liners for domestic appliances like freezers, advertising signs, display fascias, drip-trays, frames, and wells for public seating. The technique may be used to produce raised or indented lettering, guides for people with impaired vision, and other designs. Particularly interesting applications include the production of raised land-masses on half-globes for toys and teaching aids, and impression of stars and planets for models of celestial spheres.

The method is versatile and (since only simple tools and low pressures are applied) inexpensive enough for short runs. Simple formings may be produced at reasonable speeds virtually ready for sale (excess film or sheet being cut away by blanking) while more complex shapes usually need further machining to remove excess and in preparation for fitting.

The films and sheets may be transparent, translucent, or coloured and the surface finish obtained generally is comparable with that of the original—though affected by the heating and cooling, by the surface finish on the mould, and the finish on any other jigs or formers used. Material for thermoforming must be stable at the temperature applied, free from 'pin-holes' and (in contrast with shrink-wrapping) without strain or excessive orientation. (In shrink-wrapping the softening effect of heat allows a stretched film to return to its earlier unstretched state.) Rigid self-supporting shapes may be made by thermoforming but it can be applied also to softer materials like plasticized PVC. It can be adapted to fit seamless plastic linings to vessels made from other things or durable plastic surfaces to pressings from fibrous or metallic sheet—as examples, covers and housings for machinery. In such work, the article to be lined or covered serves also as the mould: it can be coloured by the plastic and given leather-like, wood-grain, marble, or other fancy finishes.

Thermoforming also is used on a large scale for the 'bubble-packs' seen often in retailing a variety of ironmongery and similar goods—in which formed transparent or transluscent plastic bubbles display an item and seal it to a backing board. In the analogous 'skin-packing' process a transparent film is formed immediately over the goods, they themselves being the mould. (Shrink-wrapping, on the other hand, has no forming stage as such; it involves the shrinking of a sleeve of film over the items, which often are bottles or cans, in a heated tunnel through which the packs are moved on a conveyor.)

Sewing

Articles may be assembled from plastics by adhesive bonding, heat sealing and welding (which are reviewed in subsequent chapters), and by a variety of physical methods of fixing including sewing with filament, thread, or wire.

Physical fixing is applied most usually to film or sheet materials but in special cases can be employed with formings or mouldings. The results generally, though no doubt adequate for the purpose, are not so strong as are obtainable by other means; usually there is a reason other than strength for selecting a physical fixing—such as for decorative effect, or when a variety of different materials is to be joined and a more conventional approach is not possible. Stitches may be used to add coloured fabric, piping, or emblems at the seam—or, indeed, for no purpose other than decoration (say, by a thread of contrasting colour or of silken sheen).

The stitches or eyelets should not be so close together that there is risk that the pieces might tear. For similar reasons, tension and strain on the join should be avoided. The fixing (clips, eyelets, thread, *etc.*) should be compatible with the articles to be joined, not harder or more abrasive. Film, non-woven, and woven fabrics all may be finished in this way. With film and non-wovens the stitches should be loose and separated fairly widely; thread should be of about the same diameter as the thickness of film or fabric. When selecting thread or other fixings for woven fabrics the gauges should be compatible with the warp and weft (not too thick, and not too fine).

When sewing by machine, low speeds and limited tension are recommended.

For some forms of protective clothing sewing may be used as primary stage of assembly before seams are welded.

Matching Other Substances

When something has been made from a familiar substance that is to be replaced by plastic it might be best to change entirely not only the method of manufacture but the object as well—to give something entirely new and appropriate in the different material. Sometimes, however, it will be essential to make the plastic items as a replica of its predecessor: as an example, this can be so when making balls for play in garden games and competitive sports. It can be comparatively easy to make one of the same size and general appearance as the original, but rather more difficult to reproduce precisely the same features of performance. For some games (as when replacing carved wood), the plastic ball should be quite inert; in others (when replacing leather or rubber) it must have the correct amount of bounce.

Therefore, when preparations are made to follow former materials, it is always necessary to examine the product in detail and to analyse all relevant aspects. To return to the sporting example—the new ball must mimic its predecessor and an important aspect of doing so is to generate the familiar sound: if, when struck, the original ball rang bell-like or sounded 'dead', so also must the plastic one. With a view to achieving this, attention should be paid to:

- (i) the presence, type and amount of 'filler' in the plastic composition
- (ii) the internal structure (whether there are voids in the material and, if so, their size, degree of uniformity, and distribution)
- (iii) the wall-thickness (whether it is of uniform thickness and density throughout)
- (iv) the integrity of the outer surface (whether it is smooth and coherent, or pitted and scratched).

The surface finish is important for resonance and any imperfections from machining or moulding (including sprue marks), should be avoided; if necessary, they should be removed in a suitable finishing process (perhaps in a lathe fitted with cup holders of suitable size and diameter). A careful visual inspection of representative samples taken after the finishing may be advisable.

CHAPTER 4

Calendered Thermoplastics

R. A. FAIRBAIRN

Brief Review of Development

Calenders were used to improve the surfaces of paper and fabrics long before they were employed to process rubber and then thermoplastics. In 1836 E. M. Chaffee patented a four-roll calender: Chaffee worked with Charles Goodyear (who developed the process of vulcanizing rubber) and the intention was to utilize the calender to produce a sheet of rubber laminated to a fabric base. The rubber industry developed rapidly and with it the design of calenders—so that just before the 1939–45 war when poly(vinyl chloride) (PVC) was introduced machinery capable of processing it into film already existed. There was development in both Germany and the United States and probably the first successful calendering of PVC was in 1935 in Germany, where in the previous year the Hermann Berstorff Company of Hanover designed the first calender specifically to process this plastic. Because of the war Germany was starved of natural rubber and industrialists there were under great pressure to develop materials that could be used as replacements for it; PVC was regarded as suitable, with the result that by 1945 the industry had become quite sophisticated in its processing.

Improvements in design continued and an important change in the 'sixties was the supply of independent drives for each roll in a calender; formerly, gearing allowed only a fixed increase in speed through the stack. The fixed increase normally was achieved by having a difference of one tooth between the gears. Independent drives permitted variable speed ratios to be used throughout the stack.

Various arrangements of rolls (or 'configurations') have been used (see Figure 37) with a view mainly to avoiding the effect of moving one roll to adjust a gap between rolls (and so affecting the next gap), and also to reduce to a

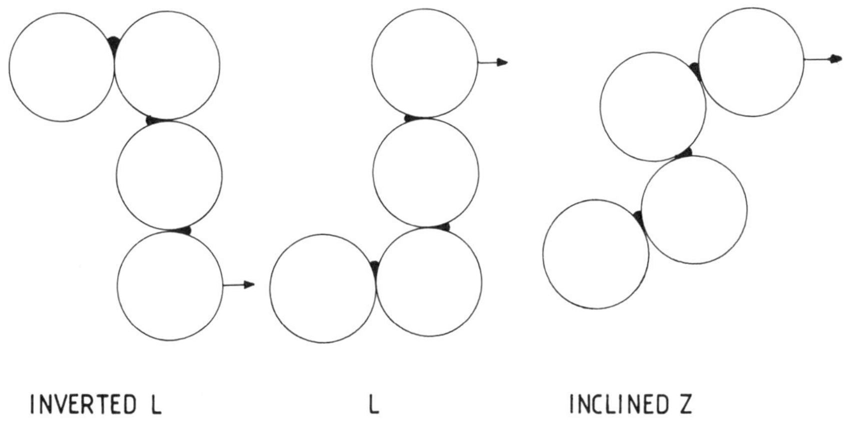

Figure 37 *Configurations for calender rolls*

minimum the transmission of the forces generated in one nip to the next. By 1938 the basic theory concerning the generation of separating forces in the calender nips had been published by Ardichvili,[1] and this work was carried further to give more accurate results, in a number of instances by the manufacturers of calenders.

It was believed that the 'Z' configuration of calender rolls achieved the isolation of one nip from the effect of changes occurring elsewhere in the stack, but this was at the cost of loss of access and reduction in the ability (because of the short web length) to prevent heat being carried from one nip to the next. It was argued that the short length between nips was an advantage in that it reduced the dwell time in the machine and so diminished the need for heat stabilizer. The Z calender was used very extensively in the rubber industry, and for a while was popular in the production of PVC—the Farrel Company being the best-known manufacturer, though others also made calenders in this form. Many such calenders are in use still but the Z configuration fell from favour in the 'sixties and it is doubtful whether any more will be built for PVC.

Poly(vinyl chloride) films are produced in two main forms—unplasticized and plasticized—and over the years different machines have been manufactured to handle the two types. When calendering unplasticized PVC there is a tendency for small particles, usually referred to as 'crumbs', to fall away from the edges of the film and from the feed nip. Such crumbs then could fall on to the finished film, where they would stick and form defects. To avoid this, producers of unplasticized film usually prefer an 'L' configuration in which the product travels up the stack and surface contamination of this kind is prevented. With plasticized PVC the problem of crumbs does not occur to any great extent and, as it is an advantage to have good access to the part of the calender where the finished film is made, an inverted L configuration is the most popular.

A further development is the Calendrette—a three-roll machine fed by an extrusion system that spreads the material evenly across the first roll (in this way trying to replace the even feed achieved between the first and second roll of the conventional four-roll calender). Such equipment was developed originally from melt-coating machines and usually was of comparatively small dimensions but in their evolution there has been a move towards the calender and some large Calendrettes have been built which have all the sophistication of the latest calenders. They are suited particularly for the production of unplasticized PVC as used in the vacuum forming industry, especially where thermoforming is on a scale large enough to justify making the required film or sheet at the manufacturing unit, and thus eliminating the calendering company as a supplier. For producing plasticized film there is a difficulty in matching the quality of finish obtained from a conventional calender, flow marks being a problem.

In the early days a calender train for plastics was a very basic affair, consisting of a line of small rollers to transport the film to an embossing nip, and a set of three or four larger cooled rollers to bring it to a crude wind-up. Probably there would be only three motors to drive the whole train and very little adjustment of speed ratios would be possible—a far cry from modern equipment, in which there may be twenty motors for a train.

Today PVC still is the commonest thermoplastic to be calendered but it is by no means the only one. Large quantities of acrilonitrile–butadiene–styrene (ABS), often blended with PVC, are calendered, and, despite the difficulties of processing, fluoro polymers are produced by calendering—though on a much smaller scale. In the 'fifties, even polyethylene was calendered, though its rheological properties had to be modified by the addition of rubbers. The techniques used for each material and product vary but the basic problems of calendering are common to all, more differences between the materials being seen in the train carrying product from the stack to the take-off position.

The Calender Rolls

Early calenders were fitted with rolls having a large internal cavity, were heated by steam and fitted with a syphon pipe to remove condensate. Such rolls gave rise to several problems—they became water-logged and often variations of wall-thickness led to eccentricity in use. In the 'sixties the cored rolls were superseded by rolls drilled for heating by means of hot water under high pressure or by hot oil—which offered the advantages of better control of temperature, improved concentricity, and the possibility of higher operating temperatures. Figure 38 illustrates the flow of water through a drilled roll.

Three materials are used for the surfaces of rolls—chilled cast iron, cast steel, and chrome plating. For the manufacture of rigid PVC and for polished unplasticized films (particularly transparent or translucent), chromium-plated rolls are standard. Cast steel has some advantages in terms of strength but, probably because of the difficulties of modifying surface profiles, has fallen from favour: by far the most popular roll surface for producing plasticized PVC with a matt finish is chilled cast iron.

A calender with plated and polished rolls can be used to give film with matt

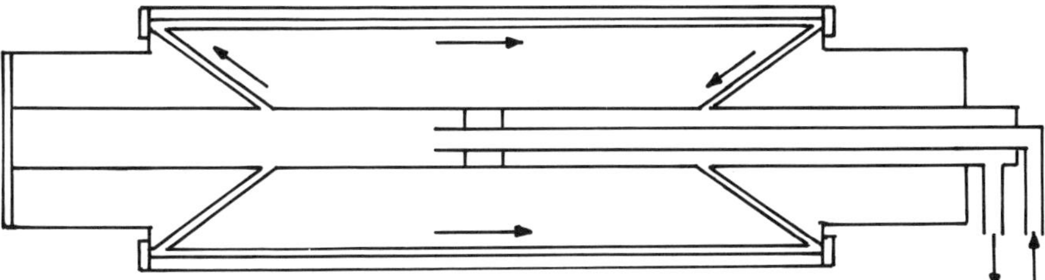

Figure 38 *Circulation of fluid as coolant in a calender roll*

finish only by in-line embossing. Frequent polishing with a lamb's wool bonnet or similar device, using chromium oxide polish, will keep the plated surfaces in good condition for a long time: stubborn stains may be removed by polishing the area concerned with a fine diamond paste.

Rolls of chilled cast iron can be used in polished state to give an acceptable glossy film but are susceptible to corrosion by fume from the PVC and their surface life is limited; after a time film produced by this means will lack the high gloss obtainable with polished chromium surfaces. The surfaces of such rolls can be matted by treatment with a suitable shot blast grit, and used to make films with matt surfaces; while the matting can be renewed easily *in situ* this method has the disadvantage that the grit will attack the surface progressively and etch out the grain boundaries in the metal to the extent that eventually the roughness of the surface will become unacceptable. The type of grit usually is selected as a compromise—while sharper grits will give more matting such a surface also will have reduced release qualities. It may be that release problems cause requirements for more lubricants, and this is counter-productive when matt surfaces are wanted. Increasing roughness leads also to greater generation of heat in the calender banks, and a need for more heat stabilizer giving rise to higher costs. For the shot blasting, grit sizes from 80 to 220 mesh are employed (mesh apertures of 0.0060 to 0.0036 inch), the grit being usually aluminium oxide or carborundum. Other finishes are obtainable by using metal peening grits or glass beads, both of which have a polishing action.

It is possible to recover rolls that have been affected unduly by grit, by honing or polishing in the calender; however, since at times rolls need to be changed because of accumulated damage to surfaces (or because of mechanical problems like failure of a bearing)—and notwithstanding the additional costs—it often is found convenient to re-grind rolls away from the calender.

The Calender Banks

Calenders work by passing the material through the gaps between successive rolls, each gap being narrower than the previous one. At a gap the excess material is sheared off the surface of the film and forms a roll of molten plastic called the 'bank'. Controlling the banks is central to the control of the calender;

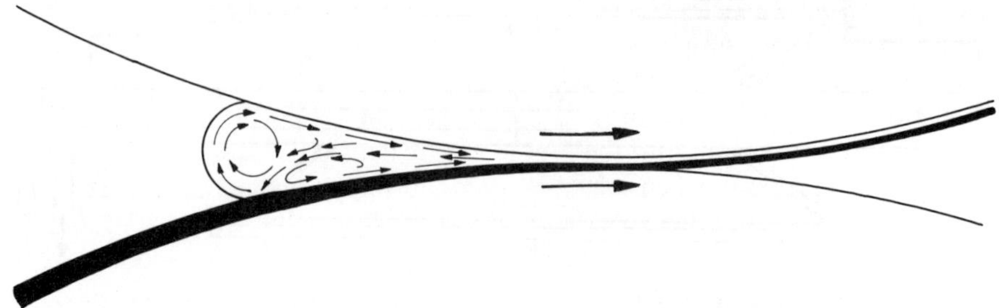

Figure 39 *Circulation of material in a calender bank (derived from G. Hatzmann—see references in text)*

the work of Hatzmann *et al.*[2] showed clearly that within the bank there was more than one current circulating, and Figure 39 is based on this.

Since only about half the material in a film enters the bank the two surfaces of the film have different histories: if it is intended to influence the back of a film it is necessary to change the bank at Roll 2/3; the front of the film is controlled by the bank at Roll 3/4.

If excess material were added to the bank without lateral transport the bank would grow in size but in fact material is transported towards the edges of a film, beyond the edges of the feed and so increasing the width of the product. Usually, after the ultimate nip, an excess width of film is trimmed away as waste: if this is not done care must be taken to make sure there is no variation in the surface finish in the areas of extended width. Occasionally a thin line of roughness will indicate the end of the width of material on Roll 3, and in the extended area flow marks are more likely to be found.

If banks are not rotating correctly there can be entrapment of air in the material—and this will give rise to marks in the surface of the film known in the trade by many names, including 'tick marks' and 'crow's feet'. The variety of different names for such faults indicates just how common they are.

Circulation within the bank depends upon the position of the currents and these are not necessarily constant across the width of a calender. Examination of a bank may reveal sluggish areas, some rotating in opposite directions, or what appears to be spiralling (the last, because of local variation in the temperature of the stock, often will result in flow marks in the film). Defects of this nature in the flow are influenced by the composition, and it may be necessary to adjust the amount of lubricant: however, if a mix is long-established the indication probably is that the conditions of calendering have drifted away from the optimum—and appropriate corrections must be made. The size of a bank can be altered by adjusting the previous nip or by changing the speed ratio of the rolls, and often this will rectify a fault.

There is a tendency for calender banks to act as a filtering system and to push hard particles out to the ends. This means that when a material is being degraded by processing to the extent that hard brown or black particles are generated, such particles will be concentrated at the very edges of the film. If the practice is to trim and recover edge material by returning it to feed the result can be the contamination of good stock.

If rolls are polished the drag at their surfaces may not be sufficient to pull material from the bank into the nip: if so, the bank will break and tongues of material form from it; when such material is absorbed back into the bank it will be cool and as a result there will be areas of surface roughness in the product. In order to overcome this it may be possible to reduce the viscosity of material in the bank by increasing the temperature at the previous rolls; alternatively, the quantity of lubricant in the composition might be reduced.

Attempts have been made to increase the number of banks by pressing a shaped blade against the surface of a roll; the idea appears very attractive (a bar or blade being cheaper to make than an extra roll) but so far has not been a success with thermoplastics. Collin[3] reported work in Germany that involved the use of bars in a variety of shapes, and in 1970 the approach was the subject of a patent.[4]

Manufacturers of floor coverings make use of the banks to produce marbled effects in the products. In essence in this method, heated granules in contrasting colours are added to a bank in a manner calculated to give minimum dispersion; in order to ensure that the effect required (and no other) is obtained it is essential to control strictly the amount of working to which the added materials are subjected.

Calender Contamination

On any calender line there are several possible sources of contamination, the first being the raw materials used. In most resins there will be a small number of discoloured grains, which are accepted up to a specified level. On occasions a manufacturer's equipment fails and produces resin with more than normal contamination of this kind. Some particles may not be discoloured but may fail to disperse and so cause small lumps in the film (known in clear films as 'fish eyes'). On the other hand, lumps may not necessarily be attributable to faults in the resin—they can arise also from poor dispersion of plasticizer, or from degradation taking place at some stage in the process.

Any component in a composition may be a source of contamination; it also can be caused by failure to keep plant clean, by accidental addition to re-worked material, by engineering problems, metallic detritus, and so forth.

Processing equipment can generate problems of this kind. As examples: if the seals of internal mixers are defective contamination can follow; when extruder feeders are used there can be localized contamination behind the gauze pack; and worn conveyor belts can release fibrous material which is very obvious in translucent or transparent film or sheet. Leakage of lubricating oil from bearings rarely causes critical problems; a more serious difficulty is condensation of plasticizer—the condensate usually gathering other contaminants before being incorporated into the product.

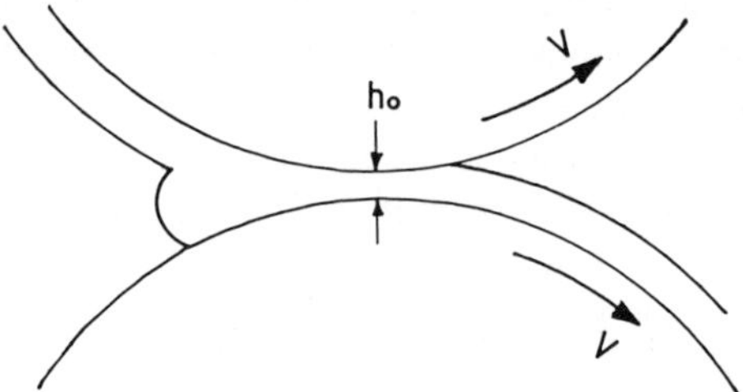

Figure 40 *Shear rate* $= 2V/h_0$

Production Speed

There is no correlation between production speed and quality of product but as speeds are increased the zone of acceptability for the control settings becomes narrower. The amounts of heat generated at the nips between rolls are related to the rates of shear; with a view to controlling the thickness of film produced the separation at a nip is maintained at constant value, and (unless corrected in some way) as speed is increased so too is temperature—roughly in proportion. Within certain limits this effect can be counteracted by reducing the temperatures of the feedstock and the rolls. Without such correction, and because of the higher temperatures being generated, there is a tendency for film produced at higher speed to be more glossy.

Figure 40 indicates how the gap between rolls and their speed of rotation are related to the shear rate.

Usually when making products of thickness greater than about 300 μm the amount of heat generated is less than that lost from the process in the normal way and the rolls must be hot enough to maintain the material at the correct temperature. Thin film usually is produced at higher speeds, and in such circumstances the heat created is in excess of requirements and roll temperatures must be lowered sufficiently to extract the excess heat. Measurements made when making plasticized PVC 100 μm thick at roll speeds of about 100 m min^{-1} showed that the material in the nip was 40 °C hotter than the roll temperatures. Bearing in mind that in general plastics are not good thermal conductors, and consequently that the transfer of heat from plastic to roll will take time—if the period in which heat may be transferred from bank to roll is not sufficient the heat will be carried instead from one bank to the next. The resulting high temperatures in the material are undesirable for a number of reasons—the time taken to degradation will be reduced, with consequent variations of colour in the finished product, and the exposed surfaces of the film will be unduly glossy.

If at the point of release from the last roll the temperature of the film rises its strength will be diminished; on the other hand, the tendency to adhere to the roll will remain. It follows that there will be stretching of the film and with the increase in surface area any roughness at the surface will be reduced and the reflectivity increased. It is a common practice when making unplasticized PVC to run the calender at slow speed for fairly thick film and when thin film is required to stretch the stock in order to achieve the gauge necessary—an approach resulting in higher production speeds for thin products. In this way the seriousness of any surface defects can be lessened, gloss will be improved, and the time taken to make small changes in thickness much reduced.

Optimization of Calender Conditions

Traditionally the setting of controls was decided by operators with reference to experience of previous successful calendering; it may be thought that constant observation linked with trial and error would give rise eventually to the setting of optimum conditions, but while experience obviously is of value it will not necessarily bring this about.

Automatic systems for controlling the thickness of the film were introduced in the 'sixties (automatic adjustments to the jack motors controlling the ultimate nip) and once this had happened it was true no longer to say that the operator had full control of the machine. There was a parallel development, as engineering controls became more accurate, in electronic computers; when the latter became available in convenient form it was logical that they should be employed in the collection of operating data and that the various units for automatic control be brought into the same system. The most important factors involved were the control of temperatures in the unit, and of tension in the web.

A progressive development of automatic control was hindered to a degree by the comparatively high initial costs of such systems and also, from the technical point of view, by a need for suitable sensors. Even so, it was necessary to investigate the parameters and to establish in an objective way whether an 'optimum' actually existed. With this in view, the conditions recorded in the course of normal production were collected and analysed—when it became apparent immediately that the information was not precisely what was required in order to ensure satisfactory processing. As examples, the temperature of heating fluid entering the rolls was recorded but not those of the material or the roll surface. It was necessary to convert the data into measurements that were significant for the product concerned. With this in view, the temperature of the film was measured at fixed points round the roll and a cooling curve established; this could be extrapolated back to the nip and the temperature of processing obtained. Since the quantity of electrical energy being used and the mass of the material being processed were known it was possible to calculate the heat generated at the nip—although this did not allow for energy lost in the transmission to the nip. However, if the increase in temperature at the nip were determined an estimate could be made of the conversion of electrical energy to heat, and applied to subsequent studies. Such a system might seem crude but in practice the results obtained were found to be helpful. Measurement of bank temperature by means of infra-red pyrometers gave similar results. Whatever approach was used (either adding to the temperature of the roll the increase calculated, or measuring the temperature of the bank) the results obtained would have a direct bearing on the processing—unlike the temperature of a

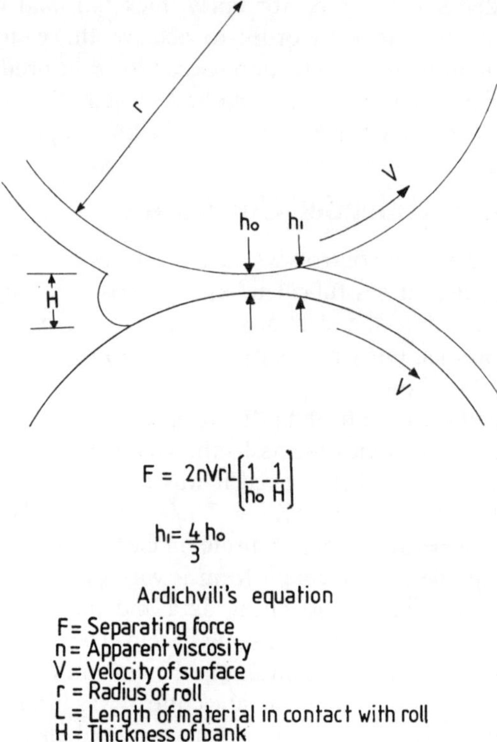

Figure 41 *Ardichvili's equation*

heating fluid, which took no account not only of heat generated but also of changes in the latter resulting from differences in formulation or running speed.

Another important calculation is that of rate of shear, which is derived from the thickness of the film in the nip and the surface speed of the rolls—due allowance being made for the expansion that takes place as the film leaves the nip.

Lastly, a measure is needed of the separating force in the nip. Normally film is required to be of the same thickness across the entire width and this is achieved by using rolls that are of larger diameter in the centre than at their edges, the final adjustments for purposes of control being made by skewing or bending rolls by applying force to their journals. It is possible to calculate the total camber applied, either from calibration of the mechanism for bending a roll or from an assessment of the effectiveness of skewing: on the assumption that the rolls become deflected in accordance with Hooke, the value thus obtained can be taken as the separating force generated.

Referring to Ardichvili's equation (Figure 41):

The roll separating force has been calculated from roll deflection (admittedly not in standard units, but this is of no importance).

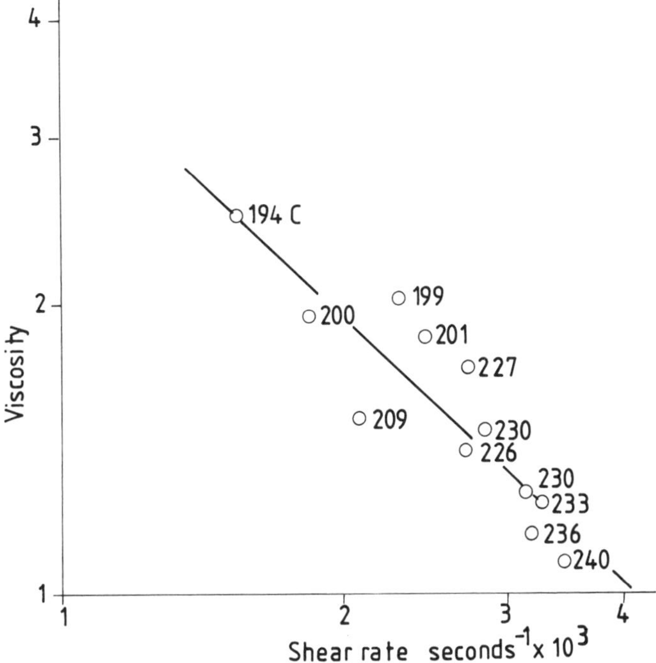

Figure 42 *Viscosity plotted against rate of shear*

Shear rate has been calculated as described earlier.

The roll radius, *r*, is constant.

Often the length of the nip, *L*, is constant at maximum width for the calender concerned but if not it can be measured.

The only unknown is the viscosity, *n*, which therefore can be calculated.

At this stage everything needed for consideration of the nip is known, and inter-relationships between parameters can be observed. The units in which measurements are expressed may leave something to be desired but the eventual results will not be affected by this.

If the viscosity of the material is plotted on a logarithmic scale against shear rate the result will approximate to a straight line as in Figure 42. When temperatures calculated in the nip are marked against the points of this plot it will be seen that the higher figures are at the higher shear rates—thus explaining the reduced viscosity. When such an approach is applied over a range of different plasticized PVC formulations the results are most interesting; it will be found that the mean lines are super-imposed, only the temperatures differing.

Results such as these suggest strongly that there is indeed an optimum way of running a calender: for each formulation, working backwards from the mean line, speeds and gap settings can be calculated and appropriate temperature settings estimated. When this has been done the conditions predicted have been found practical, although admittedly the results were no better than could be achieved by an experienced operator (and in production conditions there can be difficulty sometimes in achieving the settings desired). Such investigations should be of value: it should be possible now to pass information directly to a computer and for such data to be reviewed continually—so that deviations from the ideal become apparent and can be corrected before output is lost. However, at the time of writing this still is in the future.

Perhaps the most important area of automatic control in which no progress has been made so far is the control of the banks. In this it is hard to match the skill of the operator—the human eye can assess at a glance the size and fluidity of the banks, and an operator can make adjustments confident of their effects even if more than one bank will be involved.

Another difficult aspect is control of the point at which material leaves the roll—firstly because the environment is hostile for measurement, and secondly because any change in the tension at that stage will change also the thickness of the product.

Matting Agents

It is possible to incorporate in a composition matting agents which will give matt surface effects. This can be beneficial but considerable problems exist in the selection of such agents. In general, the basis for their effect is that they are either partly or wholly incompatible with a material and hence during processing disrupt the surface. They are sensitive to the conditions of processing, including the amount of working, and if material is re-cycled there may well be variation in the finish. For such reasons it often is more convenient to obtain consistent production of finishes required by means such as control of lubrication and the settings of the calender, rather than by a heavy reliance on additives.

Stripping Rolls

Should the surfaces of stripping rolls be matt or glossy? At first sight this question may seem to have an obvious answer, depending on the surface finish required in the product, but in practice for many of them the finish on these rolls is not of great importance. This is because, except at low speeds, the film will not be in contact with the surfaces concerned.

The presence between film and roll of a layer of air may be demonstrated easily by putting a light roller that is free to rotate in contact with a travelling film: the roll will rotate in the direction of travel but if it then is spun by hand in the opposite direction it will be seen that this has little effect on the film, and the effort necessary to cause counter-rotation is surprisingly small. Briefly, the thickness of the layer of air is controlled by the tension applied to the film, the radius of the roll, and the surface speed. The equation in Figure 43, derived from the work of Hutzenlaub,[5] expresses this.

Normal practice is to keep diameters of stripping rolls to a minimum (for example, 150 mm), and this ensures a thin air layer. The thinner it is, the more

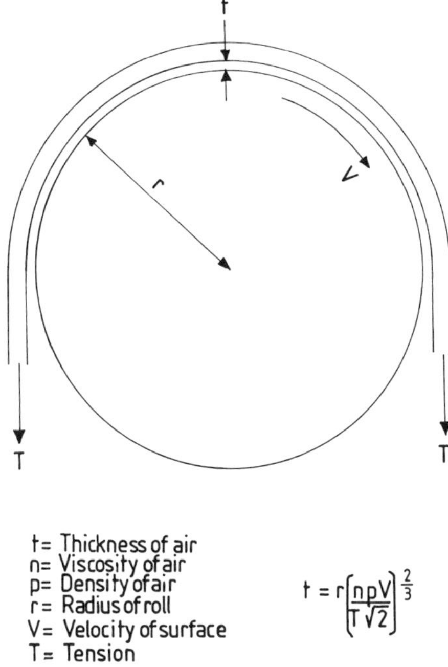

Figure 43 *Hutzenlaub equation (see reference in text)*

drive will be transmitted through the air to the film (and also such a layer will be better at transmitting heat). Rolls with rough surfaces will drive an air layer more effectively than smooth.

One problem encountered with stripping rolls is the condensation on their surfaces of plasticizer or stabilizer from the film; condensates can accumulate to such an extent that they are transferred to the film and cause unsightly marks. When plasticized films with matt surfaces are being made the condensates can be absorbed during storage back into the film (and disappear), so that provided no discoloration has taken place there is no special difficulty. However, faults of this nature are not acceptable in films with polished surfaces and in these instances the stripping rolls should be maintained at temperatures high enough to prevent condensation. Another good reason for so doing is that momentary contact between film and condensate on such a roll can be sufficient to snap the film.

While the surfacing of a stripping roll is not a critical factor, it may be noted that a polished chromium surface can be kept clean easily. Because of condensation, when highly plasticized films are being made at low speeds it may be necessary to have in contact with stripping rolls a rotating felt roller to remove any plasticizer, *etc.*

In-Line Embossing

Embossing film in line with the calender has several advantages: while giving the surface finish required it removes most of the heat remaining in the film and—since the nip of the embossing unit is the only point in the line between calender and wind-up where a film is gripped positively—it provides a single position at which fine adjustments may be made in the tension on the train.

In the simplest form a cylinder with a polished or a matt surface is brought into contact with the film—in order to improve the quality of the surface for printing, or to make matt film from a calender with polished rolls. In this process, while the upper surface of the film is modified to requirements, the reverse usually is marked by the rubber backing roller.

In many applications there is a need to maintain an embossed surface at elevated temperatures, whereas in fact it will be retained only up to the temperature at which it was applied. However, retention can be enhanced by infra-red heating of the film just before it enters the nip of the embossing unit. As PVC is a poor conductor of heat it is possible by this means, while the main mass is cool enough to keep its strength, briefly to raise the temperature of the surface to very high levels. In the course of embossing it is important to keep film in contact with the surface of the cylinder for as long as practicable, to chill the new finish and to remove as much heat as possible from the plastic as a whole. For this reason, an embossing cylinder with large diameter is very advantageous, and the film should pass round this cylinder as far as can be arranged before being stripped away. A cylinder of diameter 225 mm will process film of 100 μm thickness at up to about 80 m min^{-1}, but to emboss the same film at between 100 and 200 m min^{-1} a cylinder of diameter about 350 mm is desirable.

Embossments may be impressed in two ways, as illustrated in Figure 44. In the first, the film is pressed into the engraving on the cylinder in such a way that relatively little lateral flow takes place; hence there is little or no thinning at high points in the engraving and the pattern of the embossment will also be visible on the reverse. In the second, material is permitted to move away from points of pressure in the design so that the thickness of the finished film varies with the depth of the engraving, and the back is almost smooth.

The application for the film helps to determine which of the two types is preferred; if resistance to tearing is very important a film with constant thickness is better, but if the reverse is to be coated with adhesive it may be advantageous if it is smooth.

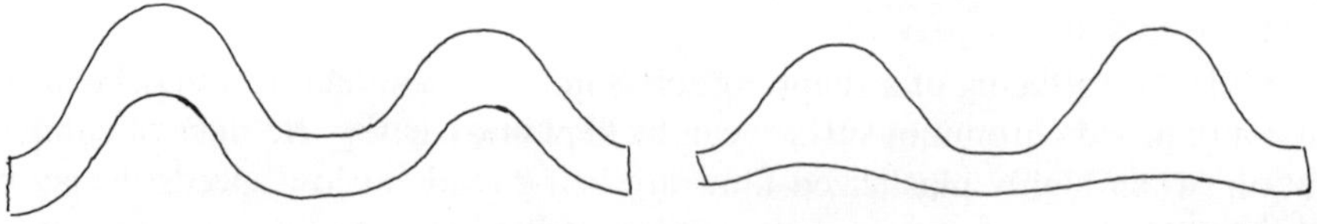

Figure 44 *Forms of emboss*

Cylinders for Embossing

A cylinder for this work must be constructed strongly enough to withstand the pressures employed both during engraving and in the actual embossing; on the other hand, the wall of the shell should be as thin as practicable in order to assist rapid transference of heat. Usually, the metal is mild steel—which is soft enough to take impressions from a hardened engraving roller but sufficiently strong to make a shell that will not distort in use.

The cylinders may be prepared by engraving but designs can be created also by etching, in much the same way as matrices for printing. Etching usually is from a photographic negative, but without super-imposition of the grid used in photogravure and similar processes to form cells that hold the ink. The steel used should be of uniform grain structure, without imperfections. In another approach, developed mainly in Japan, effects on the surfaces have been created by spraying metal.

For some types of finish, such as leather grains, the surface of a shell may be subjected to shot-blasting before chromium plating. Alternatively, slightly thicker plating may be applied, and shot-blasted afterwards. The second approach may be preferable, bearing in mind that a surface can be renewed by shot-blasting several times before it becomes necessary to remove and re-plate it.

Most cylinders are essentially hollow, with facilities for water to travel from end to end; various systems are available for the internal channels, including a ribbed central tubing designed to cause annular flow while providing support for the cylinder wall. However, the benefits of the more complicated thermal systems are somewhat less than might be expected (they achieve no more than would be gained by increasing the diameter of a cylinder). The additional cost of a more complex design in a smaller diameter has to be offset by the savings made in engraving.

The Backing Roll

Selecting the correct material and design for the backing roll are crucial factors. The most important aspects to be considered are: the composition and the thickness of the rubber coating on the roll, and the method to be used to cool the rubber.

Various parameters that influence the performance of an embossing nip were investigated by McNamee,[6] who considered the distribution of pressure through the nip and the internal temperature of the rubber coating under various conditions. The interactions that take place are complex and difficult to summarize. Factors governing the track length through the nip are: the hardness of the rubber compound, the thickness of the cover, and the closing pressure applied to the nip. The distance through the nip is influenced also by speed, since at higher speeds the rubber is distorted—in effect piling up in front of the nip.

Distance through the nip is important because time is needed for the film to distort sufficiently to accommodate the embossing, after which the distortion has to be frozen into it. Increased speed, by reason of the hysteresis of the rubber, causes the generation of more heat—and this must be removed as otherwise either the rubber may be degraded or its bond to the steel centre may fail. Another important reason for keeping the rubber cool is that if its surface becomes hot the film will cling to it rather than to the embossing cylinder, greatly reducing the dwell time of the film against the engraving and the extent to which cooling can take place—so giving embossing of inferior quality.

The rubber roll can be cooled in a number of ways: coolant may be circulated through the core protecting the bond between rubber and steel, and the surface can be cooled by running a cooled roller against it or by immersing it partially in a trough of water. The last (partial immersion in water) is very effective but before the rubber surface returns to the emboss nip it is essential to remove from it all the water—otherwise excess water will accumulate in front of the nip, cool the film, and generally reduce the efficiency of the unit. Water so trapped may cause variation of the embossment across the web; in bad cases, if water is transferred to the upper surface, glossy lines along the film will result.

The question often arises as to whether the drive should be applied to the embossing cylinder or the rubber roll but the answer seems to be that either may be chosen: because of distortion of the rubber during its rotation the steel cylinder will always have the higher surface speed—the difference between them being proportional to the nip pressure.

In general, if a softer rubber is used to cover the backing roll, that roll will operate at higher temperatures because of the greater distortion of the rubber concerned.

Defects in Embossing

Foreign bodies adhering either to the surface of the embossing cylinder or to the backing roll give faults capable of easy recognition, as they are repeated with every revolution. Similar repetitive effects may be caused by damage to the surface of an embossing cylinder—say from corrosion, or an impact.

A common cause of defects in embossing is failure of rubber covering the backing roll: sometimes, small undispersed particles become detached and leave one or more small pits in the surface, which will appear on the reverse surface of the film as raised pimples. While they may not be visible on the upper surface they may mean that the film is not acceptable (for example, if it is to be coated), and in winding-up the pimple or pimples may be imprinted on the upper surface.

Another frequent problem is failure of the bond between rubber and the steel centre; in the early stages this can lead to small cracks in the surface of the rubber, and these are reflected in the surface of the back of the film.

The surface of the rubber roll normally is renewed by re-grinding and if not done correctly this may leave a grain that will give undue roughness of the back of the film. Rubbers of different types behave differently when re-ground—polyurethane, for example, if cut at too great a speed, tends to melt and to clog the stone: the small melted granules then can adhere to the surface of the roll. If a smooth finish is needed, with some rubber compounds belt grinding can be used to advantage. Sometimes, (say, to increase the key of adhesives) a coarse stone can be used to give the rubber a rough surface. However, there can be difficulty in using a rubber roll to give a textured finish as it is hard to reproduce the same effect precisely each time, and the life of the rubber surface usually is quite short.

If there are variations in the temperature of the film as it is fed into the emboss nip the result will be variation in the depth of embossment across the width; most frequently, the edges are cooler than the centre so that towards edges the embossing is flat. This problem may be overcome by using additional heaters near the edges or by dividing the width of the heater into a number of zones; modern heaters usually are constructed with three or five zones across the width, each being controlled by a separate sensing unit.

Should the material still be hot when it leaves the embossing roll it will tend to stretch and to distort the effect. If a distortion is regular it may be acceptable but if film is released as a bow the embossing will reflect this as a curve across the width, which may be unacceptable. Distortion is particularly unacceptable with geometric and light-reflecting effects.

Creases formed in the film prior to the emboss nip will become welded into it as longitudinal faults. Such creases usually are caused by release of film from the calender roll in the form of an extreme arc, or by a total failure to control the tension between the calender and the emboss nip. Momentary contact of the film with the surface of the stripping rolls also may cause creases.

Winding the Film

Contamination of the surface of the film can be a major problem: often the static electric charge on the surface is very high—sufficient to attract any loose particles in the vicinity of the train; in summer, it can pull down small flies. For many uses, any contaminants such as these will make the film unacceptable—as examples, for coating, for printing, in packing food, or for medical applications. On the other hand, it is not easy to prevent such contamination. The cooling train and wind-up must be kept clean and free of loose material. Vigilance and regular maintenance are required: it is very difficult to guard against wear between different parts of the equipment and consequent generation of metallic dust—especially if one of the components concerned is aluminium or brass; quite large amounts of debris can be created rapidly.

For most applications creases at the wind-up will not be acceptable. The causes may be faults in the machinery or incorrect settings. Wind-up at low tension is particularly difficult to maintain—a bubble may be formed in the roll, causing the formation of transverse creases of film in the roll.

Hutzenlaub's equation[7] may be used to calculate the quantity of air entrained in a roll: in practice there always is a lot of air—usually enough to compensate for variations in thickness. After winding, the relaxation of stresses in the film will cause shrinkage, tension in the material will rise, giving an increase of pressure sufficient to expel air from the roll. Provided the film is free to move, its length will diminish and width increase: however, if a film is thinner in the centre than at the edges, or if there is a local thin area (so that the early contacts on shrinking are between thicker portions), the film may become locked at the high points and any increase in the width be accommodated by the formation of corrugations. Such corrugations do not alter the surface of the film but may detract so much from its visual appeal as to render it useless. Corrugations create problems for the fabricator, especially where automatic machinery is used.

Corrugation and other patterns of ribbing and bagginess that develop in the roll as consequences of gauge defects can be grouped together as faults of flat-lying or 'lay-flat' qualities (not to be confused with the tubular layflat film used for bag-making). It is almost impossible to express quantitatively the value of freedom from layflat problems but every user of film has in his (or her) own mind what degree of perfection is necessary in his process. Unplasticized or lightly plasticized PVC is less likely to develop unacceptable layflat patterns than are softer formulations but when such film does so it is much more difficult to use because its strength prevents the removal of these patterns by application of light tension.

The Measurement and Control of Surfaces

The maintenance of quality by measurement and control of the surfaces is one of the most complex problems; it is quite easy to make a measurement under laboratory conditions but it is often impractical to do the same during production.

It may be anticipated that some faults will be found in any film, and a specification will define to what extent faults will be tolerated in material intended for a particular purpose. Bearing in mind that the speed of the calendered film may be 100 m min^{-1}, many faults will not be visible—and it is necessary therefore either to have an automatic inspection system or a statistical plan for taking samples and making the measurements required. Normally, samples can be obtained only at the end of each production roll—which may mean that more than 1000 m have been made before a fault is detected.

Various automatic systems have been developed. One of the simplest is a test for holes in unplasticized PVC film and comprises a brush with metal 'bristles' which extends across the film and presses it gently against a metal roll. An electric charge is applied and if a hole is detected the charge passes to the backing roll and is recorded so that the number of holes found in a certain period of time is known. A difficulty with this is that no information is available about the nature of the holes—all that is known is the number detected. In an alternative approach, light is shone through the film and the number of scintillations counted: this method can be very effective, and it is possible to design a system that will differentiate between large and small holes. Such a system detects besides holes clear particles of polymer (known as 'windows') and so tends to exaggerate the number of holes, but it is possible to use it to set limits that will be acceptable to a customer.

A system employing a laser beam can be used to detect defects in the surface: the beam travels rapidly back and forth across the web and the amount of light reflected is measured. If the signal obtained is subdivided into brief periods of time it is possible not only to measure the defects but to indicate where they are on the web. The lengths and widths of defects can be measured, and the number of defects of specified dimensions counted. Experience suggests that the system can be useful in indicating the type of defect; as examples, a short, narrow fault may be a hole, while a long fault of moderate width could be dripped from condensed plasticizer; continuous faults could be creases. At present the equipment concerned is very expensive, and probably should be considered only if failure through faults will cause serious losses. It has been suggested that a system such as this might allow a reduction in manning levels but in practice so far this has not been the case.

To the best of the author's knowledge, the measurement of mattness or gloss has not been done 'on-line'. There is a system for measuring the transmission of light through a clear film, which perhaps is the nearest to this so far. Away from the calender a photo-cell with a standard light source can be employed simply to measure the amount of light reflected or scattered normally or at an angle by the surface of a sample. This system will express the mattness or gloss in numerical terms but it is preferred still to make comparisons with reference samples since this allows visual assessment of the coarseness of a finish, and of the extent of linearity.

Surface roughness may be measured with an instrument that drags a lightly loaded point across the surface of the sample to yield either a highly magnified trace of the surface or a 'centre-line average reading' (CLA). This instrument cannot be used 'on line'.

The surface energy or wettability is of particular importance when a film is to be coated or printed: it can be measured by a variety of well-established methods—for example, by determining the angle of contact of a globule of liquid resting on the surface. However, such methods will give differing results as time goes on. A figure may be obtained quickly by wetting the film in turn with a range of mixtures of alcohol and water (the properties of which are known). The mixture that is just able to wet the surface gives the value required. (For other polymers, different liquid systems are available for this test.)

A film may have a tendency to bond to itself (this is known as 'blocking') and the force necessary after a period of storage to strip it from the roll may be so great that the film will stretch or even break. Problems of this kind are most common with polished film and with heavily plasticized PVC, but also can be significant with matt surfaces. Even if accelerated methods are used the measurement of this attribute takes time (prepared samples of film are placed in an oven with a weight resting on them and after the period specified the minimum load necessary to separate the layers is measured); however, it usually is possible to control blocking by controlling the surface roughness of film.

On the other hand, in some instances—for specialized products—blocking may be wanted: usually it is brought about by modifying a formulation in such a way as to reduce the availability of lubricants at the surface, or by including additives that increase surface tack. Plainly, when such film is being made, adhesion must be kept at a level at which the film still can be stripped from the roll.

Colour is beyond the scope of this chapter but it should be noted that surface finish does alter the perceived colour: the more matt is a surface, the whiter it will appear.

References

1. G. Ardichvili, *Kautschuk*, 1938, **14**, 23.
2. G. Hatzmann, M. Herner, and G. Muller, *Kunststoffe*, 1975, **65**, 472.
3. H. Collin, *Kunststofftechnik*, 1970, **9**, 217, 241.
4. British Patent 1 303 705, 1970.
5. A. Hutzenlaub, *Kunststoffe*, 1967, **57**, 163.
6. J. P. McNamee, *Tappi*, 1965, **48**, 673; 1967, **50**, 308.
7. A. Hutzenlaub, *Kunststoffe*, 1967, **57**, 163.

CHAPTER 5

The Sealing and Welding of Thermoplastic Materials

C. HUGHES

Introduction

The development of sealing and welding techniques for thermoplastics is a continuing process: the requirements of fabrication are ever becoming more sophisticated and economic considerations provide a further constant spur for the production engineer. The first section of this chapter provides a review of the methods of sealing and welding that are in use currently, and this is followed by consideration of high-frequency welding in some detail.

General Terminology

(i) Heat Sealing

Heat sealing may be defined as the joining together of thermoplastic film by the application of heat and pressure.* In practice, although it is a general term (known also as 'thermal sealing') usually it refers to a technique in which either one or a pair of temperature-controlled elements is employed.

* ISO 472 says:
The process of bonding two or more thin layers of materials at least one of which is a thermoplastic film by heating areas in contact with each other to the temperature at which fusion of the thermoplastic film or films occurs, the bonding usually being completed by the application of pressure.

(ii) Impulse Sealing

In this method of sealing films, thin metallic strip is heated rapidly by an intermittent electrical current. It is a more sophisticated form of thermal sealing and offers certain distinct advantages.*

(iii) Hot-Gas Welding

Hot-gas welding is a method of sealing thermoplastic materials by applying a jet of hot gas which softens the materials.†

(iv) Hot-Tool Welding

In this method, heat is applied directly to the areas to be joined as a result of contact with heated metallic tools or moulds. Usually the source of heat is electrical, but if it is more convenient a gas flame also can be used.

(v) Molten-Bead Sealing

A method of sealing thermoplastic films by applying a molten bead of a like material, which is extruded continuously between the surfaces of the films before they are brought together under light pressure.

(vi) Frictional Welding

In this process, circular pieces are rotated in contact with each other, under pressure, until their surfaces melt. At this stage the rotation is brought to an end and the parts kept under pressure until the molten materials have cooled and solidified.‡

(vii) High-Frequency Dielectric Welding

In high-frequency dielectric welding thermoplastic materials are joined by applying simultaneously pressure and alternating high-frequency fields, using electrodes of suitable design and shape.§

* ISO 472 defines 'impulse sealing' and 'thermal impulse sealing' thus:
A bonding process in which the surfaces to be united are subjected to non-continuous rapid heating, pressure being maintained after heating.

†ISO 472:
A pressure welding process in which the surfaces to be united are softened by a jet of hot air or inert gas.

‡ISO 472 defines 'friction welding' or 'spin welding':
A pressure welding process in which the surfaces to be united are softened by heat generated by friction.

§ ISO 472 defines 'high-frequency welding', 'dielectric welding' and 'RF welding':
A pressure welding process in which the surfaces to be united are softened by heat produced by a high-frequency field.

(viii) High-Frequency Induction Sealing

In this process, alternating magnetic fields induce heat in a metallic insert, which is in intimate contact (at the point of joining) with the materials to be sealed.

Each of these methods is taken in turn.

Heat or Thermal Sealing

The equipment for thermal sealing usually comprises a press with at least one movable platen bar; often there are two bars, which can be adjusted to give the clamping action that is required. The bars are 'resistance heated' by means of a low-voltage heavy electric current: the degree of heating is controlled automatically by a thermostat, and often it can be pre-set.

The sealing bars should be covered by a material that will not stick to the substance being sealed: often a laminate of polytetrafluorethylene and glass fibre cloth (which may be backed with adhesive for easy application) is used for this purpose.

The heat is applied to the outer surfaces of the materials being welded, and in order to make the weld it must travel through the material to the interfaces. For many applications thermal sealing has been superseded by impulse sealing but the former process is in use still for welding thick sheets of flexible thermoplastics. The production equipment employs a heated wedge placed between the sheets, over which they are moved continuously.

Impulse Sealing

The essential difference between machines for thermal sealing and for impulse sealing is that, whereas the former employ electrode bars that are maintained at a constant temperature, the latter employ sealing bars which are heated only during the clamping and welding cycle.

The impulse sealer has either one or a pair of thin tape electrodes, usually about 0.25 inch (6.35 mm) wide, which are constructed in such a manner that the centre section of the tape has an electrical resistance somewhat higher than the edges. As a result, when an electrical impulse flows through the tape during

---

Footnote § (*continued*)

This Standard defines also:

pressure welding A method of welding depending essentially upon the use of pressure with application of heat—for example, to make thick plates or blocks from thermoplastic sheets.

spot welding A pressure welding process in which relatively small areas of the surfaces to be united are softened at spaced intervals by heat.

The Standard includes a definition too for 'solvent bonding; solvent welding' (see Chapter 6).

the welding cycle, the centre reaches a relatively high temperature. The equipment is designed so that the temperature in the centre sections of the electrodes will be above the melt temperature of the film being sealed, while at the same time the edges are below the melt temperature. Thus the film on either side of the weld is clamped under pressure, so the risks of cutting, distorting, or wrinkling it are diminished. Since the sealing electrodes are very thin they cool rapidly when the current is interrupted, so problems arising from film sticking to the electrodes are much reduced.

Sealers of this type are used extensively for making large quantities of bags of various sizes and types from thin polyethylene film, but they are not suitable for sealing thick film.

In some applications, impulse sealers are being replaced by machines which employ a continuously heated knife. In such machines, the film is supported across a roller which is coated with a resilient material such as silicone rubber (to which molten polyethylene will not adhere) and the hot knife edge pressed on to it. A strong bead edge seal is obtained.

It is necessary sometimes to manufacture film packages which are sealed adequately enough for handling and transport but which can be opened very easily. This particular requirement has been met by the use of a sealing electrode with a configuration which gives a large number of spot welds spaced at intervals.

Hot-Gas Welding

Hot-gas welding usually is carried out by means of a hand-operated torch or gun, and is similar to some methods of welding metals. A jet of hot gas (for example, air from a small compressor) is directed upon the surfaces to be welded, and at the same time a filler rod of like material is heated by the jet and laid into the weld. In the case of flat sheets, the surfaces to be welded may be prepared by a simple machining operation which mitres the edges to be joined and forms a v-channel when the two are placed together, into which the heated filler is deposited during the welding. The join between the filler and the sheets is made by localized melting and fusing. By means of this method, butt welds can be obtained with average tensile strengths greater than 80% of that of the original material. It is used extensively for fabrication from thick sheets of polyethylene or rigid vinyl.

In instances in which several similar structures are to be welded repetitively, an increased rate of production can be obtained by mounting a single torch or multiple battery on a mechanism of the pantograph type, so that several seams can be welded simultaneously. For repetitive applications, if desired, the hot-gas welding technique should be controllable by computer programme, and suitable for work with robots.

Hot-Tool Welding

Hot-tool welding can be used for most thermoplastics. In this process, the two surfaces to be joined are brought into contact with a metal plate, which usually is heated electrically and equipped with temperature control, and the contact maintained until the surfaces are seen to melt. The components then are removed from the heating plate and brought together quickly while the surfaces are still molten. A small amount of pressure is applied while cooling takes place. On occasions when the heated surfaces may become contaminated as a result of oxidation or other products of thermal degradation a small radial movement when the components are brought together helps to break through the contamination and to ensure a better weld. The thermal behaviour of certain materials is such that there is a significant risk of degradation and of staining or transfer of plastic to the tool. An example is nylon and with this the welding is carried out best using heat radiated by a tool at fairly high temperature, the radiated heat softening the surfaces to be welded. Since the components are close to the tool but not in contact with it, no material can adhere and tool cleaning is not necessary.

The costs of equipment can be quite low: precise control of temperature and timing are necessary, but if these are obtained the process is versatile and applicable for many materials and products.

The tools may take the form of metal plates or strip knives and usually are heated electrically from low-voltage transformers. A Variac transformer provides a range of adjustments and hence of control of temperature: a thermometer permits the visual checking of temperature, which in normal use would be in the range 240 to 340°C (depending upon the material or materials being joined). The heating tool, plate or knife often is nickel- or chromium-plated, or made from stainless steel—with a view to avoiding the risk of decomposition of the components being heated.

Resistance-heated retractable knives are in common use for welding flexible vinyl gaskets for seals for the doors of refrigerators and washing machine doors and panels (which are produced in large numbers). Equipment for this purpose consists basically of two metal or wood guides fixed to a bench top, on either side of a slot in the bench. Through the slot protrudes a pivoted metallic strip knife, the blade spring-loaded and capable of being retracted by means of a pedal. In operation, the two pieces of thermoplastic material (usually extrusions) are placed one against each guide and, under light pressure, are placed in contact with the sides of the heated knife blade. When the surfaces of the thermoplastic are seen to soften and begin to flow the blade is retracted and the two components brought together and fused by pressure. After a short period for cooling and hardening the welded materials are removed, the pedal lifted, and the spring-loaded heating knife returns to the high position ready for the next welding cycle.

Molten-Bead Sealing

Molten-bead sealing is suitable particularly for making large areas of polyethylene sheeting by welding several widths.

In the process an extrusion die deposits a molten bead of like material continuously along the edge of one moving length of film or sheet, and a second moving length is passed on to the edge of the first so as to entrap the bead between them; then, under pressure, a strong continuous weld is formed.

Flexible film and sheet in a wide range of thicknesses can be welded in this way, at relatively high speeds.

The bead must be hot enough to ensure that the surfaces of the films being welded are melted adequately to ensure fusion. The method is particularly suitable for use with polyethylene because of the melting behaviour of that material.

Frictional Welding

In machining, heat caused by friction often is a drawback and some means of dissipating it has to be found. However, engineers have long realized that frictional heating can be turned to advantage—such as when producing fabrications with welded joints. Circular sections of both solid and hollow design can be welded rapidly on simple and cheap equipment built on the principle of the turning lathe.

Visual checking of the components is necessary, but in general in friction welding semi-skilled labour will be adequate. The technique is best suited to making objects in long runs and when the appearance of the weld is less important than its mechanical strength. In some applications, such as the assembly of injection-moulded components, it is possible to shape the surfaces to be joined in such a way as to ensure both good welds and an attractive finish. The most satisfactory results will be obtained with equipment in which it is possible to control closely both the time of the cycle and the degree of pressure applied.

High-Frequency Dielectric Welding

High-frequency welding probably is the technique used most widely for joining both rigid and flexible poly(vinyl chloride) film and sheet. (It has been called also 'radio-frequency' or 'capacitance heating' welding.)

The method differs from most others in that initially the electrodes are at ambient temperature and that in semi-continuous production the temperature of the electrodes increases only through conduction of heat from the materials being joined. With well-designed electrodes the operating temperature seldom exceeds 40 °C.

In the process the materials to be welded are gripped under pressure and subjected to a high, rapidly alternating, voltage (at a frequency usually within the range 20 to 70 MHz, and shortly to be legislated to be frequency stabilized at 27.12 MHz ± 0.6%).

The thermoplastics that can be welded in this way are those having polar molecular structures. In such materials, induced dipole vibration occurs out of phase with the alternating electric field applied. This molecular vibration dissipates energy, giving rise to the heating and welding of the materials. A further advantage of the method is that the electrodes create a steep temperature gradient within the materials: the interfaces soften and flow under pressure, while the external surfaces remain relatively cool, and do not stick readily to the electrodes.

High-frequency welding is used primarily for fabricating flexible (plasticized) poly(vinyl chloride) film and sheet, either unsupported or with supporting cloths made from natural or synthetic materials. It can be used also to weld rigid (unplasticized) PVC surfaces, or to weld rigid to flexible, or expanded PVC. Other plastics such as acrylics, acrylonitrile–butadiene–styrene (ABS), cellulose acetate, polyamide (nylon), and polyurethane can be welded using these methods. Polyethylene film is non-polar and not amenable to high-frequency welding, although polyethylene laminates can be welded by this means to paper or to cardboard. 'Non-woven' fabrics and fabrics woven from thermoplastic filaments are being welded by high-frequency methods on an increasing scale.

Equipment for High-Frequency Welding

In essence, the equipment comprises the generator, which converts the frequencies from the mains electricity supply to high frequencies, the press mechanism, and usually some form of mechanical attachment to carry the materials towards the press—and away from it after welding is done. The high-frequency generator is designed to work with a specific size and type of press.

Figure 45 shows a high-frequency welding machine for general purpose work. Most modern machines employ a single valve oscillator working at a pre-set output frequency, the oscillator being supplied with high-voltage direct current from a transformer–rectifier power pack unit. There are full circuit-protection devices and arrangements for auto-stepped power application. On small machines, the high-frequency output is fed directly: on larger types, through a special power coupling unit. Provision for adjusting the power smoothly is in the form of a manually operated knob control (or a motorized control on larger machines), with a suitable power-level indicator. Usually the power control has the effect of tuning the press circuit to a resonant frequency close to that of the operating frequency of the generator. The closer the two frequencies become, the more efficient is the transfer of power, with consequent benefits to the cycle of heating and welding.

Electrodes with a wide range of sizes and configurations can be employed on a press of a given size. A power-level meter assists in easy re-setting when changes of electrode are made, and both the welding and the dwell times are controlled by separate auto-resetting timers.

One of the forms of mechanization encountered most frequently is the system of single or multiple sliding trays, on which lay-ups of material can be made ready for the welding. The lay-ups can be prepared rapidly, away from the press station, and while each one is welded others can be put up.

In more elaborate procedures, lay-ups can be assembled on one or more moving lines, or on rotating circular tables, where stages are added consecutively by the operators. Welding may be included at different points in the lines

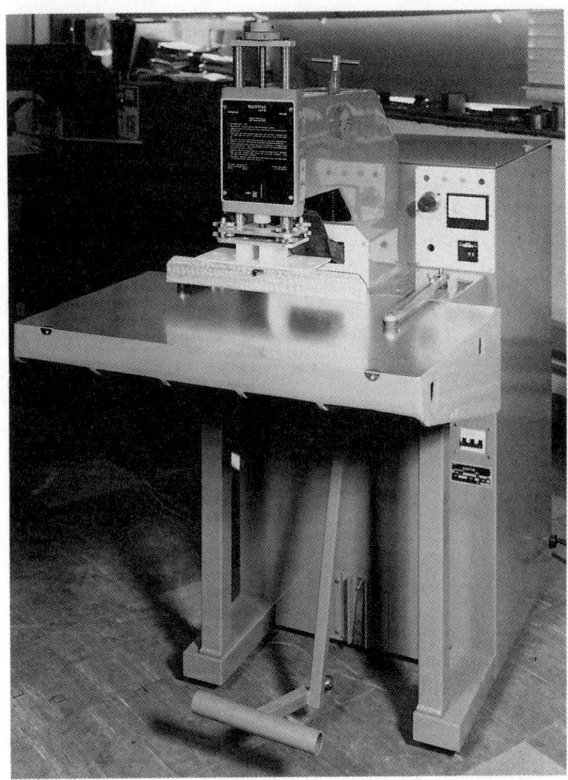

Figure 45 *Pedal-actuated machine for high-frequency welding*

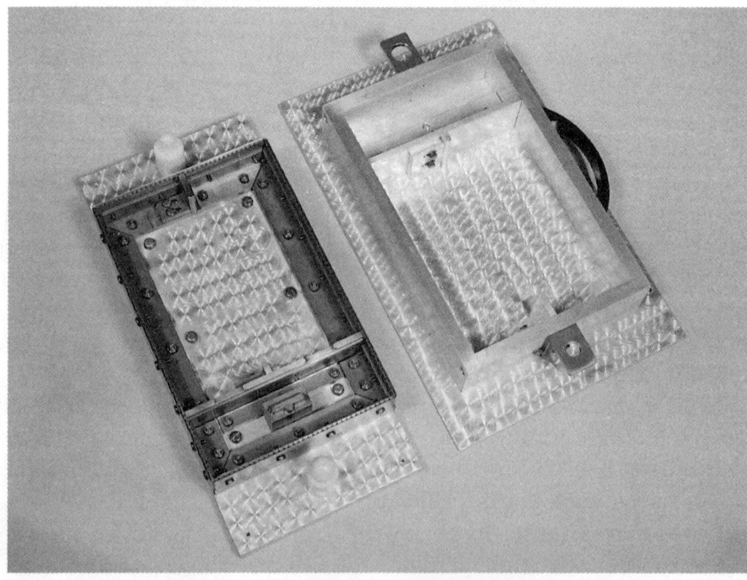

Figure 46 *Typical electrode for welding blister packs formed from plastic film*

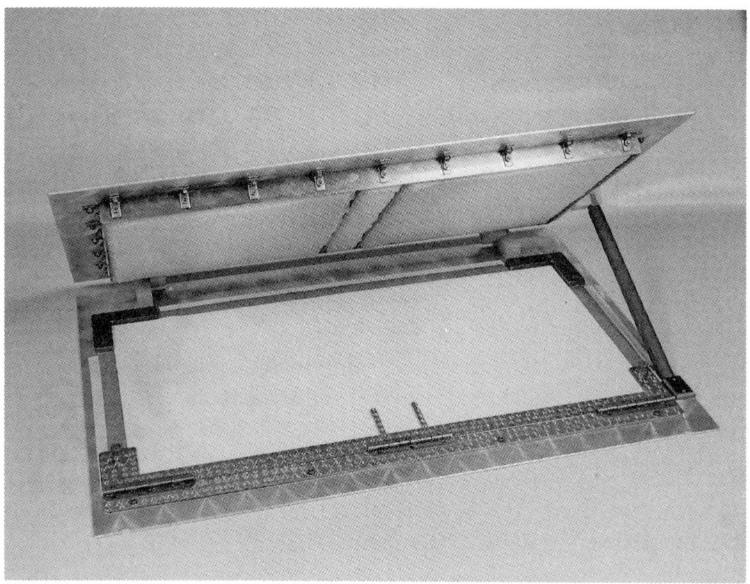

Figure 47 *Electrode for welding hinged book covers, showing locators for the work-pieces*

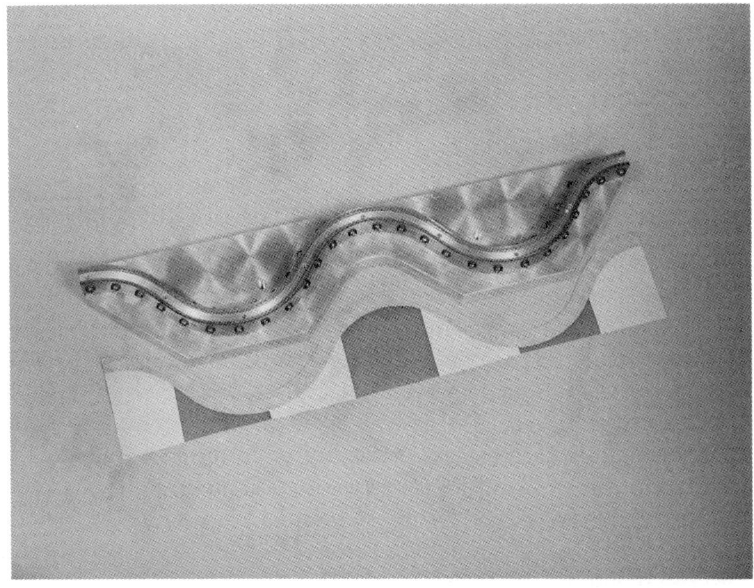

Figure 48 *Electrode for making the scalloped edges of blinds and awnings*

or tables, if the complexity of the work requires it, or as a single final task. The lines or circular tables offer opportunities for flexibility if from time to time (as usually is the case) ranges of different designs have to be assembled—perhaps including in early stages the cutting out of various components on patterns or templates. In such arrangements, cardboard stiffeners, mouldings, and other fitments can be dispensed automatically, one by one, to fall or be guided into their proper places in the lay-up (examples of goods made using equipment of this nature include book covers, ring binders, containers for liquids, and garments with buttons, fasteners, or other special fittings).

Often the machines incorporate some provision for the automatic stripping and removal of finished fabrications. 'Tear-seal' welding electrodes may be used, which not only weld the sheets but include also a cutting edge that separates each finished assembly. Further refinements include the provision of stations for printing or embossing the plastics, thermoforming units and filling devices.

High-Frequency Induction Sealing

In high-frequency induction sealing a metallic component is located at the point of joining in such a manner as not to inhibit the flow of molten material; the inset is designed to ensure a flow of material sufficient to give a seal of the strength required. The method is used mainly for the assembly of moulded components, or for the sealing of plastic bottles in instances where a tamper-evident hermetic seal is needed.

It is a technique suitable for automatic equipment: since there is no contact between the heating device and the materials to be sealed, there are none of the problems of sticking that can arise with sealing dies or electrodes. On modern machines, even with relatively thick-walled sections, sealing times can be as little as a fraction of a second. The method is ideal for work in which thermoplastics and metal components have to be joined—as in the assembly of medical syringes and the sealing of tinplate cans.

Equipment for High-Frequency Induction Sealing

Units for inducting heating are constructed like radio transmitters except that the high-frequency oscillations are concentrated into a water-cooled coil instead of into an antenna. The power level in the coil can be varied from very low to maximum, usually by adjusting a single control which works like the volume control of a radio receiver. The high-frequency power in the water-cooled coil can be regarded as equivalent to the current and voltage in the primary winding of an electrical transformer. In the high-frequency circuit, the secondary of the transformer is represented by the metallic component that is placed in or around the coil. When thermoplastic materials are being sealed, the 'secondary winding' is a continuous metal ring, wire, or disc (in practice, the insert can be almost any shape) which is located at the joint. It usually is called a 'susceptor' and is made from steel because magnetic steel can be heated easily.

Often it is necessary to apply a slight pressure at the point of sealing; the metallic insert remains embedded in the joint.

Modern equipment for continuous production normally includes high-frequency induction generators, which are switched electronically, and speeds of hundreds of heating cycles per minute are becoming commonplace.

Theory of High-Frequency Welding

(i) Principle

In conducting materials the attraction between an atomic nucleus and its associated electrons is comparatively weak, so that under the influence of a voltage field (with many metallic elements) electrons can be drawn away as a current. In insulators, on the other hand, where the attraction between nucleus and electrons is stronger, movement of electrons under similar conditions is more limited, and is known as a 'displacement current': the latter increases as higher voltages and frequencies are applied, giving rise in turn to molecular agitation and heating. It follows that the high-frequency field can be used for the heating of some non-conducting materials.

(ii) Practical Application

In practice, dielectric heating is a technique in which, when placed in a high-frequency or 'RF' field, non-conducting materials can be brought to a required temperature. There is a tendency for the molecules of the materials to vibrate in sympathy with the applied field, resulting in molecular friction that appears as heat. The extent of vibration of the molecules is related to the loss factor (F) of the material (see below), and the higher the loss factor the greater the rise in temperature.

Similarly, the greater is the voltage applied (E) and the higher the frequency applied (f), so will enhanced molecular activity and higher temperatures be achieved.

The following equation is applicable:

$$\text{Power output} = E^2 \times f \times F \times \text{constants}$$

where:

$$F = \text{loss factor} = \text{dielectric constant} \times \text{power factor.}$$

(iii) Dielectric Loss Factor

In the equation, a key figure is the loss factor—the product of the dielectric constant and the power factor, and an inherent property of the material to be heated. Some materials (such as asbestos, polyethylene, and polypropylene) have low loss factors and in practice for industrial applications, regardless of

TABLE V *Frequencies reserved for industrial, medical, and scientific equipment*

Frequency/MHz	*Tolerance*/MHz
13.56	± 0.07
27.12	± 0.16
915	13
2450	50

the voltage or frequency used, heat sufficient for sealing cannot be generated.

The loss factor of water is high, and since water is present in many products and substances this can assist in the employment of dielectric heating—particularly in baking and for drying.

Of particular interest in the present context is poly(vinyl chloride), for which a typical loss factor would be 0.03—high enough to permit heating for industrial purposes at usable frequencies. (The factor for polyethylene, on the other hand, is 0.0004—hence the difficulty in applying to it this method of heating.)

The loss factor will vary in some degree with the formulation of the material and (as the product of the dielectric constant and the power factor), will be affected by other parameters such as the frequency applied. However, in practice variations for reasons such as these are not significant.

(iv) Frequencies Used

In order to ensure sufficient molecular agitation for heating for industrial needs, frequencies in the megahertz range are necessary. Such frequencies are used also for communications and television and some years ago, by international agreement, the frequencies shown in Table V were set aside for use with industrial, medical, and scientific equipment.

Today, high-frequency generators all over the world use normally 27.12 MHz (although, as a matter of fact, for welding PVC higher frequency would be preferable and some old equipment at 50 MHz, without precise control over tolerance, would give more effective results than modern generators).

In an instance where a weld is required, say, over a length of two metres or more the voltage may diminish to the extent that the strength of the weld varies along the length. It is possible sometimes to correct such a fault by altering the arrangements for feeding, while another solution is to employ a lower frequency. Hence, 13.56 MHz may be found in instances when welding over lengths in excess of two metres is being carried on.

The other frequencies shown in Table V—microwave frequencies— involve the use of magnetrons. Such units are available at power levels of 1 to 5 kW at 2450 MHz but only two proven magnetrons, with power levels of 30 kW and 50 kW, are available at 915 MHz.

(v) Harmonic Frequencies

Besides the fundamental frequency, generators give radiation at harmonic frequencies, and these (depending upon the fundamental frequency and the sweep band of the harmonics) could cause interference with communication or television channels.

In order to ensure that equipment satisfies the specified limits it may be necessary to include line filters to carry extraneous frequencies to earth, or to screen the press platen with a metallic hood (so that no radiation can pass beyond the hood). In extreme instances of such a difficulty, it may be necessary to enclose the equipment in a suitably screened room.

(vi) Design of Generator

High-frequency power for industrial use is generated by a standard oscillator circuit operating under Class C conditions (that is, modified Colpitts, modified Hartley, tuned anode/tuned grid), using industrial triode valves, usually with ceramic envelopes and air- or water-cooled anodes.

The tank circuit may be in the form of discrete capacitors or inductance coils but more usually it is of aluminium cavity construction, with the valve enclosed by inherent inductive and capacitative elements. The last-mentioned arrangement has the advantages of being partly self-screening and of minimizing the possibility of parasitic oscillations.

(vii) Possible Operating Difficulties

Air-borne Interference. A condition of authorization under the Wireless Telegraphy Act is that radiation at frequencies such as 27.12 MHz, and within approved tolerances, must not interfere with communications for government services such as ambulances. Since these services do not use the reserved frequencies there should be no difficulty—except perhaps for harmonics. If interference should arise for this reason action must be taken to control or to eliminate it, perhaps even for a time ceasing to work until remedial measures are completed. Such measures involve the effective screening of pre-heaters, hoods, *etc.*, and the careful maintenance of screening and testing units.

Mains-borne Interference. When welding equipment is being used there is a possibility that both fundamental and harmonic currents might flow back into the mains electricity supply; however, this can be a more serious problem at the frequencies used for induction heating as distinct from those for dielectric welding.

Exposure of Operators Just as it is undesirable to be exposed to excessive ultra-violet radiation in sunlight, so also is it undesirable to be exposed excessively to high-frequency radiation.

In Britain, the Health and Safety Executive issued in 1990 the recommendation that users of such equipment monitor the generators and tooling regularly and make quite sure that no operator was receiving continuously 10 mW cm^{-2}. Meters for measuring the radiation are available commercially, and calculations from readings obtained can give figures for continuous exposure.

While the human body is equipped to regulate and to dissipate heat it is understood that organs such as the lenses of the eyes and the male *testes* are sensitive to excessive heating, and hence could be damaged if appropriate care were not taken.

Materials Used in High-Frequency Welding

(i) Flexible Calendered Poly(vinyl chloride) Film or Sheet

Typical formulations of vinyl film, sheet, or sheeting, containing appreciable amounts of plasticizer, are suitable for welding by this means—and such materials are encountered most frequently in practice.

However, when flexible vinyls such as these are printed—particularly with areas of ink containing a conductor like carbon black—difficulty can arise from arcs at the surface. Similar problems can be found with metallic print, and with metallized surfaces (which normally cannot be welded): lacquered surfaces too can impair or prevent welding.

Laminates of flexible vinyl films are used frequently in the manufacture of air beds, buoyancy items, inflatable toys, paddling pools, and so forth, and such laminates usually can be welded without difficulty.

(ii) Rigid Poly(vinyl chloride) Film

Vinyls with low plasticizer contents also can be welded, but normally pre-heated electrodes and electrodes equipped to stabilize temperature are necessary. Similar vinyl materials can be welded, or rigid and flexible. Problems may be experienced with printed, lacquered, or metallized surfaces, just as with the flexible films. Flock-coated items can be welded, but not (of course) flock face to flock face.

(iii) Cellulose Acetate Film

Cellulose acetate film can be welded in this way, subject to constraints similar to those for rigid poly(vinyl chloride) film.

(iv) Flexible Polyurethane Film and Coated Fabrics

Materials such as these, used in inflatable life jackets, boats, rafts, and other buoyancy equipment, can be welded by means of appropriate techniques.

(v) Poly(ethylene terephthalate) Film

Amorphous poly(ethylene terephthalate) is not suited to welding by these means: the material may be modified with glycol to make formings for packaging, and this modified material can be tear-sealed.

TABLE VI *Typical pressures required for high-frequency welding. The figures given are per square inch or per square metre of the face of the welding tool*

Material	Type of Machine		
	Small (*up to* 4 kW)	*Medium* (5 *to* 10 kW)	*Large* (*over* 10 kW)
Flexible poly(vinyl chloride) film (*e.g.* 2 × 0.5 mm)			
lb in^{-2}	30 to 100[a]	100 to 150	100 to 150
kg m^{-2}	210.9 to 703.1	703.1 to 1 054 650	703.1 to 1 054 650
Padded lay-ups such as seating, with cover of fabric-supported material			
lb in^{-2}	minimum: 100[b]	100 to 250	150 to 250[c]
kg m^{-2}	minimum: 703.1	703.1 to 1 757 750	1 054 650 to 1 757 750
Rigid sheet or sheeting; more difficult work in general			
lb in^{-2}	minimum: 100[b]	[d]	[d]
kg m^{-2}	minimum: 703.1		

[a] If sufficient power is available, lower pressures may be possible.
[b] The capacity of the press rather than of the generator may limit the size of weld.
[c] For some difficult work, pressures in excess of this may be needed.
[d] Items of work must be considered individually.

As noted earlier, polyethylene usually is considered unsuitable. Occasionally, for specific uses, satisfactory welding of polyethylene has been achieved by employing barrier lay-ups through which the degree of heating required could be obtained.

In an industry in which new materials, formulations and presentations are being brought forward all the time it is not possible to compile a comprehensive list of products that are, or may be, susceptible to high-frequency welding. The information given above is intended as a guide, but if it appears that this form of welding would be a useful method of fabrication for a new material or product it would be a wise approach at an early stage to have practical trials.

Indications of the pressures required for welding materials of different kinds are given in Table VI.

Types of Weld, Electrodes, Jigs, Dies, etc.

The efficiency of a manufacturing line and maintaining quality in its output depend very largely on the design and suitability of electrodes and other ancillaries. (Care in setting up also is important—ensuring the electrodes are level, that the correct power, pressure, and dwell times are observed, and so forth.) It is important to purchase appropriate and reliable machinery, but good machinery can be undermined if electrodes and ancillaries are not as they should be.

Most electrodes are made from rule sections of tool brass: they can be made by the users of equipment, and adapted by them for the methods and angles of fitting required for their various machines. More complicated shapes may be made by cutting or milling from solid brass bar.

A typical electrode for tear-seal work would be hinged on a top plate, the hinges insulated (for example, by polypropylene blocks); a lay-up then is placed on the bottom plate, the unit closed and fitted under the head for welding. Hinged systems such as this have many applications—as in making pre-cut covers for books, where card or board inserts must be placed as close to the electrode as possible, to ensure a close fit after welding.

In a tear-seal electrode, the profile allows sealing and cutting to take place at the same time. The tear or cutting part of the profile normally would be at the outer edge and raised above the welding part, to approximately 50% of the total thickness of the material.

By this means, homogeneous materials can be sealed and separated in a single operation; unfortunately, for obvious reasons, the method is not suitable for sealing and cutting vinyls supported on knitted or woven fabrics.

Similarly, fabric-supported materials can be welded vinyl sides together, and plain welds can be made over stitched seams in such materials when, as in the protective garment trade, it is required to make the seams waterproof.

Simple electrodes may be engraved with patterns of stitching or other designs (including trade names, logotypes, *etc.*) and in the course of welding also emboss permanently the design in the area of the weld.

More elaborate decorative effects may be obtained by taking from a hand-tooled or stitched master product a liquid silicon replica, the cured silicon matrix being used in conjunction with the electrode to reproduce the original finish on vinyl. (This method of embossment is suitable for products such as belts, footwear, handbags, luggage, and cases for stationery.)

'Half-tooled' and other patterns and finishes analogous to work in leathers, embroidered fabrics, or even ivory, may be obtained by applique welding—where typically a flexible vinyl of a different colour is superimposed and welded over the substrate, material not forming part of the design being removed. Very attractive results can be obtained, using contrasting or complementary colours and finishes, but the technique also can be more expensive than ordinary tear-seal work because the cut pieces (as examples, the enclosed portions of letters like 'a', 'b', 'o', and 'p'), may have to be taken away by tweezers.

Electrodes, jigs, and dies required for work of this nature also would be more complex than those for simple straight welds—hence, they are more exacting and expensive to prepare; the extra cost involved has to be justified by the quantity to be made (that is, the cost can be carried by being spread over many units), or the individual products must be of high value.

At times facilities for setting-up and welding electrodes are required for work in 'three dimensions'—as in the covering of the backs of kitchen chairs, where the weld in the flexible poly(vinyl chloride) material has to follow both the profile and the curve of the inner plywood shape. Similar needs are encountered for the manufacture of goods such as inflatable buoyancy chambers for boats and life rafts, where again quite elaborate electrodes and ancillary facilities can be appropriate.

Barrier Materials

Where layers of unequal thickness are to be welded it is necessary, in order to ensure that heating and welding take place precisely where required, to include in the lay-up a suitable barrier material (in such a way as to compensate for the thinner film or sheet, and to direct the high-frequency radiation to its inner surface).

Barriers are essential too in all tear-seal work, to prevent electrical contact between the cutting surface and the electrode or jig: the welding machines are fitted with 'arc-anticipating units', which will operate in the absence of a suitable barrier.

Various standard products and some proprietary materials are available for use as electrical or thermal barriers, including:

Glass-fibre fabric coated with polytetrafluorethylene
Natural and silicone rubbers (silicone with fabric backing)
Poly(ethylene terephthalate) film with paper backing
Varnished cotton.

It must be remembered that under the conditions of use most barrier materials will be heated in some degree, and appropriate allowance made for this.

Selected Applications

(i) Inflatables

The employment of polyurethane materials for life-saving equipment and other inflatable goods has been an important change in this field. High-frequency welding is recognized to be the only way to produce complete seals for products of this nature, including boats, buoyancy bags, escape chutes, escape units for submarines, fenders, floating booms (as containment for spillages of oil or other substances), life jackets, life rafts (of capacities ranging from one to 30 people), and lifting bags (for floatation and salvage).

Other important but perhaps less rigorous applications in the field of leisure include air beds; air-filled structures as enclosures for exhibitions, sports arenas, and so forth; balloons for advertising and other purposes; beach balls; inflatable arm-bands, cushions, rings, *etc*., as swimming aids; and inflatable surf-boards.

(ii) Medical and Surgical Goods

High-frequency welding is very suited to the manufacture of a wide range of products in the fields of medicine and surgery—as containers for solutions, body fluids, catheters, tubes for tracheotomy, and so forth. The completeness of the seal once again is a most important factor, including the ability to weld in place joints and connections as needed.

In general, the design of production lines for medical goods should allow a higher degree of flexibility than with large-volume consumer items, since orders can be smaller and there may be considerable variations of size within any one range.

(iii) Packaging

High-frequency welding has a part to play in this field, including the following.

Sachets. Extruded flexible vinyl 'lay-flat' (tubing, folded flat along the length) is clamped, filled, and sealed in sections, each of the capacity required—to give discrete sachets of preparations such as bath foams, hair lacquers, and shampoos.

'Blister' Packs. This type of pack can be produced by heat-sealing rigid transparent or translucent film to pre-coated card but vinyl thermoformings can be welded more quickly and the welding technique is more versatile.

Transparent Boxes. Boxes can be made in translucent or transparent acetate or rigid vinyl film, with creasing by application of heat and welds at the final tear-seal.

Wallets and Displays. The welding technique is applicable to a great variety of special fabrications, for the display of goods or in advertising.

(iv) Protective Clothing

The fabrication of protective clothing is another important field, in which there are increasingly special requirements. An indication of these is as follows.

Motor Cycling. One- or two-piece protective suits normally are made from fabric-supported flexible vinyl or from polyurethane materials. A typical garment would be sewn but with welds over the seams to ensure that it is waterproof. With polyurethane, welding alone often is quite satisfactory. In some parts of the garments, patches are over-welded as re-inforcement.

Civil Engineering, Mining. Waterproof clothing for employees in civil engineering and mining projects comprises normally separate jacket and trousers, both of which are welded from supported flexible vinyl.

Ambulance, Police, and Postal Services. Clothing for these purposes is typically as for civil engineering but often panels of light-reflecting or fluorescent material are incorporated by welding.

Special Hazards. Disposable clothing for hazardous work (as examples, in decontamination, or in atomic radiation at low levels) may be made from unsupported vinyls by welding.

Sports. Goods such as anoraks, cagouls, and golfing jackets and trousers often are made by welding from lightweight polyurethanes laminated to knitted nylon fabrics.

CHAPTER 6

Adhesives for Plastics Fabrication

P. J. C. COUNSELL

Introduction

The use of adhesives to bond plastics to each other and to other substances—such as metals, paper, and wood—is long-established and may well pre-date even the simpler techniques for joining these materials by applying heat in various ways.

Adhesives offer both advantages and disadvantages in comparison with other methods of assembly, some of the most important of which are presented in summary form at the conclusion of this chapter. (The chapter as a whole comprises an introductory section, a summary of theoretical considerations underlying the effective use of this approach, a review of the types and typical compositions of adhesives suitable for bonding plastics ('solvent-based', 'water-based', 'hot-melt', and 'reactive'), requirements and methods for treating the surfaces in preparation for bonding, and a number of examples of the successful use of adhesives.)

The origins of the plastics industry go back well over a century to the early exploitation of nitrocellulose but the manufacture and use of plastics in a large way came later with the development of petro-chemicals and the resulting ready availability of the many precursors required. Most of the polymers that are of commercial importance now were introduced in the period between the two world wars, or in the years immediately following. It seems reasonable to anticipate that the polymers sold in quantity at present are likely to remain significant for some time in the future; newer materials may be added but changes perhaps will be linked mainly to the modification of existing types for more specific purposes and techniques of manufacture.

In comparison with the large tonnages of plastics consumed in various ways the quantities of adhesives may seem rather small, but this is understandable bearing in mind that only a few grams of adhesive will be needed to support assemblies such as an aircraft or an article of furniture.

Since the amounts required are likely to be limited, only a few polymers have been developed specifically for use as adhesives; more usually, the manufacturers have accepted polymers from large-scale production and modified them as necessary by compounding. In essence, work of this nature is based on formulation and mixing, and a very wide range of resins, ancillary chemicals, plasticizers, fillers, and so forth, is employed in order to create products which have the balance of properties required for each purpose.

It is worth noting that many of the manufacturers have close connections with the surface coatings industry, and there is much common ground between the formulation of surface coatings and of adhesives. Much knowledge (on matters such as the properties of formulations, appropriate methods of application, and qualities like resistance to environmental conditions) is derived from practical experience over a long period of time, and thus has a value that often is difficult to protect; because of this, manufacturers usually are not willing to disclose detailed information.

The preparation of adhesives for plastics presents some special problems (the surfaces of plastics are especially difficult to unite in this way) but many of the difficulties have been overcome and bonding is a method to be considered on equal terms with the welding or fastening of components by mechanical means. Bonding with adhesives is the method to be preferred when offering the prospect of satisfactory results at lower costs per unit.

Theoretical Aspects of Adhesion

Adhesion is a common daily occurrence that we recognize without difficulty, but unfortunately recognition is much easier than explaining in detail how it comes about. Why indeed should there be so strong an attraction between the molecules that it becomes difficult to separate the surfaces or bodies concerned? Summarized below are three theories or explanations that have been advanced over the years, often with much compelling argument and data in substantiation. Very briefly, it is said that:

1. Adhesion may be attributed to the effect of attractive forces between molecules at the interface, giving rise to chemical and physical adsorption of the materials concerned
2. Atoms at the interface inter-diffuse, so the materials become entangled and the transition between the separate bodies or surfaces concerned becomes less distinct
3. Electrical charges created at the interfaces lead to electro-static attraction between the bodies or surfaces.

In the view of the present author, the first suggestion above—chemical and physical adsorption—is the most persuasive; if it is correct that the basis is intimate molecular contact between adhesive and substrate then for good results an adhesive must wet the substrate effectively. For anyone who has dealt with adhesives in a practical way the importance of wetting is quite clear.

Materials such as polyolefin and polyfluorinated hydrocarbon plastics present special difficulties for the technologists concerned because of the problems of wetting. (With these, bulky methyl or fluoroalkyl groups at the surfaces prevent close contact between the molecules.)

The second theory—inter-diffusion—would seem applicable to the particular case of the adhesion of like surfaces (known also as 'auto-adhesion'). Certainly it occurs in the welding of plastics, and in instances such as tyre manufacture when unvulcanized rubbers adhere spontaneously. On the other hand, it is unlikely that much inter-diffusion takes place when materials dissimilar chemically are brought together. It is well-known that many plastics are incompatible and if mixed will, over a period of time, tend to separate—forming eventually discrete layers. Behaviour of this kind is precisely the reverse of inter-diffusion.

The third suggestion—the electro-static theory—nowadays attracts little support. It was proposed originally when electrical discharges were observed when bonds were parted, but this feature can be explained by the mere separation of an electrical charge when rupture takes place. Tribo-electric effects giving rise to static charges occur when many materials are subjected to friction or to other mechanical operations, and it is tempting to extend this observation to embrace the energy required to break a bond between adhering materials. Of course, specific instances do exist of electro-static attraction giving rise to adhesion (as in office photocopying, where charged particles of toner ('ink') are attracted to the surface of paper carrying an opposite charge, and then fused subsequently to the paper by heat). However, the general application of electro-static theory as an explanation for adhesion must be called into question.

Adhesives Suitable for Bonding Plastics

When adhesives are used they must be in fluid condition during application, so that they can be transferred easily to the substrate from the containers in which they are stored, and caused to flow and to wet effectively the surfaces to be bonded.

Later the viscosity of the adhesive must increase so that eventually it takes on solid form: the transition from liquid to solid is brought about in various ways, as follows.

Solvent-Based Adhesives. In these formulations the polymers comprising the adhesive proper are dissolved in volatile solvents, which evaporate in the course of the transition.

Water-Based. In these adhesives, the polymers are in aqueous dispersions or solutions from which water evaporates to effect the transition; in the case of dispersions, the particles coalesce to form a continuous film; with solutions, the film is formed from polymer coming out of solution.

Hot-Melt. ISO 472 defines 'hot-melt adhesive':

> A thermoplastic adhesive that is applied in a molten state and forms a bond on cooling to a solid state.

Essentially, the transition is effected by the cooling of molten polymers.

Reactive. in this type, chemical changes take place in the formulation, as a result of initiating factors such as the application of heat, the mixing of two active components, exposure to ultra-violet radiation, or the absorption from the atmosphere of water vapour.

Film Adhesive. A further type which is of importance commercially is the 'film adhesive' which exists usually as pre-formed shapes on suitable substrate materials (often, siliconized release paper). When being applied, such adhesives are activated by heat or pressure and so converted into a semi-fluid condition, when they can wet the surface to be bonded. The 'pressure-sensitive' adhesives are an example of this type—the application of finger pressure alone being sufficient to form the bond.

However, the final strength of the bond obtainable with such adhesives is limited as in general they do not possess high cohesive strength. Other disadvantages include poor strength at elevated temperatures, and indifferent resistance in the long term to sustained loads ('creep').

Polymer systems and formulations for the principal types of adhesives are reviewed in more detail in the sections following.

(i) Solvent-Based Adhesives

This group is very versatile and capable of performing well in a wide variety of applications. The adhesives concerned are based on solutions of elastomers or resins (natural or synthetic) in organic solvents, and further ingredients including:

Fillers—to reinforce the polymer and to reduce the overall cost
Modifying resins—to increase the 'tack' of the adhesive and, sometimes, to reinforce it
Plasticizers
Stabilizers and anti-oxidants—to help prolong the service life of the adhesive
Wetting agents—to improve stability in storage prior to application and, sometimes, to improve the ability of an adhesive to bond more intractable surfaces
Pigments (if required)—to impart colour to the eventual dry film of adhesive.

The nature of the solvent plays an important part in establishing the properties of the adhesive: as examples, the volatility of the solvent influences the speed of drying and the time during which adhesive-coated surfaces are capable of forming a bond when brought into contact.

Nearly all the organic solvents available commonly are flammable and the volatility under ambient conditions governs the volume and speed at which, when an adhesive is being used, ignitable vapours are released. The most volatile 'quick-drying' solvents present particular requirements for forced ventilation and for other precautions, such as the elimination of any possible source of ignition.

In general the chlorinated hydrocarbon solvents are non-flammable and at first sight might be considered more suitable for use in adhesives, but such materials also have disadvantages:

The flammable solvents will dissolve a wider range of polymers and resins

Flammable solvents also offer a wider range of volatilities

There are serious doubts in relation to the hazard to health that chlorinated solvents could pose (and concern too about the damage believed to be caused by their use to the ozone layer in the upper atmosphere).

For these reasons, the use within the European Community of many of the most well-known of these solvents is to be ended within ten years.

Concern as to possible effects of solvents on the environment no doubt is influencing the extent to which solvent-based adhesives are used. In many parts of the world air quality is affected adversely by features such as exhaust emissions from motor vehicles, and further restrictions on these and other emissions (like volatile solvents) must be anticipated. 'Global warming' is said to be taking place because of increased volumes in the atmosphere of 'greenhouse gases'—the most important of which is carbon dioxide, a common product of the oxidation of organic materials. Chlorofluorohydrocarbons ('CFCs') and methane also are of significance—and hence some of the solvents used hitherto, as possible contributors to 'global warming', may be thought undesirable.

Such doubts about the future are encouraging the gradual replacement of solvent-based adhesives by other types.

(ii) Polymer Systems for Solvent-Based Adhesives

Acrylonitrile–Butadiene Elastomers. These polymers, the so-called 'nitrile rubbers', are used dissolved in ketone or other highly polar solvents. When they are compounded with thermosetting phenolic resins it is possible to obtain good resistance to elevated temperatures.

Typical applications include the bonding of vinyls, of expanded polyurethane, nylon, vinylidene chloride copolymers, and phenolic resin mouldings.

Polychloroprene Elastomers. Compositions based on polychloroprene have been the mainstay of the 'contact' adhesives (in which coalescence of film to form an instant, tough bond takes place when surfaces coated with the adhesive are brought together under light pressure). These types almost always are compounded with magnesium and zinc oxides, which serve two main functions—to provide an acid-acceptor mechanism for traces of hydrochloric acid arising from degradation of polymer and a degree of vulcanization of the film (thus improving its physical properties and resistance to heat).

Polychloroprene adhesives offer good resistance to acids, alkalis, oils and water, and are used widely for bonding rigid thermoplastics such as ABS and

PVC. They are tolerant of difficulties in application and often will perform well even if applied and used in conditions that are not ideal.

Polyurethane Elastomers. Linear polyurethane polymers dissolved in ketone solvents provide a very versatile group of adhesives. Good results are obtained without fillers or other ancillary components in the formulation (indeed, the polymers concerned often are not compatible with the modifying resins and plasticizers that have been used for some time in other types of adhesive).

Polyurethanes show particularly good results when used for bonding plasticized poly(vinyl chloride) compositions and good results also with other plastics and rubbers, such as the thermoplastic styrene–butadiene–styrene ('SBS') copolymers.

These advantages have led to the wide use of these adhesives in making shoes, where to a large extent they have replaced other types.

Styrene–Butadiene Copolymer Elastomers. 'SBR' elastomers are employed in low-cost contact adhesives suitable for less-demanding applications—such as when exposure to elevated temperature is not likely, and when a bond of moderate strength is adequate. They can be dissolved in aliphatic hydrocarbon solvents and used to bond solvent-sensitive substrates like expanded polystyrene.

(iii) Water-Based Adhesives

Products comprising hydrophilic polymers dissolved in water are well-known and used widely as adhesives but are of little general significance for bonding plastics. The present chapter is concerned only with products based on polymer dispersions, which consist of small discrete particles of diameter about one micron (1 μm, or 10^{-3} mm) suspended in a continuous water phase. In most instances a protective colloid is present at the interface between the particles of polymer and the water and this helps to stabilize the dispersion and prevent premature coalescence of particles. Dispersions such as these are known as 'oil-in-water' types. With them, the molar mass of the polymer species comprising the dispersed particles does not affect the viscosity and so polymers of high molecular weight can be applied in this way.

It is possible for the dispersion to be the other way round—a continuous polymer phase in which droplets of water are dispersed. In this type, called 'water-in-oil', the polymer phase controls the viscosity and usually such dispersions are more viscous.

The solids contents of these dispersions generally are of the order of 50% to 60%, occasionally even higher, so that one may deposit from a certain thickness of wet adhesive quite a thick dry film of polymer. Solvent-based adhesives generally have lower solids contents (20% to 30%), and since the shrinkage after drying of water-based dispersions also is lower such adhesives can be formulated to fill gaps between materials as well as to bond them.

As water is non-flammable so inherently are the dispersions; they are without the environmental fears and risks associated with volatile solvents and there may be an economic advantage too in omitting a range of solvents that is derived from petrochemicals.

However, adhesives of this type also have disadvantages. The high surface tensions may make it difficult for them to wet hydrophobic surfaces like those of plastics (although in this the incorporation of wetting agents or small amounts of solvent can be of help). Usually considerably more energy is required to dry water-based adhesives; often such adhesives perform best when one or more of the materials to be bonded is porous and capable of adsorbing water from the dispersion. Some dispersions do not have good stability in the presence of ions of multivalent elements, resistance to freezing or to excessive agitation. When such conditions are likely to be encountered, the addition of stabilizers may be necessary. It should be kept in mind that the stabilizers are hydrophilic protective colloids and that the presence of small amounts of these in the dried adhesives will affect adversely their resistance to water.

(iv) Polymer Systems for Water-Based (Dispersion) Adhesives

Acrylics. The acrylic polymers are versatile and can range from soft and 'tacky' through to tough 'rubber-like' substances. As a group they give dry films offering good stability on exposure to ultra-violet and visible light—important qualities for uses in which visual appeal is required. Their resistance to heat is not particularly good, though adequate in many cases for bonding materials such as poly(vinyl chloride) compositions and expanded plastics.

Polychloroprene. Polychloroprene dispersions have a range of qualities similar to those of solvent-based polychloroprene adhesives and a similar range of uses. As an example, the bonding of vinyl materials with phenolic resin/paper decorative laminates often is carried out with these products. It is necessary to incorporate acid-acceptor dispersions of metallic oxides, and the dispersions in general do not provide such long open times as solvent-based polychloroprene adhesives.

Styrene–Butadiene Rubber Latices. SBR latex is used mainly when the requirements are fairly modest and low cost is necessary. An example is in fixing vinyl-based floor coverings to wooden or concrete sub-floors—in which products comprising heavily filled styrene–butadiene rubber latices give adequate results.

Vinyl Acetate–Ethylene. Materials based on copolymers of vinyl acetate and ethylene exhibit a good balance of properties. As the ratio of vinyl acetate to ethylene is increased the copolymers become softer, more hydrophilic, and offer enhanced adhesion. However, at the highest levels of vinyl acetate the films become hard again, showing good adhesion to substrates such as wood and paper but with resistance to water somewhat inferior to that of the intermediate copolymers.

Dispersions of vinyl acetate–ethylene copolymers have been used with success in bonding plasticized and unplasticized poly(vinyl chloride).

(v) Hot-Melt Adhesives

Essentially, hot-melt adhesives are blends of polymers that are melted and kept at an elevated temperature until applied to a cooler substrate: the second substrate is brought into contact immediately with the molten film and afterwards the adhesive cools rapidly, a bond being formed almost instantaneously. The temperature of the molten adhesive generally is in the region 160 to 200 °C, so that its viscosity is low enough for application—and this means that the substrates to be bonded must be capable of withstanding the effect of adhesive at this temperature (this can cause difficulties with some thermoplastics, especially expanded materials).

Such adhesives are required to remain stable at elevated temperatures over periods of several hours (during normal daily operation of production equipment) and formulations that decompose under such conditions or in which pronounced changes of viscosity occur are not suitable. In order to achieve the bond strength required the adhesive must wet properly the substrates as soon as it is applied—so the temperature of the substrates can be important (if too cold they may absorb heat, cause cooling of the adhesive prematurely—before the surfaces are wetted—and impede bonding).

For successful use of such adhesives, not only must the formulation be suitable but the system for application also must be efficient. The molten adhesive must be maintained in a fluid condition and at the correct temperature for application: over-heating could cause degradation, and if it is too cold the viscosity of an adhesive might be too high. The hot fluid is pumped to the substrate through a suitable nozzle but if the distance between the reservoir and the nozzle is appreciable there may be cooling and a need for heated hoses. Sometimes, molten adhesive is transferred to a coater of the heated-roller type, where again the control of temperature is important. There is a wide variety of equipment for application and every manufacturer of hot-melt adhesives will advise on systems that are suitable for use with them.

The most significant limitation on the usefulness of these adhesives is poor strength at elevated temperatures. Most compositions of this nature are not suitable for use at temperatures in excess of 50 to 60 °C, and show also relatively poor resistance to sustained loads (that is, resistance to 'creep'), even at room temperature.

However, techniques have been developed for formulating hot-melt adhesives from modified polyurethanes containing labile isocyanate groups and these are capable of reacting with moisture in the atmosphere to give cross-linking—the overall resistance of the system to heat thus being improved. It is likely that in the future further developments along these lines will be seen.

(vi) Polymer Systems for Hot-Melt Adhesives

Many thermoplastic materials are available commercially and it might seem at first sight that all of them could be considered for hot-melt systems. However, relatively few of them offer the suitable combination of melt viscosity, long-term melt stability, wetting action, and mechanical properties in the solid state.

Bitumen. Bitumen is a black thermoplastic obtainable from the refining of crude oil. Its composition is somewhat variable, depending upon the type of crude, and it tends to stain light-coloured substrates. However, the relatively low cost of the material makes it attractive for some applications and it may be compounded with other thermoplastics—like the polyolefins.

Hot-melt adhesives based on bitumen show good adhesion to many materials and are resistant to water and weather. One important commercial use in which plastics are involved is the construction of road signs and other street furniture, where these adhesives are used to bond polyethylene sleeves to steel tubes.

The addition of thermoplastics to modify road surface compounds based on bitumen has been reported, the object being to improve the flexibility of the compounds and resistance to skids.

Butyl Rubber. Butyl rubber is used in conjunction with other thermoplastic hydrocarbon rubbers (for example, polyisobutylene) to make pressure-sensitive hot-melt adhesives. Such formulations include also tackifying resins and oils. Adhesives of low viscosity are produced, and these may be applied by the 'hot-melt spray technique' (see page 106).

Ethylene Copolymers. Ethylene copolymers probably are the most important materials in hot-melt formulations. Ethylene–vinyl acetate and ethylene–ethyl acrylate polymers are very versatile and available in a wide range of grades offering different co-monomer contents and viscosities. The melts are stable and compatible with various modifying resins, waxes, extenders, and fillers. Adhesion to many substrates is good—including the polyolefin plastics, which are difficult to bond with most other types of adhesive unless the surfaces are pre-treated.

The range of applications includes: sealing cartons, bonding packaging films and laminates, fabrications for the motor industry, and applying veneers to chipboard and plywood.

Polyamides. Polyamides for this purpose are derived from the reaction between dimer acids and short-chain polyamines: they range from brittle, resinous solids with sharp melting points to more rubbery and flexible substances with wider spreads of melting temperatures. Materials of the former group give rather brittle bonds but offer somewhat higher service temperatures than are possible with the more flexible types. Polyamides have good resistance to oils and to solvents but their resistance to water is less satisfactory than for less polar materials. They also are more prone than other hot-melt systems to thermal degradation.

Styrene Copolymers. The so-called 'thermoplastic rubbers' based on styrene–butadiene–styrene and styrene–isoprene–styrene block copolymers can be used for hot-melt adhesives, particularly when extended with tackifying resins and oils. They can be made into pressure-sensitive adhesives, as melts with low viscosity—being applied from fine spinnerets which are oscillated to make a pattern on the substrate, only one surface of which has to be coated. In such a system, the fine filaments of adhesive coming from the spinnerets cool very rapidly, so there can be no damage to substrates that are sensitive to heat. Permanent tack means that the adhesives have very long open times, and coated articles may be stored for a period before being bonded to the second surface.

(viii) Reactive Adhesives

Several different types of reactive adhesive exist and are used, the properties and applications depending in each instance on the chemistry of the setting reaction. Since the latter accompanies the change from liquid to solid form there is a limitation in all such systems in terms of the time available after reaction has started. (The period in which the adhesive can be applied in these circumstances is known as the 'pot life'.) During the reaction the lengths of the molecular chains increase progressively, giving in general first an increase in the viscosity, followed by gelation, by setting to a solid condition, and lastly by extension and/or cross-linking of the chains to reach the final form. The last stage may require many hours or even days for completion, so the adhesive will not become fully effective until some time has elapsed.

Setting is governed by the kinetics and the Arrhenius equation, and accelerates with increased temperature (a convenient 'rule of thumb' is that the setting time is reduced by half for each 10°C increase: if 24 hours are required for an adhesive to set at 25°C, this may be reduced to $1\frac{1}{2}$ hours at 65°C, and to about three minutes at 115°C).

(ix) Cure Mechanisms for Single-Part Reactive Systems

In single-part reactive systems the formulation contains all the components necessary to give cure but this does not happen under normal conditions, curing being initiated by one or more of the following.

Heating. The reaction is started by heating the formulation to about 150°C: at first the viscosity of the mixture declines but after reaction has commenced viscosity rises progressively, leading to gelation and hardening. At the relatively high temperature the final stages take place rapidly and often total heating times of 15 to 30 minutes are adequate.

Examples of systems cured in this way are:

Epoxides cured with dicyandiamide

Butadiene rubbers cured with sulphur/rubber accelerators.

Because of the temperatures applied in curing, such adhesives are not suited to use with most thermoplastics.

Since curing at higher temperatures gives more extensive cross-linking, such adhesives usually offer enhanced resistance to heat.

Taking up Moisture from Atmosphere. Adhesives that react by taking up atmospheric moisture remain fluid while in storage but absorb water after being applied and this initiates cross-linking, gelation, and hardening.

The best examples are polyurethanes containing water-labile free isocyanate groups; these perform well on duties such as bonding polyurethane surfaces for sporting events.

Another well-known type is the RTV ('room-temperature vulcanizing') silicone sealants, in which adsorbed water causes hydrolysis of acetoxy groups on silicon atoms within the chemical structure, creating sites for cross-linking and the consequent hardening of the sealant to a rubbery state.

These adhesives must be packed so as to prevent ingress of moisture before application, and to ensure they are in satisfactory condition after distribution and storage. This means in turn, in some instances, that before the packs are closed air must be displaced from them by applying, say, dry nitrogen gas under pressure. When containers of the latter kind have been opened and part of the contents used the material remaining may be de-stabilized and gel quite quickly.

Irradiation by Ultra-Violet or Visible Light. Systems of this nature comprise materials such as acrylic resins in which a photo-sensitizer (for example, benzophenone) is incorporated. On exposure to ultra-violet radiation the photo-sensitizer dissociates into free radicals which then catalyse polymerization of the acrylic material. The reaction is controllable and takes place within a few seconds; the ultra-violet radiation may be directed and focused so as to obtain cure precisely where needed. Products like these are used to retain coatings, wiring, and other components in the assembly of electronic goods.

The main disadvantage is that at least one of the substrates must be transparent to the radiation initiating cure. This requirement prevents also the filling or pigmenting of such materials. Nevertheless, many applications are being developed—bonding glass, transparent plastics, jewellery, and optical fibres.

Presence of Water on Surfaces to be Bonded. Substances in the Earth's atmosphere adsorb water from it, the water in many instances being bound tenaciously to the surface. This provides the basis for curing of the cyanoacrylate adhesives, which takes place on contact with surfaces covered with such a water layer.

The products essentially are esters of α-cyanoacrylic acid, predominantly methyl and ethyl:

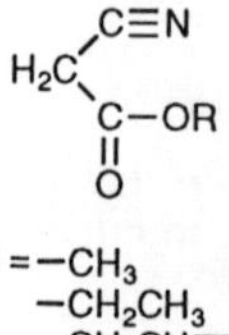

$$H_2C{=}C(C{\equiv}N)-C({=}O)-OR$$

$$R = -CH_3,\ -CH_2CH_3,\ -CH_2CH{=}CH_2,\ -CH_2CH_2CH_2CH_3$$

The cyanoacrylate molecule has the electron-withdrawing groups, cyano and ester moieties, attached to the same carbon. They activate the carbon atom to nucleophilic attack from the moisture on the surface of the substrate, and this results at room temperature in an anionic polymerization with rapid growth of the chains and very short setting times (a few seconds).

The most rapid polymerization takes place on nucleophilic surfaces (that is, basic surfaces containing alkaline components) and the reaction on acidic surfaces can be slow. To obviate this a basic activator (for example, an amine/solvent blend) may be necessary on acidic or porous surfaces like wood, and this should give satisfactory bonds within convenient periods of time.

Cyanoacrylates can be used to bond many materials, including most thermoplastics and even the more difficult ones like polyethylene, polypropylene, and ethylene–propylene–diene ('EPDM') rubber. The best results are obtained with close contact and narrow bonds (some formulations have limited ability to bridge large or irregular gaps between the surfaces).

Many cyanoacrylates give rather rigid bonds which may in consequence be brittle and not able to resist peeling forces. However, in recent years more flexible toughened grades have been developed for applications in which resistance to peeling is required.

Since they are thermoplastic the cyanoacrylate adhesives have limited resistance to heat; they do not resist moisture, and can be softened by highly polar solvents like ketones. They are expensive but since only a small quantity is necessary to form a bond, their overall economy in use is good.

(x) Cure Mechanisms for Two-Part Systems

In two-part systems the adhesive is supplied as two materials in separate containers and to initiate setting the two must be mixed immediately before application. Mixed material either has to be used within the pot life of the system or must be discarded; application equipment or tools used for the work either must be disposable or must be cleaned with a suitable solvent immediately after the cycle.

With most systems of this nature it is important that the two parts be mixed in the correct proportions (that is stoichiometrically) so that the reactions are completed and an excess of neither component is left in the hardened product. The most usual ratio of mixing is equal proportions of both parts (1:1) but differing ratios are by no means uncommon—especially for systems in which chain polymerization initiated by free radicals takes place (when relatively small amounts of free radical initiators will start the setting).

Often, to prepare for work requiring only a small amount of material, it is convenient to draw the appropriate weights or volumes of the two components from the containers into a suitable disposable cup, then to mix and apply to the substrates by a method appropriate for the viscosity of the mixture. If the scale warrants it is preferable to use application equipment designed for the purpose, in which the components may be extruded as necessary from twin reservoirs through a suitable static mixing nozzle and then to the substrate. It is possible to obtain for this work applicator guns with dual pistons, operated either by hand or by pneumatic pressure. Alternatively, for production on a very large scale, equipment may be obtained with large pressurized tanks serving feed lines to the mixing nozzle or nozzles.

Polymer systems suitable for two-part adhesives include the following.

Acrylics. Many acrylic monomers are capable of being formulated into adhesives. Methyl or lower alkyl methacrylates often are used, although they have a strong odour which frequently is said to be objectionable.

Monomers with lower odour (such as hydroxyethyl methacrylate and tetrahydrofurfuryl methacrylate) are available, but in many ways the bonds obtained with the more odorous materials are better.

Generally, acrylic adhesives are cured with a two-part redox system—one part of which (the 'initiator') is in the base component, and the second (the 'curative') in the accelerator. Most often the former is a hydroperoxide, while the curative is a reaction product of aniline and n-butyraldehyde.

In application the base component, containing the initiator, is spread on one of the two surfaces of the assembly and the accelerator (with curative) is put on the other surface. Mixing of the two components takes place when the surfaces are brought into contact, with the help of light rubbing, and the subsequent polymerization and setting reaction take place within a few seconds.

Acrylics such as these (the so-called 'second generation acrylics') contain a rubbery polymer intended to toughen the matrix; in a patented process, chlorosulphonated polyethylene was grafted for this purpose to the acrylic materials.

As a general comment, acrylic adhesives are tough materials with good strength in both peel and shear and are reputed to be capable of bonding even to surfaces that are slightly contaminated (such as by oil). The strength of the bonds obtained increases rapidly at room temperature. Such qualities have given rise to large-scale applications in a variety of industries bonding plastics such as ABS, polystyrene, polycarbonate, poly(methyl methacrylate), phenolic resins, and rigid vinyls.

Epoxides. Two-part epoxide adhesives were introduced soon after the second World War and are the type used most commonly for structural work. They have been investigated extensively and very wide ranges of resins, modifiers, curing agents, and other additives have been suggested for them (a complete summary is beyond the scope of the present chapter).

Epoxide adhesives are based on precursors containing the three-membered, oxygen-containing oxirane group:

This group is reactive towards nucleophilic and electrophilic species, giving rise to opening of the oxirane ring and to further reaction between the molecular chains.

With these adhesives the setting reaction proceeds without significant change in volume, so that the final assembly suffers little stress arising from shrinkage.

There is much use of modifiers of the basic system, including accelerators to increase the speed at which curing takes place, toughening agents intended to give enhanced resistance to peel forces, and fillers to alter the rheology and to reinforce the set materials.

For curing at room temperature, systems based on primary and secondary amines and polyamides are used extensively. More reactive products are obtained with aliphatic chains rather than aromatic amines, which have lower basicity. Sulphur compounds like mercaptans react rapidly with epoxide systems and will provide a basis for rapid-curing—but the bonds resulting tend to be rather brittle and also to have the unpleasant odour of the mercaptan.

Many such materials will cure at room temperature but a moderate heating cycle reduces greatly the time taken for curing and can give products with enhanced resistance to elevated temperatures or to contact with chemicals.

These adhesives bond well with most plastics, other than the polyolefins and fluorinated polymers.

Polyurethanes. Polyurethanes were discovered in the 'thirties by Otto Bayer and have proved to be very versatile. It is possible by modifying the chemistry of the precursors to make products ranging from flexible expanded materials to hard, rigid resins—and similar scope is available also in designing polyurethanes for adhesives.

The chemistry is based on the reaction of isocyanate and hydroxyl groups:

$$R^1\text{—N=C=O} \quad + \quad \text{HO—}R^2 \longrightarrow R^1\text{—NH—}\overset{\text{O}}{\overset{\|}{\text{C}}}\text{—O}R^2$$

where R^1 and R^2 form part of the molecular chains.

The isocyanate group is very reactive and will condense with other active compounds containing hydrogen, like amines and thiols. Products resulting from such reactions, respectively, are the ureas and thiocarbamates. With amines the rate of reaction is somewhat faster than with hydroxyl compounds, while that with thiols is slower. However, a wide variety of catalyst systems (many based on amines and organometallic compounds) is available for the acceleration of isocyanate reactions.

Isocyanates also can react with water, to give initially carbamic acid—which decomposes immediately into carbon dioxide and an amine.

Two-part polyurethane adhesives can be formulated to be tough, elastomeric materials capable of giving good bonds under both peel and shear loadings: they are particularly good for bonding dissimilar substrates such as metals to plastics, or to glass, perform well under cooler conditions, and show acceptable resistance to temperatures up to about 100°C. (Hybrid materials containing a proportion of polyamine precursor for reaction with isocyanate groups will withstand higher temperatures, but usually this is at the expense of flexibility in the set product.)

Some polyurethanes—those based on polyester polyols—are subject to hydrolysis at elevated temperatures: polyether polyols are superior in this.

Structural adhesives based on polyurethanes are used widely in the transport industry, especially for bonding plastics reinforced with glass or other fibres. They are used also as sealants in glazing for motor vehicles—most notably following the introduction in recent years of front and rear windows bonded as elements in the construction.

Pre-Treatment of Plastics Prior to Bonding

A variety of techniques is available for joining plastics to themselves or to other materials—mechanical fastening, welding, and bonding with adhesives—and all are used extensively. Each method has both advantages and drawbacks; for plastics, bonding might be said to be preferable when they are to be joined with dissimilar materials.

Reference was made earlier to difficulties of wetting plastics surfaces, and to problems in this regard compared with hydrophilic materials like glass, metals, and timber. With some plastics very effective bonds are possible, but as a group, mainly for this reason, they are amongst the most difficult to join by adhesives.

The paragraphs below review several methods that are applied with a view to improving this position.

(i) Removal of Contamination

It is desirable always to remove from surfaces to be bonded all forms of contamination—general dust and dirt, flakes of paint or rust, grease, moisture, oils, and plant materials like fibres, pollen, or straw. This can be done in various ways—by washing, cleaning with solvents, applying vacuum suction or air pressure, and light abrasion, as appropriate. However, plastics require special consideration because often aids like anti-static and mould-release agents are added to the compositions in order to change the behaviour of the surface, and these either may migrate to the surfaces during processing or may be applied deliberately at certain stages. The presence, types, and amounts of such agents always should be considered when trying to obtain satisfactory bonding of plastics.

(ii) Mechanical Roughening

Several methods are available for roughening the surfaces to be bonded—ranging from simple rubbing with suitable cloth, paper, or abrasive in powder form, to more severe treatments like mechanical grinding or shot-blasting.

The most suitable in any particular instance is best established by trials in advance. Sometimes, fairly severe abrasion (by increasing the surface area available) is the most successful. On the other hand, excessive roughening may weaken the substrate and so give inferior results. With fast-setting adhesives like the cyanoacrylates, microscopic bubbles of air may be trapped between adhesives and substrate and so impair bonds.

(iii) Priming

Primers of low viscosity, usually dispersions or solutions, may be applied in thin layers to the substrates and generally are allowed to dry before the adhesive proper is spread. Such primers are intended to wet the surfaces thoroughly, and should be compatible with the relevant adhesive. In some instances they offer the further advantage of sealing porous substrates and so making possible more consistent bonds.

Examples of materials used in such primers include:
Acrylic dispersions (for polystyrene)
Chlorinated rubber in solvent solution (suitable for general purposes)
Phenolic resins in solvent (for nylon)
Polychloroprene in solvent (general purposes)
Polyurethanes with isocyanate functionality (for polyurethane adhesives)
Chlorinated polypropylene in solvent (for polyolefins)
Chlorinated vinyl acrylic dispersions (for epoxide adhesives)
Silanes in solvent (for glass).

(iv) Chemical Treatments

The usual chemical treatments comprise aggressive materials (which must be used with due care) and these attack the surfaces to give an oxidized layer with higher polarity and enhanced wettability. The surfaces before and after application of such treatment may be compared with the help of electron microscopy, when often increased roughening of surfaces also can be detected.

Chemical treatments include:
Chlorinated isocyanurate compound in ethyl acetate solvent* (for rubbers, especially thermoplastic styrene–butadiene–styrene rubbers)
Chromic acid (for all plastics, but particularly polyolefins)
Acidified potassium permanganate (for all plastics)
Sodium in naphthalene/tetrahydrofuran (for all fluorinated plastics).

With the last example in particular the time of treatment is of critical importance. The reagent reacts with fluorine atoms in the composition, leaving a surface rich in carbon to which the adhesive may bond. There is a tendency for the surface to darken under the treatment.

(v) Flaming and Corona Discharge

The application of flame or corona discharges is thought to be effective mainly by oxidizing the surface, so giving a polar, wettable layer. Such treatments are most appropriate for treating large areas swiftly, as in the manufacture of plastic film.

In flame treatment, an oxidizing flame is deployed so as to impinge on the surface for a very short time (thus avoiding melting the plastic). As time is

* Satreat—a trade mark of the Shoe & Allied Trades Research Association.

allowed to elapse afterwards the effectiveness of the treatment diminishes and because of this it is important to apply the adhesive as soon as possible after flaming. This approach is used quite extensively to treat polyolefin plastics.

Electrical corona discharge may be carried out in air or in other gases. For plastics based on olefins the best results are obtained in dry air but in treating nylon greater success was reported with discharge in a nitrogen environment.

Corona discharge is a popular method for applications on a large scale and because it can be carried out quickly in dry conditions; in general it is less hazardous than chemical treatment and leaves no visible sign on the surfaces treated. On the other hand, there is some evidence that it is less effective in promoting adhesion to plastics than treatment with a substance like chromic acid.

Examples of Bonding Plastics with Adhesives

(i) Cyanoacrylates

An outline of the chemistry of cyanoacrylate adhesives was given earlier in this chapter. They have been found to give good results with a wide variety of plastics, especially:

Acrylonitrile–butadiene–styrene
Phenolics
Polycarbonate
Polyesters
Poly(methyl methacrylate)
Polystyrene
Polyurethanes
Styrene–acrylonitrile.

In many cases a light sanding of the substrates in advance of application was found to improve the strength of the bonds.

For some plastics, surface treatments were necessary—for example, chromic acid treatment for polypropylene. When fillers were present in the plastics the results with cyanoacrylates often were better (this was apparent in particular with polyethylene).

In most instances the resistance of cyanoacrylate bonds to humid aging (at 40°C, saturated humidity) was found to be quite good; when there was a reduction of strength on exposure to these conditions no further diminution occurred within three months.

(ii) Hot-Melt Spray Adhesives

The general principle of application of this newer type of adhesive is as described earlier, melting taking place in a closed tank at 170°C and the material being pumped through heated lines to a heated nozzle for transfer to the substrate. The difference is in the design of the nozzle, which may either give a fine spray of tiny droplets of adhesive or fine threads of the molten material,

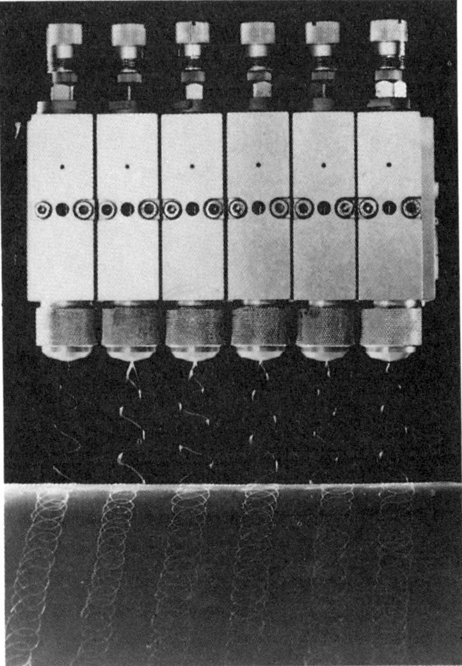

Figure 49 *Arrangements of nozzles for hot-melt spray of adhesives*

and these can be arranged to give patterns of application as illustrated in Figure 49.

Equipment available offers opportunities for widths of coatings ranging from a few millimetres to one metre or more. Since the adhesive is applied in fine spray or threads there is much less chance of damage to substrates sensitive to heat. The open time is quite lengthy, so adhesive may be applied over large areas before bonding is done—and as there are no solvents there are advantages with regard to the health of operators, safety, and the risk of fire. Such systems are suitable too for application by robots.

Hot-melt spray adhesives will bond successfully carpets, fabrics, laminating panels, metals (both painted and unpainted), and both polyethylene and polypropylene film and sheet. Two important commercial uses at the time of writing are: bonding sleeves of low-density polyethylene to ductile iron pipes; and bonding sound-deadening materials and carpets to the floors of motor cars.

(iii) Acrylic Adhesives for Bonding Boat Hulls

In recent years a new type of thermosetting polymer based on polydicyclopentadiene has been developed and marketed.* Usually this is amber in colour and it is made by a reaction injection moulding process employing various catalysts:

* Metton—a trade mark of Hercules, Inc.

Figure 50 *Catamaran with adhesive-bonded hulls*

its properties render it suitable for producing structural components of boats and motor vehicles.

An example of use for this material is the construction of hulls for a novel type of catamaran—a vessel designed to be transported easily on the roof rack of a car (and for which in consequence the total weight must be kept down). The frame and mast assembly are made from aluminium tubing, bonded with an epoxide adhesive; the floats are made in two halves from Metton, and fixed in the course of manufacture using a two-part acrylic adhesive, rubber-toughened, with a very fast setting time (setting hard in nine minutes). The two components for the adhesive are packed in dual cartridges and injected on the line of the bond by a special dual-piston gun with a static mixing nozzle. This method of fixing the floats is economical and capable of withstanding the stresses involved both in transporting and sailing the vessel (see Figure 50).

(iv) Bonding Grit to Polypropylene Walking Boards

The range of uses for polyolefin plastics, and for polypropylene in particular, has been expanded greatly in recent years: one application of interest in this context comprised large mouldings (approximately 1 m by 0.5 m in area) designed to interlock and to be placed on football grounds or other sports fields

TABLE VII *Results of shear tests of bonds with and without primer*

Identification of specimen	*Adhesion strength of grit and adhesive*/MPa
Without primer	0.1
With primer (chlorinated ethylene–vinyl acetate)	4.1

to prevent crowds and passers-by causing damage to the turf. In view of its durability, strength and lightness in weight, polypropylene was considered very suitable for the purpose but was deficient in the sense that when wet (for example, from rain) the surfaces of the mouldings were slippery. In an attempt to overcome this, the possibility of bonding fine grit to the mouldings was investigated.

The grit comprised particles ranging from 1 to 2 mm in diameter and the initial studies were into ways of retaining these in a layer of wet adhesive: epoxide and moisture-curing polyurethane adhesives were tried. However, adhesion to the substrate was poor (it was found possible to remove easily the entire film of adhesive and grit).

The use of primer solutions was investigated—chlorinated polypropylene and chlorinated ethylene–vinyl acetate, both in toluene: these gave improvements in adhesion, the second being the superior. Such a primer, in conjunction with a moisture-curing polyurethane adhesive, gave a product where it was possible to remove small particles of grit and adhesive only by prising from the surface with a knife.

Quantitative confirmation of this outcome was obtained by shear tests, for which specimens were prepared both with and without the primer, a steel strip being bonded over the grit by an epoxide adhesive and the test carried out after curing of the second adhesive. Adhesion strengths recorded for the grit and adhesive were as shown in Table VII.

(v) Installing Synthetic Surfaces for Sport

For a variety of reasons, demand arose for synthetic surfaces on which games could be played or displays and physical exercises presented. In some instances, existing playing fields, bowling greens, and tennis courts were inadequate—perhaps occupied too fully and not always available when wanted. Grass surfaces could become worn, saturated with moisture, or eroded in places, so that time had to be set aside for their recovery. For some sports and pastimes there was also a requirement for all-the-year-round and all-weather facilities.

These needs led in turn to investigation of ways of preparing synthetic surfaces so that in behaviour they would be very similar to natural lawn or turf. The factors examined in this connection included the heights to which sports balls would rebound, the resistance to balls rolled over the surface ('ball drag'), and the absorption of impact. All were found to depend largely on the method of construction of the synthetic surface, including its underlayers—the correct

Figure 51 *Applying adhesive for a synthetic surface for sports events*

specification of which would be at least as important for the outcome as the specification of the upper or top layer.

Most surfaces of this kind now are laid on a sub-base of hard core, tarmac, or concrete in which has been installed adequate drainage for surface water: over this is spread a shock-absorbing layer of porous rubber crumb, made from shredded tyres bonded together as flexible planar material of thickness 5 to 25 mm and supplied in the form of continuous rolls. The rubber layer may be laid loose or pegged, and sometimes is bonded to the sub-base with a moisture-curing polyurethane adhesive.

The entire installation is finished with a carpet of synthetic grass, which comprises usually a woven polypropylene lattice in which tufts of green polypropylene fibre or filament are fixed. The lattice is sealed on the under-side with a bonded film—which keeps the tufts in place and may be made from polyethylene, polyurethane, rubber latex, or vinyl plastisol.

There are alternative methods of applying the carpet to the shock-absorbing layer:

(1) *laid loose, with seams bonded*: a broad tape is placed under the seam ('seaming tape'—a strong so-called 'geo-textile' with a polyolefin laminated to the underside), adhesive applied to the tape by means of a spreader, and the two lines of carpet rolled into position. Figure 51 illustrates application of the adhesive.

Originally for this work two-part polyurethane adhesives (100% solids) were used, mixed immediately beforehand, but material of this type is sensitive during setting to contact with water and rainfall soon after application was found to cause weakness. In consequence, moisture-curing polyurethane adhesives now are preferred; such systems can be applied under most weather conditions, even at temperatures down to 5°C.

During laying, the carpet is weighted with sand and afterwards this is brushed into and among the grass-like filaments, where it forms a permanent part of the surface and helps to determine its behaviour.

Figure 52 *Employment of a heat-activated adhesive to fix the crown in a protective helmet*

(2) *bonded-down*: a moisture-curing polyurethane solvent adhesive is used to bond the shock-absorbing layer to the sub-base, and the same adhesive spread in bands on the surface of this layer—the strips of carpet then being unrolled into place. In this method, sand is not employed for weighting.

(vi) Assembling Military Protective Helmets

A new approach was proposed for making effective helmets which could replace the former British army steel helmet. Essentially the new helmet used modified phenolic resins reinforced with nylon, and the 'crown cap' inside was thermoformed from polyethylene. Formerly the crown cap was attached to the steel by rivets—not an appropriate method for fixing polyethylene to reinforced plastics. Instead a method was developed with a hot-melt adhesive based on ethylene–vinyl acetate copolymers cast as film on release paper. For assembly, the cast film is cut in advance to match the intricate shape required and activated by heat to bond under light pressure; subsequently, a further heat activation is employed to fix the crown cap in place (Figure 52 illustrates this).

Helmets made in this way are tested in laboratories approved by the Ministry of Defence, the bonds being required to withstand exposure to extreme heat and cold, ballistic and drop tests—so simulating the rigours experienced in the field.

Protective helmets like these are being used not only by the British military but also by other armed forces overseas, as well as for police riot-protection duties in Britain and other countries in Europe.

Some Advantages of Using Adhesives

In summary, to conclude:

(i) adhesives may be used to unite dissimilar materials for which welding is not practicable

(ii) most adhesives can be employed at ambient or moderately higher temperatures, whereas for welding high temperatures usually are necessary

(iii) the stress in adhesive bonds is distributed throughout the area of the join—unlike methods of mechanical fastening in which stress will be concentrated where bolts, rivets or screws are fitted

(iv) structures made with adhesives often are lighter in weight than mechanical assemblies

(v) the line of a bond can provide a seal against the ingress of air, moisture, or other contamination

(vi) adhesives can assist the absorption of vibration within an assembly, and provide other features (like electrical insulation)

(vii) normally, adhesive bonds should not distort substrates, nor (unlike welding) leave joins which may require smoothing subsequently, or concealment in other ways

(viii) generally in adhesive bonding it is necessary to have access only to one side of each substrate, which is not the case with other methods of joining

(ix) the rate of production often can be increased, especially if fast-setting adhesives (like hot-melts, cyanoacrylates, and second-generation acrylics) are employed.

Disadvantages

On the other hand:

(i) no single adhesive system is capable of bonding all materials satisfactorily under all conditions (it is necessary to choose a suitable system from the wide range available)

(ii) for the best results it usually is necessary to prepare surfaces before application of the adhesive.

(iii) a period of time must be allowed before an adhesive bond will reach its maximum strength; the full loading of assemblies in advance of this must be avoided

(iv) different systems offer different properties; many have limitations such as in resistance to heat, to humidity, or to other external conditions: hence, care is necessary in the selection and testing of adhesives for particular uses

(v) the testing of assemblies may require destruction of the bonds—making it difficult to standardize quality control procedures

(vi) normally, bonded assemblies are not demountable.

Acknowledgments. The author wishes to acknowledge the assistance of colleagues at Evode in obtaining the material on which this chapter is based. Thanks are due also to Apex Leisure Investments Limited for the illustration of the Alikat catamaran.* Lastly, the author thanks the directors of Evode Group PLC for permission to publish the chapter.

* Alikat—a trade mark of Apex Leisure Investments Limited.

CHAPTER 7

Decorative Laminates

P. ALLEN and M. F. KEMP

Introduction

'Decorative laminate' is defined in ISO 472 but in common usage has come to mean sheet materials consisting of decorative surface papers impregnated with melamine resin and consolidated under heat and pressure with plies of core paper permeated with phenolic resin. In a wider sense the term can be applied to many associated products—including: laminates in solid colour; laminates with facings such as metal foils, textiles, or wood veneers; polyester laminates; direct faced boards; and composite boards comprising thin laminates bonded to substrates of various kinds.

Products of this nature based on melamine resins were developed in the 'thirties but it was not until the immediate post-war years that they became established commercially and were used widely in applications such as construction, furniture, ships, and other transport vehicles.

They offer many technical advantages (which may be summarized in terms such as 'durability' and 'ease of maintenance') but probably the most important is that they are pre-finished and after fabrication need no painting or other protective finish, like lacquering. The sheets are supplied ready for fabrication and installation, and normal hand tools and wood-working machinery can be used for cutting, sawing, routering, and so forth. Thus they are well-suited to most methods of assembly—from fixing on-site and fabrication in small workshops to fully automatic lines of production.

The properties and characteristics of laminates depend not only on the types of resins and papers used in their manufacture but also on thickness, surface finish, colour and—to some extent—on the kind of substrate to which a laminate is bonded. However, the main attributes generally can be summarized as follows:

(i) a wide range of colours, patterns and textures
(ii) ease of handling and fabrication
(iii) durability (good resistance to wear, scratching, and impact)
(iv) cleanliness and hygiene (resistant to staining)
(v) resistant to chemicals, heat, and moisture
(vi) stability of colours.

When describing in more detail the characteristics and methods of manufacture of laminated sheet materials of different types it is helpful to use the classifications below, each of which is covered by a British Standard:

High-pressure laminates
Decorative continuous laminates
Direct-faced boards
Composite boards.

—and to take each of these in turn.

High-Pressure Laminates

General

A definition of high-pressure laminate ('HPL') is given in BS EN 438: 1991[1] as follows:

> A sheet consisting of layers of fibrous sheet material (for example, paper) impregnated with thermosetting resins and bonded together by means of heat and a pressure of not less than 7 MPa, the outer layer or layers on one or both sides having decorative colours or designs.

Such sheets are made in thicknesses ranging from 0.6 to 30.0 mm. Normally, products of a thin-veneer type range from 0.6 to 1.5 mm in thickness and have a decorative surface on one face only, the reverse being sanded to facilitate bonding to a substrate. Sheets from 2.0 to 30.0 mm are known as 'Compact laminates': between 2.0 and 5.00 mm they are available either with one or both faces decorative and can be used, provided they are supported adequately, without being bonded to a substrate. Laminates thicker than 5.0 mm invariably are double-faced and are self-supporting.

The sheets are supplied in sizes appropriate for the requirements of the main users—Table VIII gives examples. The commercial ranges include three basic

TABLE VIII *Examples of sizes and types of use for high-pressure laminates*

Type of use	*Sheet size*/mm
Manufacture of:	
cubicles	3660 by 1525
doors	2150 by 950
kitchen worktops	3050 by 1320, and 3660 by 1320
wall panels	3050 by 1220

types of HPL sheet—Standard, Flame-Retardant, and Postforming—each of which is available in several grades. The basic types are designated 'S', 'F', and 'P', respectively, and in specifications this type reference is prefixed by two letters the first of which denotes whether Vertical, Horizontal, or Compact grade ('V', 'H', or 'C') and the second whether General Purpose or Heavy Duty ('G' or 'D'). Laminates of Vertical grade typically would be in the range 0.6 to 0.75 mm thick; Horizontal 0.8 to 1.5 mm, and Compact from 2.0 to 30.00 mm. Heavy Duty grades would offer superior wear and resistance to scratching compared with General Purpose.

A laminate designated 'HDS' would be a horizontal heavy duty standard type; 'VGP' would be vertical general purpose postforming; and 'CGF' compact general purpose flame-retardant. However, BS EN 438 permits the use of an alternative system in which the letter S, F, or P is followed by a three-digit code defining durability in terms of resistance to wear, impact, and scratching: using this system, an HGS laminate would be designated 'S333' and a VGP 'P222'.

Table IX gives typical constructions of high-pressure laminates and the main components are considered in further detail below.

Thermosetting Resins Used in Decorative Laminates

Melamine–Formaldehyde Resins

Melamine-formaldehyde resins are used in the surfaces of most decorative laminates and have excellent properties for this purpose in respects such as:

(i) hardness, as exemplified by resistance to abrasion and to scratches
(ii) clarity of the protective surface layer
(iii) resistance to chemicals, staining, and moisture
(iv) resistance to heat
(v) light stability.

Melamine, a white powder, was discovered and identified by Liebig in 1834 but commercial manufacture came only in 1939, by Cyanamid Company of America with dicyandiamide as raw material. Melamine is 2,4,6-triamino-1,3,5-triazine with a structure as shown in Figure 53. On reaction with formaldehyde in aqueous solution the melamine powder dissolves rapidly on heating to form various methylol melamines, as in Figure 54. After further heating and the elimination of water the methylol melamines condense to form resinous polymers.

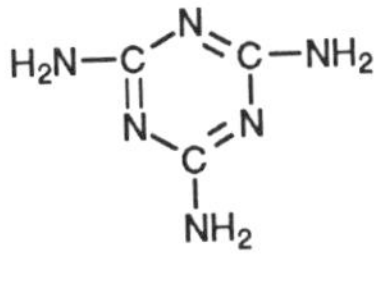

Figure 53 *Melamine*

TABLE IX *Typical constructions of high-pressure laminates*

Component	*Resin*	*Paper*	*Designation (grade) of laminate) and number of plies*					
			HGS	*HGP*	*VGP*	*VGS*	20 mm *CGS*	30 mm *CGS*
Overlay	melamine–formaldehyde	α-cellulose	1	1	1	1	2	2
Print	melamine–formaldehyde	printed pigmented	1	1	1	1	2	2
Core	phenol–formaldehyde	kraft	4 to 6	4 to 5	3 to 4	3 to 4	90 to 120	135 to 180
Release	—	release-coated	1	1	1	1	—	—

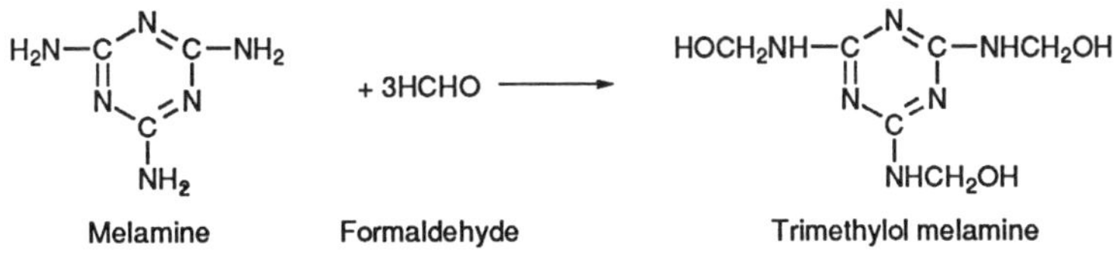

H_2N—C(N)—NH_2 ... NH_2 + 3HCHO ⟶ $HOCH_2NH$—C—$NHCH_2OH$ ... $NHCH_2OH$

Melamine Formaldehyde Trimethylol melamine

Figure 54 *Methylol melamines*

The type of polymer obtained depends on factors such as the pH and temperature of reaction, the ratio of melamine to formaldehyde, and the type of catalyst employed. For decorative laminates, melamine–formaldehyde is prepared by reacting melamine in stainless steel kettles under reflux, alkaline conditions with 37% to 46% formaldehyde in aqueous solution. The reaction temperatures used vary from 80 to 100 °C and are maintained until the condensation has reached the desired end point—that is, reacted sufficiently but still water-soluble. The end point is checked by measurements of viscosity, cure time, and water tolerance. Depending on the type of laminate to be produced, other constituents (surfactants, plasticizers, release and anti-foam agents) normally are added to the base resin before impregnation of the surface papers. It is common practice also at this stage to adjust the pH by adding acid catalysts.

The resins are thermosetting and polymerization is advanced by the application of heat. A useful feature of the reaction, exhibited also by phenolic resins, is that by withdrawing the source of heat the polymerization can be held at any stage. Thus, impregnated papers can be dried and held in the partly cured state, final curing taking place later during pressing.

Melamine–formaldehyde resins can be purchased in the form of spray-dried powder, which dissolves readily in hot water.

Phenol–formaldehyde Resins

Phenol–formaldehyde resins are used in the core assembly and in conjunction with the kraft paper give the characteristic brown colour to the back of the laminate. Features of these resins making them suitable for the purpose include:

(i) good resistance to water
(ii) good dimensional stability
(iii) excellent rigidity
(iv) good resistance to impact.

Resins for decorative laminates are made by reacting phenol with formaldehyde under alkaline conditions. Normally, reaction would be in a stainless steel vessel fitted with heating jacket, condenser, and mechanical stirrer. The raw materials are reacted with caustic soda under reflux conditions for approximately one hour, after which cooling and discharge take place. The result is a slightly viscous water-soluble resin, brown in colour, which has a considerable shelf-life.

As with the melamine–formaldehyde resins, additives such as flame retardants (depending upon the type of laminate to be produced) can be added before the impregnation stage. The resin then can be used in impregnating kraft core papers.

Manufacturing Details of HGS Laminates

Raw Materials

Overlay Paper. Overlay paper is α-cellulose paper of high purity, with substance weight in the range 20 to 80 g m^{-2}. An essential feature is the refractive index, which is virtually identical with that of melamine–formaldehyde resin: hence, after lamination—when all the fibres of the paper are wetted and consolidated with resin devoid of air—the overlay becomes transparent. It forms a durable, hard, clear layer to protect the decorative print layer below.

The qualities required in an overlay paper for satisfactory impregnation and lamination can be expressed as:

(i) a rapid and copious absorption of resin
(ii) to prevent tearing during impregnation, a specified level of wet tensile strength
(iii) to prevent advancement or retarding the cure of the resin, a pH close to neutral.

Print Sheet. The print sheet is paper of high purity and substance weight in the range 60 to 140 g m^{-2}, printed to give the effect or pattern desired, and capable of being impregnated with melamine–formaldehyde resin. The background paper is pigmented to achieve the base colour required, using pigments that resist fading both in the intermediate processing and in the final laminate. Normally the amounts of pigments used are between 10% and 40%, and they must be heavy enough to give the degree of opacity required.

Properties required in the paper may be summarized:

(i) the surface to be printed must be smooth in order to accept an even deposit of ink
(ii) it must absorb resin readily
(iii) its tensile strength when wet must be high enough to withstand the stresses of impregnation.

The printing inks must be formulated with pigments that are compatible with the resins, capable of withstanding the high pressures and temperatures encountered during lamination, and offer good resistance to fading in the finished laminate. Resins used in the ink base must be capable of fixing the ink sufficiently to hold the print during impregnation, and yet either must cross-link during lamination or allow penetration of the melamine–formaldehyde resin to take place.

Various printing techniques may be used. Gravure is the most common, and is employed for the greater part of printing for laminates. Screen printing, which uses heavy deposits of ink, often is used for special artwork—and in such cases it is essential to obtain a good bond between the ink system and the melamine–formaldehyde resin. Photographs and other images may be reproduced on a large scale by a special technique involving jets of ink controlled by computer.

Plain Colour Papers. When no pattern is required papers in plain colours are employed. Normally they are of heavier substance weight—between 80 and 220 g m^{-2}—and as with the print papers they must be:

(i) colour-matched, using pigments unaffected by the chemicals used in processing and resistant to fading in the finished laminate
(ii) sufficiently absorbent to allow penetration by the resins
(iii) strong enough when wet to withstand the stresses of impregnation.

Colour-Matching. It is essential in producing all types of laminates to ensure tight control of colour (acceptable matching, sheet-to-sheet, should be possible even between batches made several years apart). It is not possible to match papers either before or after impregnation with the precision required and hence all plain colour and printed papers are colour-matched after lamination. In addition to visual judgment it is normal to use spectrophotometers and associated computers to decide colour acceptance (or otherwise).

Core Papers. These commonly are kraft papers of substance weight between 80 and 260 g m^{-2}, usually for impregnation with phenol–formaldehyde resins.

This component (the core paper) contributes the greater part of the strength of the laminate, including such qualities as dimensional stability and resistance to impact, to tearing, and to water.

Separator or Release Sheets. Separator or release sheets are used to free the reverse side of a laminate either from pressure padding material or the adjacent laminate; normally they are paper of the glassine type, coated with release agents.

Impregnation. Apart from the release sheets all the raw material papers described above are required to be impregnated with a liquid resin—either melamine–formaldehyde for the surfaces or phenol–formaldehyde for the core. This stage of the process, essentially, involves filling the paper with the appropriate amount of the liquid resin, driving off solvents, and advancing the cure until a handleable sheet of semi-cured impregnated paper is obtained.

The machines used (known as 'treating machines') consist of three main sections:

The wet end—where in an unwind stand the reel of paper is supported and fed under suitable tension controls into a bath of resin at controlled temperature. In many cases, instead of total immersion, the paper is wetted on one side only—to allow the resin to penetrate, force air out, and give a fully saturated sheet. To allow time enough for the required penetration a 'skying' system of variable rollers is used, as shown in Figure 55; this transports the web upwards and then down again before a final immersion in the resin, followed by metering.

Metering of the resin can be done in a number of ways:

(i) by means of 'percentage rollers'—rollers with an adjustable gap between them

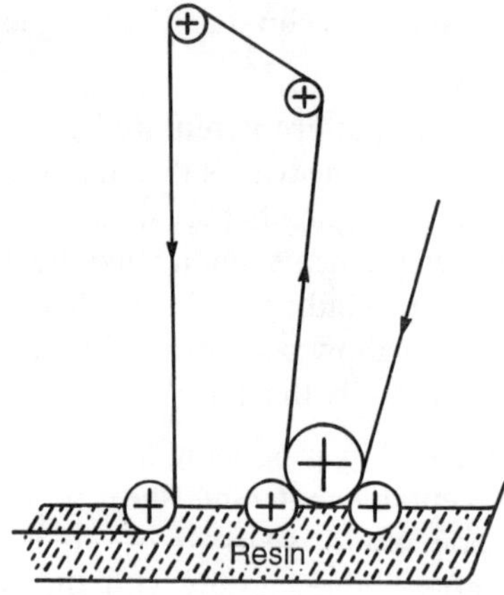

Figure 55 *Diagrammatic representation: system for 'skying' paper when wetting with resin*

(ii) by various systems of roller coating, where gap and speed adjustment of a pair of rollers is used to control the take-up of resin from the bath; a third roller then transfers the resin to the paper web
(iii) by 'dip and scrape'—immersion followed by scraping away of excess resin: this may be used as an elementary method of application, such as for certain types of core materials.

Figure 56 illustrates different methods of metering.

The second stage is drying, the web being carried through an oven with hot air as the heating medium. Drying must take place without any physical interference (such as by conveyor bars) to the resin layers on either side and to accomplish this the web usually is supported entirely by jets of hot air blown from nozzles above and below: the result is a sinusoidal path, with good drying characteristics (see Figure 57). In the last zone of the oven cold air is introduced to bring down the temperature of the web.

After it leaves the oven the web is reeled or cut into sheet form. The treated stock is checked to ensure that the correct amount of resin has been applied, and that drying and curing to the degree required have been obtained. This is done usually by measuring resin content and volatiles; in some instances a small heated press may be used to measure the amount of resin flow.

Control of the resin content may be achieved by adjusting the metering system while flow and volatiles usually are controlled by varying the speed to allow the web less or more time in the drying oven.

Pressing. The impregnated sheets then are assembled in sequence, as illustrated in Figure 58, and laminated under heat and pressure.

The pressures employed are between 7 and 14 MPa; common temperatures are up to 150 °C, for a cycle of approximately 90 minutes in total. Most of the curing occurs between 130 and 145 °C, where normally the temperature will be held for twenty to thirty minutes.

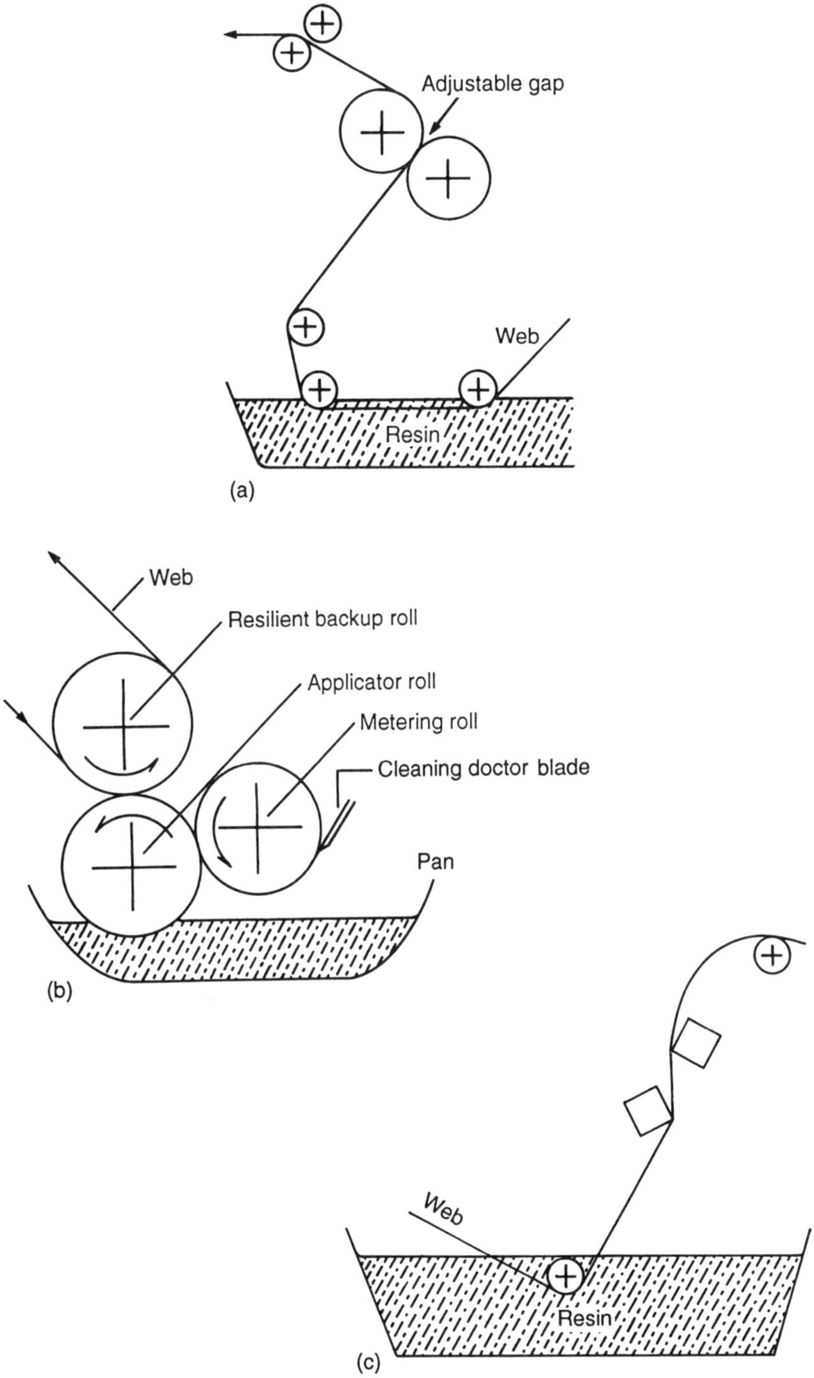

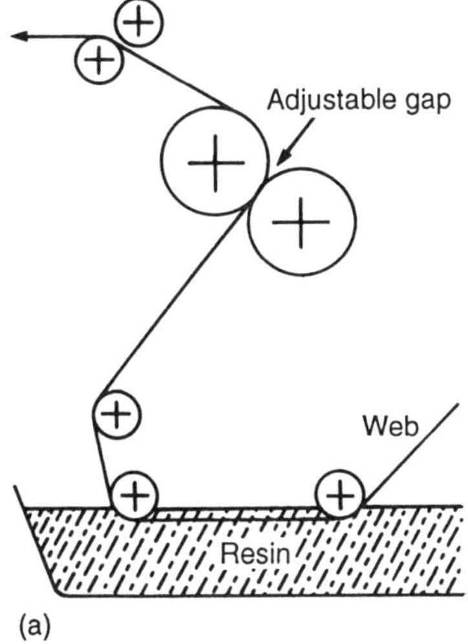

Figure 56 *Methods of metering:* (a) *Percentage roller metering;* (b) *Coating system with metering roll;* (c) *Dipping and scraping*

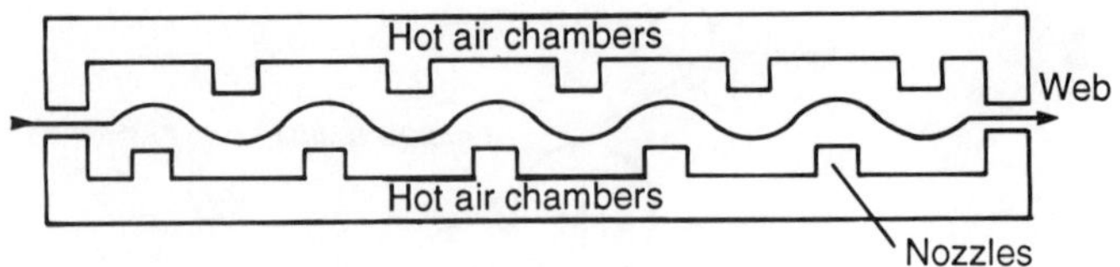

Figure 57 *Sinusoidal path of web through drying oven*

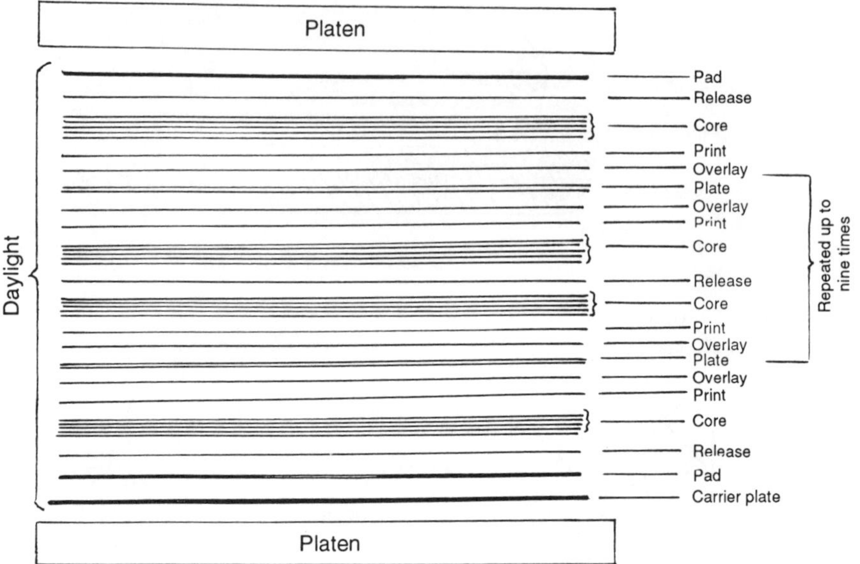

Figure 58 *Diagrammatic representation: lamination in a multi-daylight press*

Materials for a press are assembled in 'daylights'—the distance between the open platens of a multi-platen press. Numbers of up to twenty laminates may be pressed in a daylight, separation between them being by means of metal or laminated plastic plates. Pads consisting of kraft paper, or of some other suitable material, are used to even out the distribution of pressure. Figure 59 shows a high-pressure multi-daylight laminating press. The platens are heated by means of hot water under high pressure, or by steam; a modern press might have up to 24 daylights.

Plates. In the pressing, the surface finish required in the laminate is transferred from that of a metal or laminated plastic plate. Such plates used to impart the different types of finish include:

(i) Plane surfaces; these are obtained from hardened steel plates with surfaces polished to give a high gloss, or alternatively blasted or ground with abrasive particles to give a matt finish. (Plates such as these can be used in conjunction with embossed release papers to produce other textures.)

(ii) Light textures; these are taken from hardened steel plates that have been etched by photographic techniques and finished suitably by chemical or

Figure 59 *High-pressure multi-daylight laminating press*
(Photo Formica Limited)

electrolytic polishing. (In order to achieve release from the steel plate, a release agent is incorporated in the surface resin.)

(iii) Heavy and structured textures; usually these are produced not from metal but from plates comprising phenolic laminates, the texture required being transferred from a suitable master. (Release from the laminate plate is obtained by means of a layer of aluminium foil backed with paper, or with release papers coated suitably.)

Finishing. After the pressing cycle the laminates are separated and passed to final finishing, where suitable cutters are used to trim all four edges and the reverse sides sanded to give the finish needed subsequently for adhesive bonding. Following the sanding, type and batch identifications are printed on the reverse or face of the laminate, or it may be identified by means of a removable adhesive label.

Other HPL Products

In addition to the S, F, and P types of high-pressure laminate considered so far there are several variants worthy of special mention—some admittedly outside the scope of BS EN 438 but which nevertheless could be described as high-pressure decorative laminates, as follows.

Solid-Colour Laminates. Such laminates are of construction similar to the normal HPL except that the core layers are decor papers impregnated with melamine resin (rather than phenolic kraft). Fabricated panels faced and edged with these products have the appearance of solid, homogeneous material (since no dark phenolic edge is visible where the edging strip meets the surface). The uniformity of colour throughout the thickness also gives these laminates good wear qualities, as there is no contrasting core layer which might show through should the surface become abraded or scratched.

Inter-leaving different colours, chamfering, engraving, routering, and sand blasting may be applied to give unusual decorative effects with solid-colour laminates.

Special laminates are made with a contrasting colour below the top surface, for use in engraving. Normally they are of thickness 1.0 to 3.00 mm and can be used to make engraved signs, instrument fascia panels, and so forth.

A recent variant of laminates of this nature is surfaced with two or more transparent overlay sheets, each of which carries a printed pattern that forms part of the whole design: the result is to give depth to the design, and a three-dimensional effect.

Metal-Faced Laminates. In these products, instead of a melamine decor surface paper, aluminium or copper foils are bonded in the press to a phenolic core.

Polished, matt, and brushed finishes may be obtained and, in addition, other decorative effects from embossed, oxidized, and etched foils. The surface may be lacquered, anodized, or chromium-plated to prevent oxidation, and can be coloured to simulate other metals like brass and bronze.

In general such laminates are less durable than melamine surfaces and therefore not normally recommended for hard wear in applications such as counter tops and working areas.

Wood Veneers. Wood veneered laminates are available in two forms, which may be described as 'natural' and 'surfaced'.

The natural product comprises veneers of real wood which have been sorted and joined edge-to-edge (for example, by stitching), and bonded under heat and pressure to layers of kraft paper impregnated with phenol–formaldehyde resin. A barrier layer immediately below the veneer is impregnated with melamine–formaldehyde resin and prevents upward migration of the darker phenolic resin.

The surface is a sanded natural wood, which may be stained, polished, or lacquered as desired. The properties of the surface are similar to those of the natural product, except that as a result of densification during manufacture the veneers on these products are harder.

The surfaced laminate is of similar construction but to improve durability is pressed with an overlay impregnated with melamine–formaldehyde.

Some of the properties of wood veneer laminates are inferior to those of conventional HPL (as examples, colour-fastness, resistance to heat and impact) but the surfaced wood veneers will meet the requirements of the furniture Standard BS 6250 Part 3[2] for horizontal surfaces in severe use.

Since it is natural wood the colour of the laminate may vary from sheet to sheet (and may change on aging) but these products combine the appeal of real wood with enhanced durability, ease of handling, and fabrication.

Textiles. Laminates can be constructed in a manner similar to that for surfaced wood veneers but, in place of the wood, with the textile finish of an incorporated linen or hessian fabric.

Exterior Grades. Most decorative laminates are intended only for use indoors but on occasion (such as when there is a degree of protection against the elements, and shade from direct sunlight) they may be used with success outside for applications such as panels in doorways of retail shops. Interior grades will keep their integrity out-of-doors but over a period of time there is a tendency for their surfaces to whiten and colours also may fade (the whitening is attributed to a deterioration in the bond between the resin and paper fibres). The extent of the change of colour and the speed at which it takes place depend on the extent of exposure to ultra-violet light and on the stability of the pigments used.

In laminates intended for use out-of-doors papers with superior colour-fastness may be used and normally their construction includes an overlay sheet of ultra-violet absorbing thermoplastic film—which helps to reduce fading of the colour, and whitening.

Thick compact exterior grade laminates are self-supporting and suitable for applications such as 'infill' panels or the decorative cladding of exterior walls. They are available with surfaces of anodized aluminium and coloured lacquer-coated foil as well as normal melamine.

The suitability of a laminate for outdoor use may be tested by accelerated methods. Tests available include: colour-fastness, resistance to corrosion, resistance to frost, and the absorption of water.

Anti-Static Laminates. The conventional laminates are relatively good electrical insulators and in consequence in dry conditions can accumulate static electric charges; these can present problems—as when discharges take place suddenly in the course of manufacture or assembly of electronic components. It is particularly important for applications such as computer furniture and benches intended for electronic assembly that static charges will be dissipated at controlled rates.

Anti-static laminates are made for requirements like these and their construction includes one or more conductive layers (for example, carbon-impregnated paper). The working surface concerned must be grounded ('earthed') by means of a suitable terminal, and with cable or other conduit to provide a path for the discharge.

Electrical requirements for such an installation would be specified in terms such as: surface resistivity, resistance to ground, and rate of decay of voltage.

Flooring Grades. HPL was introduced originally for flooring some 25 years ago, mainly for light duties as in control rooms, computer suites, and so forth. At that time the resistance to wear (as measured by the Taber abrader), though adequate for the applications concerned, was only a little better than that of general purpose laminate. However, improvements of the order of five- to ten-fold have been achieved since then by incorporating mineral fillers in the surface paper and/or the resin—so that HPL flooring now can be used not only in domestic halls, kitchens, and so on but also in commercial buildings like hotels, offices, and shops.

For such uses the laminate normally is bonded to a fibreboard or chipboard substrate (typically, 6 mm thick), which is backed with a compensating laminate to prevent bowing, then cut into square tiles or planks that are tongued and grooved.

Special Art-Work and Graphics. If required special designs can be incorporated in S, F, P, and C laminates by replacing the usual decor paper with a melamine-impregnated sheet carrying the design. Examples include: abstract effects, diagrams, emblems, lettering, logotypes, maps, murals, photographs, plans, signs, *etc.* A variety of techniques may be used to transfer a design to the decor sheet:

(i) air-brushing
(ii) laser-scanning of a photographic transparency, with computer-controlled jet printing in colour
(iii) painting by hand
(iv) photographic graining (using variations in density of grain to give tones)
(v) silk-screen printing.

Whatever method is employed the inks or paints must be compatible with the process (otherwise defects such as blistering, blurring of images, or delamination, can occur).

Decorative Continuous Laminate

The British Standard for decorative continuous laminate ('DCL') is BS 7332: 1990,[3] which gives the following definition:

> A continuously manufactured sheet consisting of layers of fibrous sheet material (for example, paper) impregnated with phenolic and/or aminoplastic and/or polyester thermosetting resins, assembled and bonded together by means of heat and pressure, the outer layer or layers on one side having decorative colours or designs.

In the continuous process, pre-impregnated surface and core papers are drawn from reels and fed between two endless steel belts which rotate on large heated rollers. The layers of paper are pressed together as a result of pressure applied to the backs of the steel belts: the resins soften, flow, and cure under the effects of heat and pressure to form a continuous laminate which after cooling either can be reeled or cut by guillotine to sheets of the length required. (Figure 60 shows a continuous laminating press.)

The pressure applied is normally in the range 3 to 5 MPa; temperatures are similar to those used for HPL but with machine speeds of 5 to 30 m min^{-1} the cure times for resins are much shorter.

The thickness of such laminates ranges typically from 0.4 to 0.9 mm and they are available in Standard, Flame-Retardant, and Postforming qualities. The last is the most usual—the thinner postforming laminates being suitable for bends of very tight radius.

Figure 60 *Continuous laminating press*
(Photo Grecon)

Laminates faced with melamine can have cores based on either melamine or phenolic resins; when bonded to supporting substrates they are suitable for both vertical and horizontal applications (wall panels, working surfaces, *etc.*). Light surface textures can be produced by pressing against textured papers, or by using in the press textured steel belts. As an alternative to sanding the reverse of the laminate a parchment backing can be applied, which also will facilitate bonding with adhesives. Parchment offers an advantage with very thin laminates, in which irregularities resulting from sanding might show through.

Continuous laminates are manufactured also using polyester resins in place of melamine; however, such laminates are less durable and best suited to light duties such as shelving, vertical surfaces, and edging.

Direct-Faced Boards

Direct-faced boards, as the name suggests, are made by bonding directly to the supporting substrate one or two layers of decor paper impregnated with resin, using a single pressing operation to achieve the bond. (This is in contrast with the composite board—see below—for which there are two pressing operations.)

Depending upon the requirements of the product, direct-faced boards may be produced at high or low pressures on static or continuous presses. One of the best-known types is melamine-faced chipboard ('MFC'), which is covered by

BS 7331: 1990:[4] that Standard specifies six grades of MFC for indoor use, classified by mechanical properties and by resistance to wear and to moisture.

Direct-faced fibreboard has been on sale for more than 25 years but as yet there is no British Standard for it.

For applications such as cladding fire-rated bulkheads and doors in ships and other vessels decor papers have been bonded directly to thin steel sheet. However, for these purposes steel coated with PVC and thin continuous laminate bonded to steel now are used more commonly.

Composite Boards

Composite boards are made by bonding decorative laminate of thickness normally 0.4 to 1.5 mm to one or both faces of a substrate, using a suitable adhesive. The most common method of manufacture is flat pressing, which (depending upon the type of adhesive) may be carried out hot or cold. If the adhesive is of the spray-coated contact type, mechanical nip rollers may be employed instead of flat pressing.

For most constructions and applications it is essential to face the reverse of the composite board with a balancing laminate similar in construction and moisture content to the decorative face laminate (failure to do this results invariably in differential movement of the faces and problems with bowing of the board).

Composite boards and panels faced with decorative laminates are specified in BS 4965: 1991.[5] This Standard gives the types of substrates, decorative laminates, and adhesives that may be used in combination to obtain various levels of performance. The requirements are classified in the following terms:

(i) resistance to moisture of the substrate and adhesive
(ii) performance rating of the surface laminate
(iii) fire rating of the composite board.

Substrates include chipboard, plywood, medium-density fibreboard, hardboard, fully glued blockboard, cement-bonded particle board, mineral board, and vermiculite board.

For special applications core materials outside the scope of BS 4965 frequently are used—as examples:

(i) honeycomb cores, made from aluminium foil or kraft paper (either resin-impregnated or not); typical uses include panels for caravans, and the ceilings of railway coaches
(ii) metals such as aluminium or steel—for panels in marine, off-shore, and transport applications
(iii) expanded plastics such as phenolic, polystyrene, or polyurethane, in board form or foamed *in situ*—for panels where lightness in weight and good thermal insulation are required.

Technical Properties of Laminates

A total of some 24 properties are specified in the current British Standards for decorative laminate products, as follows:

Appearance
Dimensional tolerances
Dimensional stability
Flatness
Resistance to:
- Wear
- Boiling water
- Water vapour
- Dry heat
- Cracking
- Impact
- Scratches
- Staining
- Cigarettes

Light-fastness
Fire performance
Compact laminate only:
- Resistance to moisture
- Flexural strength
- Flexural modulus
- Tensile strength

Postforming laminate only:
- Postformability
- Blister resistance

Melamine-faced chipboard only:
- Resistance to wet heat
- Surface soundness

Composite boards only:
- Adhesion of laminate to core.

Most of the test methods involved were developed for decorative laminates, with the intention of reflecting requirements in performance of the various types of product.

BS 4965 for composite boards and panels includes very few tests for physical properties but specifies precisely the individual components for each type of composite construction, referring to the appropriate standards.

As was made clear in the Introduction to this chapter, the characteristics and properties of laminates depend on a variety of factors: before going on to consider the special qualities of flame-retardant and postforming materials a number of the more important general features are reviewed as follows.

Surface Construction

The resistance to wear of most plain colour laminates is typically about 700 revolutions in the Taber test—adequate for most applications without need for a protective overlay.

On the other hand, the resistance of laminates with printed decor papers depends very much on the substance weight of the overlay paper used to protect the ink layers. There is a tendency, unfortunately, for an overlay to impair the definition of print and brightness of colour—which can be undesirable, or, particularly with more exotic designs such as metallized and pearlescent effects, even unacceptable.

Printed laminates with no overlay are intended only for very light duty applications; they have a wear resistance (Taber test) of around 80 to 100 revolutions, compared with 400 to 600 revolutions achieved normally by horizontal grade print laminates.

Thickness

Mechanical strength, rigidity, and many other properties are dependent on or related to thickness, one of the more important of them being dimensional stability.

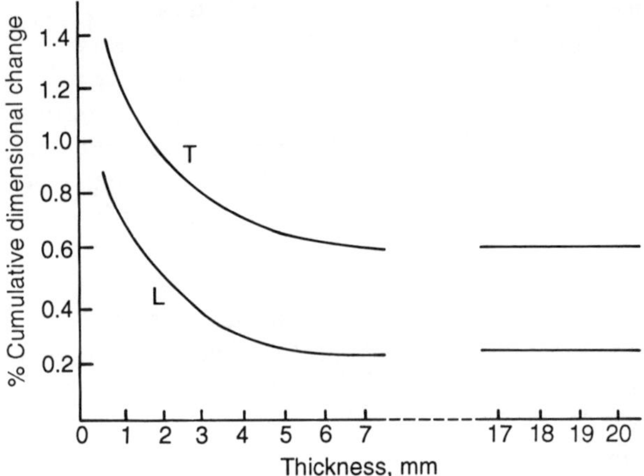

Figure 61 *Illustration from BS 438 Part 1 of the relationship between thickness and dimensional movement in both longitudinal and transverse directions*

In dry conditions decorative laminates will shrink, and in damp they will expand; unless suitable precautions are taken the associated movements can give rise to stress cracking of laminates and the bowing of composite boards. Since the phenol–formaldehyde resins are more stable in this respect than melamine–formaldehyde, laminates with phenolic kraft cores have dimensional stability better than those with melamine core papers—and thick laminates incorporating many plies of phenolic core paper are more stable than thin laminates with fewer plies.

Orientation of the paper fibres also is a factor: because of this, moisture-related movement across the width of the laminate can be about twice as much as along the length of the sheet.

Figure 61 from BS EN 438 Part 1 illustrates the relationship between thickness and dimensional movement in both longitudinal and transverse directions. The curves show the maximum dimensional movement permitted between extremes of humidity—from oven-dry to saturation. In many applications such extremes may never be encountered in practice but with a view to avoiding problems after installation it is prudent to follow the recommendations of the manufacturer with regard to pre-conditioning, counter-veneering, and so on.

Resistance to impact is a characteristic of the composite board rather than of the face laminate, dependent both on the thickness of the laminate and the type and quality of the substrate. Laminates of thickness less than 0.8 mm have little inherent resistance to impact and a substrate of good quality, such as chipboard, plywood, or MDF, is required to provide the necessary support. Post-formed worktops faced with very thin laminates can suffer damage from impact on the curved edge, where the structure of the underlying core is more open and provides less-uniform support than the fine face of the chipboard.

A relatively thick laminate (1.2 to 1.5 mm) is required to provide adequate

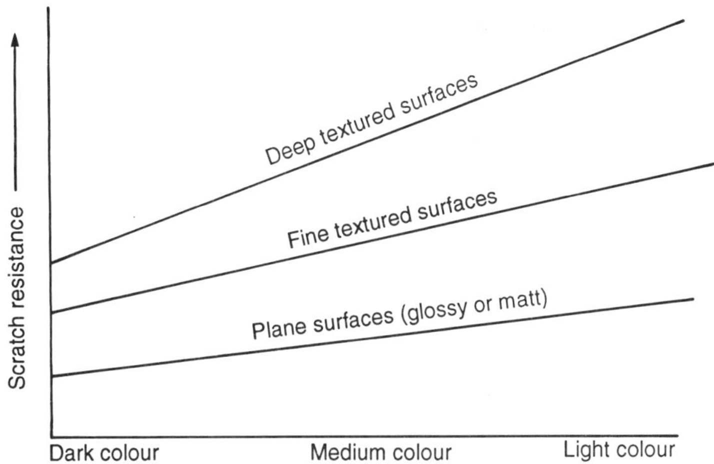

Figure 62 *Effects of surface finish and colour on resistance to scratching*

impact resistance with honeycomb substrates: alternatively, such substrates can be given a skin of thin plywood or hardboard before application of the decorative laminate.

Surface Finish and Colour

The relationship between the hardness of a surface and its resistance to scratching is evident (and, by analogy, polyester laminates can be scratched more easily than melamine)—but scratch resistance is influenced also by surface finish and colour. In general, scuff and scratch marks are more obvious on plane finishes (whether glossy or matt) and on dark colours than on textured finishes and light colours. Figure 62 illustrates this relationship.

While textured laminates offer superior resistance to scratching they are less easy to wipe clean than plane surfaces. This can be an important consideration in applications such as bench tops for microbiological and nuclear laboratories, where ease of decontamination is a primary requirement.

Fire Performance

Melamine–formaldehyde resins are inherently flame-retardant and normally, when bonded to chipboard, to plywood, or MDF, HGS and VGS laminates will achieve Class 2 in the spread of flame test (BS 476 Part 7[6]).

Class 1 can be achieved by bonding to a non-combustible substrate such as calcium silicate board; on this basis maritime authorities such as the Department of Transport and Lloyd's Register of Shipping have approved S laminates for applications such as the cladding of bulkheads in ships.

In F grade laminates fire properties are improved further by incorporating additives like inorganic phosphates in the core resin or paper. Thus, when

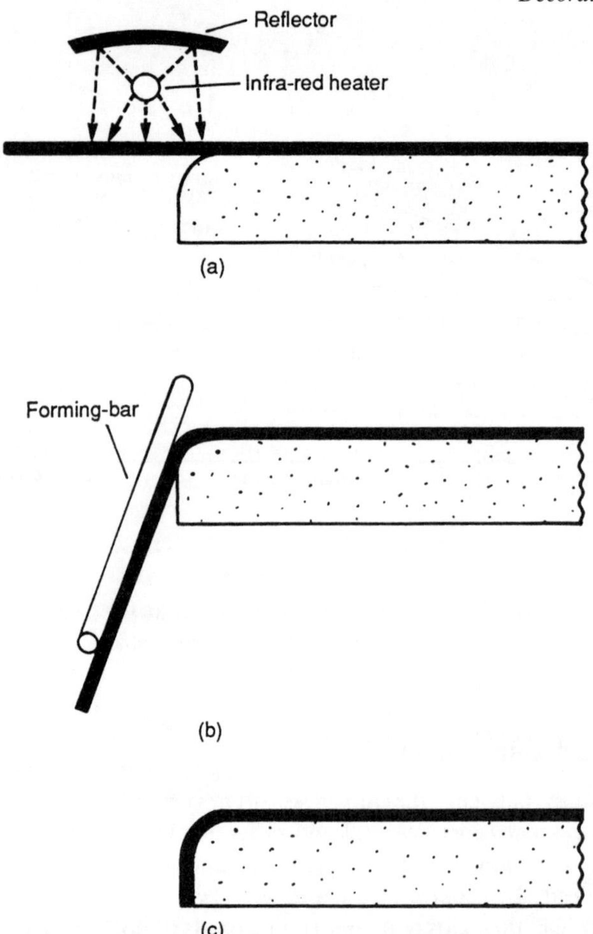

Figure 63 *Stages in postforming:* (a) *Profile at edge;* (b) *At the temperature of forming, the laminate is shaped around the profile and bonded simultaneously;* (c) *Surplus material is trimmed away to give a smooth finish*

bonded to suitable flame-retardant substrates, such laminates can achieve Class O performance as defined in United Kingdom building regulations.[7]

The fire resistance of a composite board when tested to BS 476 Parts 20 to 23[8] is a measure of the ability of the panel to prevent the passage of flame and heat for a certain period of time (such as a half hour, 1 hour, *etc.*). It is a function of the substrate used, the face laminate having little or no bearing on the outcome of the test.

Recently flame-retardant laminates with low levels of smoke emission have been developed to comply with requirements for certain applications in transport (for example, BS 6853[9]). When bonded to non-combustible substrates these laminates will give smoke densities of about half those produced by the normal F laminates.

Speaking generally, the fire properties of laminates based on polyester resins are inferior to those of the melamine–formaldehyde materials.

Postforming

The process of postforming consists at its simplest of applying localized heat to the surface of a P laminate which has been bonded previously to a substrate with a radiused edge profile (see Figure 63).

As the temperature of forming is reached the laminate overhanging the profiled edge becomes pliable and can be formed around, simultaneously being bonded to, the radiused profile. Surplus laminate then is trimmed away, to give a finished panel with a smooth radiused edge.

Postforming laminates, unlike some thermoplastic sheet materials, cannot be formed into compound curves—but with appropriate techniques complex multiple internal and external bends ('S', 'U', *etc.*) can be achieved easily.

At one time formability was obtained by under-curing the laminate at the pressing stage, but this had several disadvantages:

(i) there were difficulties in controlling precisely the degree of under-curing
(ii) a limited shelf-life (since formability deteriorated over time)
(iii) reduced resistance to heat, moisture, staining, and scratching.

However, improvements in plasticized resin systems have obviated to a large extent the need for under-curing and overcome such deficiencies.

For prototype work and small production runs static forming machines usually are employed (in these the panel remains stationary, while, under manual or automatic control, the machine heats and forms the laminate).

Items such as working surfaces for kitchens and other panels required in large numbers are fabricated on large machines which are automated fully and in which the postforming operation is continuous. On such machines the composite panel is carried by conveyor from the pressing stage and passes through a battery of infra-red heaters to activate the adhesive and soften the laminate. Then, as the panel is carried through on the conveyor, fixed bars and rollers form it around a profile. Two edges can be formed simultaneously at speeds of about 15 m min^{-1}. Figure 64 shows a machine of this type.

As a general rule, thin laminates can be formed round smaller radii than thicker ones. The minimum radius achievable for an HGP laminate 0.9 mm thick is around 10 mm—much depending upon the technique of forming used and the element of skill applied.

Flame-retardant postforming laminates are slightly less pliable and for these the minimum radius recommended normally is 15 mm. On the other hand, polyester laminates are more flexible than melamine and very tight bends can be obtained in these without cracking or crazing of the surface.

In recent years postforming compact laminates (typically, 2 to 10 mm thick) have been developed, using within the laminate thermoplastic films to act as slip layers. These products require special techniques for forming.

Figure 64 *Automated machine for postforming*
(Photo Orama Fabrications Limited)

A Comparison of Properties

As has been indicated, laminates are constructed in a variety of ways for differing uses and because of this it is not possible to make general comparisons between the qualities of different types.

It should be noted also that the methods of test given in different British Standards are not necessarily equivalent—for example, there is no correlation between the tests for resistance to scratching in BS EN 438 and in BS 7332.

However, it is possible to compare the durability of some of the more common products by taking typical values for resistance to impact, scratching, and wear derived from appropriate test methods in BS EN 438 Part 2 (see Table X).

International Harmonization

There are British Standards as shown for each of the four main groups of products considered in this chapter but High Pressure Laminate is the only type for which European (EN 438: 1991) and international (ISO 4586: 1987) standards exist.

It will be permissible to apply the CE mark to sheets of High Pressure Laminate in accordance with the regulations contained in the European Construction Products Directive; however, until European fire test methods and specifications are harmonized fire certification by individual countries still will be necessary.

A European Standard for Continuous Laminate is being developed within CEN Technical Committee 249 but at the time of writing there have been no proposals to harmonize standards for Direct-faced or Composite Boards.

TABLE X *Comparison of properties. (In all cases typical figures are given)*

Type of laminate	*Impact*/ Newtons	*Resistance to Scratching*/ Newtons	*Wear*/ revolutions
Heavy duty high-pressure laminate, 1.2 mm	25 to 35	3.0 to 7.0	1000 to 3000
Horizontal general purpose laminate, 0.9 mm	20 to 25	2.0 to 5.0	400 to 800
Light duty laminate (non-overlay print), 0.6 mm	15 to 20	2.0 to 4.0	80 to 120
High-wear melamine-faced chipboard	10 to 20	2.0 to 4.0	300 to 700
General purpose melamine-faced chipboard (non-overlay print)	5 to 20	2.0 to 4.0	80 to 120
Polyester continuous laminate (print), 0.7 mm	15 to 20	1.0 to 2.5	80 to 120

References

1. BS EN 438: 1991 Decorative high-pressure laminates (HPL)—Sheets based on thermosetting resins, Part 1: Specifications; Part 2: Determination of properties.
2. BS 6250 Part 3: 1991 Domestic and contract furniture: Specification for performance requirements for cabinet furniture.
3. BS 7332: 1990 Decorative continuous laminates (DCL) based on thermosetting resins.
4. BS 7331: 1990 Direct-surfaced wood chipboard based on thermosetting resins.
5. BS 4965: 1991 Decorative laminated plastics sheet veneered boards and panels.
6. BS 476 Part 7: 1987 Fire tests on building materials and structures: Method for classification of the surface spread of flame of products.
7. The Building Regulations 1991 Approved Document B (Published by HMSO).
8. BS 476 Parts 20 to 23: 1987 Fire tests on building materials and structures: Methods for determination of the fire resistance of elements of construction.
9. BS 6853: 1987 Fire precautions in the design and construction of railway passenger rolling stock.

CHAPTER 8

Mouldings—Their Surface and Finish

A. WHELAN

Materials Used for Moulding and Extrusion

The distinction is by no means as clear-cut as once it was but still it is convenient to classify plastics either as 'thermoplastic' or 'thermosetting'—terms defined in ISO 472 as follows:

> *thermosplastic:* Capable of being softened repeatedly by heating and hardened by cooling through a temperature range characteristic of the plastic and, in the softened state, of being shaped repeatedly by flow into articles by moulding, extrusion, or forming.
>
> *thermosetting:* Capable of being changed into a substantially infusible and insoluble product when cured by heat or by other means such as radiation, catalysts, *etc.*

The decorative laminates described in the previous chapter are made with selected thermosetting resins; while resins of this type can be moulded and extruded by methods similar to those outlined in the present and the next chapter the materials employed for these processes predominantly are thermoplastic. Many such plastics can be moulded and extruded under suitable conditions, the most important in terms of quantities used being those that combine properties satisfactory for the purpose with convenience in processing–especially the polyolefins (polyethylene and polypropylene), poly(vinyl chloride), and styrene polymers and blends. Other plastics with special qualities, such as better resistance to chemical attack, heat, impact, and wear, also are used—including acetals (polyformaldehyde or polyoxymethylene), polyamides, polycarbonates, thermoplastic polyesters like poly(ethylene terephthalate) and poly(butylene terephthalate), and modified poly(phenylene oxide), which is known also as 'modified polyphenylene ether'. Many of the materials in this group are less easy to process than those used in larger quantity—for reasons such as hygroscopic behaviour, a need to apply higher temperatures, or for more critical control of the temperatures—but the properties required of the mouldings justify a disadvantage of this nature.

A comparatively new group of materials—'thermoplastic elastomers' or 'thermoplastic rubbers'—combines the ease of processing of thermoplastics with qualities of traditional vulcanized rubbers, especially elasticity. Because of convenience in processing there is much interest too in blends of plastics with elastomers, which may be modified by the inclusion of filler or glass fibre. As an example, a rubber-like material that can be processed as a thermoplastic can be made by blending and melt-mixing an ethylene–propylene rubber with polypropylene. The use of such blends may be helpful when there are needs to reclaim and re-process material, and in order to obtain products with qualities intermediate between those of the main components of the blends.

Selection of Grade

A plastic may be identified conveniently by a common name such as 'polyethylene' but each of the many plastics and rubbers of commerce comprises in effect a family of materials with differences between grades such as in molecular weight and in the distribution of molecular weight. The molecular structure is important for processing and for the finished product, and further modifications in thermal and other properties may be obtained by including small quantities of co-monomers and additives of various kinds.

The user first must select the type of plastic to be used, then the grade appropriate for the purpose—bearing in mind the specification and any special requirements in the mouldings, like transparency, translucency, gloss or matt finish. The suppliers usually are able to supply information to assist the selection, either verbally or in writing, and may be able to help with demonstrations and tests before production is started. On occasion, when large quantities of material are likely to be required, a grade may be developed specially for a particular purpose.

As a rule, when the moulding or component is to be subjected to quite severe mechanical stress, a grade with high viscosity (stiff-flowing) will be most suitable; such grades are of the highest molecular weights and exhibit superior mechanical properties. However, sometimes guidance of this general nature cannot be followed as (because of moulding behaviour) the products obtained show unacceptable strains or poor surface quality. It is preferable too to use easier-flowing grades (that is, with lower molecular weight) when injection moulding articles with thin walls, so as to ensure rapid and complete filling of moulds. Grades of this kind will give moulded components with very smooth surfaces.

Additives

Material may be purchased in finished from ready for moulding or extrusion or essentially as a raw material which may be blended or compounded immediately before use. In either case additives for different purposes are likely to be present. Behaviour during processing and the properties of the products (including appearance) can be changed dramatically by the conditions and by the presence or absence of such additives. Some types (as examples, plasticizers and heat stabilizers) are essential with certain formulations but not required in others. With some materials a formulation for moulding or extrusion may be comparatively simple but for others [notably poly(vinyl chloride)[1]] it may be quite complex. It is not appropriate here to take the various types in detail but brief reference is made below to the components in formulations which have direct effects on appearance and surface finish.

Colorants

The colorants may be classified in two main groups:

(1) dyes—soluble colouring systems that give translucent colours

(2) pigments—insoluble substances giving opaque colours.

The second group is used most widely, and within this the colorants used most frequently are white and black.

The polymer compound and components moulded from it usually are coloured uniformly throughout (that is, the colours of the surface and the interior are the same). However, with a view to reducing costs of colouring and of compounding, methods have been developed for colouring only the surface, by sandwich moulding or by co-extrusion of product and surface layer.

Until fairly recently most plastics were purchased fully compounded in the colour required for moulding or extrusion, but with a view to offering ranges of colours more cheaply and conveniently it is quite usual now for machines to be fed with blends of 'natural' thermoplastic and masterbatch—a convenient wax-like base containing appropriate concentrations of colorant and other additives (such as anti-static agent). The supply of masterbatches has become an important specialized business.

Jazzy, marbled, or tortoiseshell effects can be produced in moulded components by using starting materials in a selection of different colours, or by using two or more masterbatches. Such effects are attractive for certain purposes but it is important to remember that when preparing the melt intensive mixing must not take place: machines are made specially to ensure that the pattern required in a moulding can be reproduced precisely.

Agents against Aging

Exposure to heat and to light will affect plastics, and most formulations include agents to give protection against such changes, both during processing and subsequently. In this connection some of the most common chemical reactions are oxidation, attack by ozone, and dehydrochlorination. When heated, polyolefins degrade readily by oxidation and for this reason polyethylene and polypropylene customarily are associated with anti-oxidants. PVC degrades more readily by dehydrochlorination, so the additives necessary in these formulations are those that restrict this form of change. Attack by ozone is a feature of rubbery materials and may be controlled with anti-ozonants. Most polymers are prone to degradation by ultra-violet light and hence the inclusion of stabilizers against this.

An early sign of degradation is a slight change of colour—for example, yellowing. In finished components some areas may yellow and others shielded against exposure keep much of the original colour. The patterns resulting might be unsightly and unacceptable—as also particularly with yellowing of transparent or translucent items (like light fittings); opaque colours are affected by this form of degradation and here too from time to time there may be complaints, difficulties in matching, and rejection.

Lubricants

The main functions of a lubricant are to stop the material adhering to processing equipment and to ease flow during moulding or extrusion. Such agents may be classified as 'external' or 'internal', a good example of the former being stearic acid added to PVC to prevent sticking. The primary purpose of the internal lubricant is to ease flow, and this may be exemplified by glyceryl monostearate in unplasticized PVC. Without the inclusion of small amounts of such lubricants it would not be practicable to make certain items in this material.

Lubricants have the effect of improving the gloss and appeal of mouldings since they help to bring the material into close contact with the metal finish of the mould and facilitate transfer of that finish to the plastic. An 'easy flow' resin such as polystyrene for making thin-walled containers would be based on polymer of low molecular weight containing an internal lubricant such as butyl stearate or liquid paraffin; the composition, in the form of small uniform pellets, then would be coated with an external lubricant—say, zinc stearate or a wax—and when fed to the injection moulding machine would melt easily, giving material of low viscosity and products with high gloss surfaces.

Lubricants sometimes are added to reduce the effects of friction in use: an example would be silicone oil—which gives mouldings with high gloss and when these mouldings are transported in bulk helps to reduce surface damage caused by rubbing between then.

Fillers and Reinforcements

The modification of plastics with inert fillers of various kinds is long-established, particularly with thermosetting materials. The use of fillers and reinforcements now also is frequent practice with thermoplastics, not only for making items in the field of engineering[2] but also for the more common products.

The term 'filler' may be unfortunate and a little misleading since often an important purpose of this additive is to modify the properties of the polymer and make them more suited to an application. Many substances are used in the form of fibres, which improve the rigidity of the composition (a frequent weakness in thermoplastics)—one of the most typical of these being glass fibre, with its advantages of being inert and available in standard form.

The use of fillers can create difficulties over the quality of finishes: often their incorporation causes roughening of the surface, particularly when fillers are in forms such as long fibres or large particles. Hence, when a finish of good quality is required it is preferable if possible to use filler in small particles and there may be benefits from treating it beforehand—the inclusion of a lubricant will help to improve dispersion, hamper agglomeration of the particles, and give mouldings with better surfaces.

Abrasion during processing reduces the length of fibres (to below 1 mm) but even so materials filled with fibres nearly always have comparatively poor matt surfaces. The use of expensive highly polished moulds is not very helpful and a more suitable approach to this problem is a mould with a finely textured surface, like a leather grain. Many designs are available and are good especially for applications like motor vehicles, where reflective surfaces could be a hazard. Another way of masking the effect of filler on the quality of a surface is to apply by sandwich moulding an unfilled coating over the filled plastic.

Reclaimed Material

Many injection moulding machines are run on feedstock comprising a mixture of new and reclaimed resin (speaking generally, this is possible only with thermoplastics). Reclaimed material includes feed systems from previous operations, unwanted and rejected components, and its incorporation in suitable proportion often is an important factor in controlling costs. The material is collected and re-ground to a suitable size, and must be mixed with the new in a definite, established ratio (which must be maintained) to ensure that the properties of mouldings, including surface finish, remain consistent. Great care is necessary to ensure that the re-ground stock is clean, dry, and of regular particle size. If dirty, the appearance of the products will be affected and damage to the mould or machine may occur. The presence of water or absorbed moisture will be deleterious, affecting the clarity of transparent material and giving streaking of surfaces of both transparent and opaque products. Should the particle size not be controlled within specified limits the feed will be irregular, leading to variations in the size and weight of the articles and to other defects compelling rejection.

It is important to ensure that the new material in a feedstock contains appropriate additives, particularly stabilizers, to accommodate satisfactorily the amount of re-ground being added. (Otherwise degradation, changes of colour, and so forth, may occur.)

Matting and Anti-Blocking Agents

Various agents are available the addition of which gives matt or 'silk' finish in the product. Such additives have the effect also of reducing 'blocking' (the tendency of a material to stick to itself) of film and sheet. Blocking can be controlled by altering the surface in such a way that (as an illustration) microscopical protrusions help to separate the layers of film. In polyethylene this can be achieved by incorporating inorganic materials such as kieselguhr of fine particle size; the first addition reduces the tendency for blocking and further amounts lead to a more matt finish (typical additions range from one half to three parts per hundred of resin).

For use with PVC, polymeric additives based on acrylic resins are available. It is thought that these form a layer on the surface of the PVC that reduces gloss without much impairing clarity—a feature useful in applications such as non-reflective coverings for documents.

Rubber

For many years rubbers have been included in plastics compositions to modify and improve mechanical properties. A well-known instance is polystyrene, the addition to which of about 10% rubber changes this essentially brittle substance to one with an impact strength sufficient for many uses not practicable otherwise. More recently it has been found helpful for certain types of work to add rubber to polypropylene; this gives an improvement in impact strength, particularly at low temperatures, and can enhance the appearance of the product—because although gloss is reduced the additive limits shrinkage after moulding and sinking of the surface. Up to 30% of ethylene–propylene rubber can be added, with the material resulting still being effectively rigid. Mixtures of this nature are seen frequently as fenders for motor vehicles and as bumper guards on the sides of cars.

Blends like these are based on rubber which has not been vulcanized or cross-linked; useful thermoplastics also can be obtained if particles of cross-linked rubber are dispersed in the matrix—this giving superior rubber-like qualities.

As with fillers the inclusion of a rubbery polymer generally reduces the gloss on the surface of finished products; essentially, this is because of the elasticity of the rubber and the tendency of the particles to deform during processing and later (after shaping) to protrude at the surface. The effect may be ameliorated in various ways: with extruded polystyrene sheet containing rubber it is possible to apply to the surface a glossy film of unmodified polystyrene, or to pass the extruded sheet through a series of highly polished rolls.

Other Impact Modifiers

Impact strength can be improved by other methods of modification—for example, by biaxial orientation of the material during or immediately after moulding. When transparent or translucent bottles are required in unplasticized poly(vinyl chloride) strength can be improved by including up to about 10% methyl–butadiene–styrene copolymers; the bottles resulting retain a good finish.

Plasticizers and Extenders

The use of plasticizers also has a long history; in commerce they are associated most usually with the cellulosic plastics and poly(vinyl chloride). Many different plasticizers are available, offering a wide range of properties, and by modification in this way PVC can be produced in a great variety of forms—ranging from the unplasticized (such as in rigid drain-pipes) to plasticized film for inflatable goods like boats and paddling pools. The inclusion of suitable plasticizers ensures that the material flows more easily during processing and finished products are softer and more flexible. Transparent articles with very high gloss can be obtained in plasticized PVC.

Extenders are referred to sometimes as 'liquid fillers'; their most common use is in plasticized PVC, where chlorinated waxes used in this way serve also as flame retardants.

Flame Retardants

Suitable flame retardants often are included to reduce the flammability of products: compounds containing halogens have been used for this purpose but unfortunately in a fire they can lead to the generation of fume and smoke; for this reason, among other substances that may be preferred is aluminium trihydrate, which decomposes on being heated strongly to give large amounts of water. However, the greatest benefit is obtained when this is added in large amounts—sufficient to reduce also the flow of the plastic during processing; as temperatures approach 200 °C processing becomes difficult and in consequence the appearance of the products may be much impaired.

Blowing Agents

Many products are based on expanded plastics—that is, cellular materials made either by introducing gas (usually nitrogen) during processing or including in the formulation a chemical blowing agent. Such agents generate gas when required and a group used widely for this purpose is the azo compounds, which decompose over a fairly narrow range of temperature—at, for example, the melt processing temperatures associated with the material concerned.

The use of blowing agents often results in products with lower gloss and inferior finish, although techniques are available for producing expanded materials with high gloss surfaces.

Nucleating Agents

Metal salts of organic acids (for example, sodium benzoate) are used as nucleating agents, while many other materials will function similarly: they are added in order to increase the crystallization in certain materials, notably polyamides, and have the effects of reducing the size of crystal structures but inducing the formation of more of them within a shorter period of time. More rapid crystallization implies faster solidification and can help to raise significantly the rates of output in processes like injection moulding; reduction in size of the crystal structures also leads to improvements in the appearance and clarity of products.

Anti-Static Agents

Many formulations include suitable anti-static agents, which migrate gradually to the surface and so reduce the surface resistivity. Dispersal of static electric charges by this means is beneficial over the long term for the appearance of the articles (since dust is less likely to collect on them) but the presence of additives of this nature can necessitate some changes in moulding practice.

Importance of Dispersion

In all cases to give the best results additives must be dispersed well and to this end, melt-mixing by means of a compounding extruder can be very helpful—giving continuous output typically in the form of rods that can be cut subsequently to form pellets of regular size. However, in some instances lower levels of dispersion can be tolerated, in order perhaps to reduce costs or to give flexibility in operations.

Melt Processing

In processes such as the moulding or extrusion of thermoplastics the compound is heated until it softens—usually not to a liquid but to a sticky material of high viscosity that will not flow quickly under its own weight. After heating in this way it is shaped and allowed by cooling to set in the form required. Shaping can be continuous, as in calendering or extrusion, or discrete as groups of components or injection mouldings. The heating is more efficient if the material is stirred or sheared—as occurs for example in an extruder, where melt passes along the length of the screw, or screws, rotating inside the barrel.

The materials fed to the machine must be clean and dry; in moulding they may be as fine powder, re-ground, or uniform pellets; generally higher and more consistent outputs are achieved when the last are used. In all cases it is essential in order to maintain quality and keep to specification that the feed remains consistent throughout.

Since in general they are poor conductors of heat plastics require large inputs of energy as heat to bring them to the working temperatures: the amounts of heat needed differ between materials—essentially because of their molecular structures and crystallinity—as illustrated in Table XI.

Changes taking place in the composition in the course of processing usually are undesirable; if materials are not dry various problems can arise from evaporation and condensation of water vapour; if oxygen is present there may be oxidation, shown first by altered colour and followed by variations in properties; the melt may be overheated—leading in the absence of air to

TABLE XI *Heat contents of some injection moulding materials*

Material	*Temperature/°C*			*Specific heat/*	*Heat to be removed/*
	of Melt	*of Mould*	*Difference*	J kg^{-1} K^{-1}	J g^{-1}
Acrylate–styrene–acrylonitrile	260	60	200	2010	402
Acrylonitrile–butadiene–styrene	240	60	180	2050	369
Butadiene–styrene	220	35	185	1968	364
Cellulose acetate	210	50	160	1700	272
Cellulose acetate butyrate	210	50	160	1700	272
Polyamide 6	250	80	170	3060	520
Polyamide 6.6	280	80	200	3075	615
Polyamides 11, 12	260	60	200	2440	488
Polycarbonate	300	90	210	1750	368
Poly(ether ether ketone)	370	165	205	1340	275
Polyethylene					
low density	210	30	180	3180	572
high density	240	20	220	3640	801
Poly(ethylene terephthalate)	240	60	180	1570	283
Poly(methyl methacrylate)	260	60	200	1900	380
Polyoxymethylene	205	90	115	3000	345
Poly(phenylene oxide)	280	80	200	2120	434
Poly(phenylene sulphide)	320	135	185	2080	385
Polypropylene	260	20	240	2790	670
Polystyrene	220	20	200	1970	394
toughened with rubber	240	20	220	1970	433
Styrene–acrylonitrile	240	60	180	1968	354

degradation, often with the generation of gas that might be dangerous; contaminants like dust also can cause difficulties during this stage.

Not only do the temperatures at which different thermoplastics are processed differ widely, but so also does the behaviour of material at those temperatures. To take one example, unplasticized PVC, even when stabilized, can only be held for a few minutes at its processing temperature of 175°C (347°F). Polysulphones, on the other hand, require melt temperatures in the region of 400°C

(752°F). To prevent degradation it is necessary to take account not only of the stability or instability of a compound but also of questions such as the period of time at the temperature, whether oxygen is present or the surrounding atmosphere is inert, and what materials are in contact with the plastic. As an example of the last, copper causes rapid degradation of polypropylene, and therefore copper cleaning pads should not be used on equipment for processing this material.

Aside from the possibility of hazards from substances evolved during decomposition (unless shown otherwise it is prudent to regard such compounds as harmful)—the fumes, gases, and condensates concerned can mar the products, giving unsatisfactory surfaces, stains, and so forth.

For consistent production to a specified standard control of the viscosity of the melt also is required: as noted above, the viscosity is affected by the molecular weight of the polymer, its distribution, by formulation, and by the heat applied; as temperature is increased the viscosity is diminished and to ensure flow as necessary throughout a run it is essential that such aspects be specified in advance and maintained within the limits.

Orientation

Some manufacturing processes for plastics require, in addition to shaping, orientation of the molecules, the products concerned not being satisfactory without this. Examples include polyolefin films, which are stretched while still hot either in a longitudinal or in longitudinal and transverse directions in order to obtain thin sections with adequate strength (or, on the other hand, for ease of tearing in one direction). The larger blow mouldings usually are oriented to make thin sections strong enough, while normally filaments extruded for weaving would be oriented longitudinally.

However, molecular orientation takes place in any case in some degree in any process involving melt flow and is 'frozen' into the objects when they cool and the material solidifies. Such orientation can cause difficulties (like local brittleness and a tendency for the objects to distort), and will be especially troublesome if in the course of production the pattern of orientation changes. Orientation and strains in transparent objects can be studied conveniently using a source of polarized light (when they can be seen in different colours), but is more difficult with opaque items.

Design of Moulds

It is not appropriate here to consider in detail the design of moulds for the various processes but the basic requirements can be summarized conveniently:

(i) a mould should be made from a suitable material in such a way as to ensure easy flow to the cavities with the minimum of waste

(ii) all tolerances should take account of the thermal properties of the materials to be moulded, including their rates of cooling and shrinkage on cooling

(iii) it should satisfy all the thermal and mechanical demands of the work, and resist the materials used, all with margin enough to guarantee the safety of operators
(iv) fitting and use should be as convenient as possible, bearing in mind that in practice it may be operated by robots
(v) the surface of the mould should be finished in such a way as to transfer that finish to the surfaces of mouldings.

When melt is forced or released into a mould it replaces air in the cavity and provision has to be made to allow that air to escape quickly: this provision is known as 'mould venting'. If fast filling of the mould is required effective venting is especially important and without it the trapped air will prevent plastic from making close contact with the surface and lead to unsatisfactory finish. (It should be remembered that compression raises rapidly the temperature of trapped air and in extreme cases, if that air cannot disperse, the melt may burn.) Various methods of venting are used—such as through the ejection pins or slots machined along the parting line of the mould. In vacuum venting, air is removed by applying vacuum to the cavity; the mould may be designed so that on being closed the faces seal against a rubber gasket, vacuum being applied to a manifold or channel. This method is particularly useful when filling is fast and when volatiles are being emitted (as in the injection moulding of rubber).

Cooling

In most melt processes the large amounts of heat required to bring the material to the temperature for shaping also have to be dissipated quickly and conveniently afterwards; some of the most common problems are associated with inadequate or poor systems for cooling, or poor control of the systems.

Not only may the rate of cooling be slow it also may vary from one part of the moulding to another, and this can have marked effects on the crystalline morphology, molecular orientation, shrinkage—and consequently on the qualities and appearance of products. If, for example, the rate of crystallization differs in different parts of a moulding the sizes of crystals also will differ, and its clarity will be impaired.

It is a physical law that materials shrink on cooling and with plastics this may occur both while cooling in the mould and after ejection. It is essential therefore to take account of this factor when designing the mould and selecting the conditions (so that, as examples, the dimensions of articles or internal volumes of containers comply with specification): with this in view it is important to know in advance the shrinkage anticipated; that of an amorphous thermoplastic like polystyrene is different entirely from a semi-crystalline material such as high density polyethylene—typically 0.6%, compared with up to 4% for the polyethylene. When polymer molecules crystallize they pack more closely than in the amorphous state, and hence the shrinkage is greater.

In mouldings with thick sections and extruded profiles the rates of cooling may be comparatively low and crystallinty high in consequence. For this reason it can be difficult to mould by injection polyolefins in thick sections, the material shrinking so much that sink marks become visible at the surface: special techniques may be necessary to overcome this.

Injection Moulding

In essence in injection moulding the two halves of a mould are clamped together and the material, after being softened by heat, is injected into the mould cavity. It is a repetitive technique capable, when operated correctly, of making large numbers of identical complex mouldings at high speed; it is used extensively for thermoplastics, most types of which can be processed in this way, and with slight modifications can be used also for thermosetting plastics and rubbers. When moulding thermoplastics, in order to obtain setting, the mould is kept cooler than the melt; with thermosets, on the other hand, the barrel of the machine is cooler than the mould and the mould (to ensure curing) hotter than the material.

So as to fill the mould at the speed required, high pressure is applied to the melt; to compensate for shrinkage, it usually is necessary to force in extra material in a second stage and to prevent over-filling at that time a lower second-stage or hold pressure may be used.

When the component has cooled sufficiently and set in the shape of the cavity the mould is opened and it is ejected, the cycle then being repeated. Provided good practice is followed it is possible to make items even of complex and intricate shape with dimensional accuracy. Mouldings so made often require little or no finishing—but if required their appearance may be altered or enhanced by polishing, machining, metallizing, plating, printing, and so forth. Small components especially can be manufactured in large numbers quickly and economically; large mouldings also can be made but the forces required for clamping the moulds are considerable and it can be helpful instead to employ a modified method such as injection compression moulding, or to mould using material capable of expansion.

To produce mouldings with high surface gloss the finishes of the cavities and cores of moulds must be excellent and free from faults. Since the temperature affects the rate of output and the finish on mouldings, precise control of this also is very important.

As an example of good practice in these respects when moulding acrylate–styrene acrylonitrile (which gives poor gloss at low temperatures and uneven colour at high):

(i) the material fed to the machine must be dry, free from contamination, and of consistent composition

(ii) since high injection speeds are recommended, mould venting must be good: typical depths and widths of vents would be 0.025 to 0.05 mm and 4 to 6 mm, respectively

(iii) mould temperature should be 60 °C (140 °F).

Engineering for plastics is subject to continuous improvement and it is relevant at this point to mention two aspects of electronic control in particular that may be used to assist and improve results.

'Multi-live-feed moulding' is effected by a device that can be fitted between the injection cylinder and the nozzle and contains hydraulic pistons controlled by computer. The pistons are activated during the hold period to force the melt to mingle at joins (formed, for example, when flowing round a core or pin), and so gives significant improvements in strengths at joins as well as superior finish.

Pressure in a cavity may be controlled by fitting appropriate transducers to sense when the mould is full and bring about switching to secondary holding pressure; such transducers may be linked with computer programmes determining the amount of material packed in to compensate for shrinkage—and so give more uniform output.

Sandwich Moulding

Sandwich moulding is a form of injection moulding employing more than one material, one normally surrounding or encapsulating the other. This approach can give cellular materials with rigid outer skins, or (say) a substrate largely of re-cycled plastic with a surface layer of new material. The machines are equipped with two injection units, one for each material, which may be operated jointly or in sequence.

The sandwich process extends the scope of injection moulding and (as an instance of this) makes it possible to provide fibre-filled mouldings with surface finish of very high gloss. It is applied to produce reflectors for headlights in motor vehicles—where the substrate is heavily filled and resistant to distortion by heat yet the surface is mirror-like.

Expanded Plastics

Injection methods can be employed for expanded plastics—including the type known as 'structural foam' in which outer layers are more dense than the inner. With these, a suitable blowing agent is incorporated in the formulation and decomposes at the moulding temperature to form gas; in the outer layers the gas escapes before the material sets but inside it remains trapped—to give substance within of lower density. The melt containing gas must be transferred at high speed (before the gas escapes) but, as it flows easily and the components usually are of relatively large cross-section, high pressures may not be needed.

Mouldings made in this way have relatively rough surfaces (caused by the escaping gas) but in some applications this can be turned to advantage. If the components are intended as replacements for wood they can be grained and printed after moulding and even (if desired) have 'worm holes' included.

Gas Injection Moulding

During gas injection moulding a gas such as nitrogen is injected into the melt in such a way that it stays in discrete pockets in the thicker sections; and as the material cools these pockets of gas help to keep it in contact with the walls of the mould. Moulding of polypropylene in thick section may be done by this means, to obtain surfaces of standard high enough for electroplating (as in moulded replacements for metallic bathroom fittings).

Injection Compression Moulding

The pressures required for injection at high speeds can range up to 245 MN m^{-2} (35 000 lb in^{-2}), so that high clamping pressures are needed to stop the mould opening during filling. This in turn means that equipment must be substantial (and costly) in order to withstand repeatedly the pressures involved.

One way of limiting the need for high clamping pressure is to design the mould in such a way that it can open slightly without melt escaping: in practice this means that a precise quantity of melt is injected into a lightly closed mould, which then parts or 'breathes'. When the transfer of melt is completed, power is applied to close and to clamp the mould—just as in compression moulding. Cooling of the melt takes place in the usual way.

This process—injection compression moulding—offers other advantages besides lower clamping forces: compression of the hot melt gives excellent surface finish and lower pressure at injection reduces orientation in the melt, so improving dimensional stability. The method is used currently to make compact discs of high quality from polycarbonate.

Reaction Injection Moulding

Reaction injection moulding is associated with polyurethanes and is different entirely from moulding techniques used commonly for thermoplastics in that the starting materials usually are liquids at room temperature. As the name implies, a chemical reaction takes place in the course of moulding and complex polyurethane items can be made in one step from mixtures comprising polyol, polyisocyanate, and a suitable promoter of the reaction.

The ingredients are pumped to a mixing head in the form of fine streams of liquid which atomize and mix, and then into the mould. If a blowing agent such as methylene chloride also is included the mould is filled only partially to allow the mixture to expand on reaction and fill the cavity.

Low clamping pressures only are needed, and this in turn means that quite large products can be made on inexpensive plant. By varying the raw materials it is possible to produce either rigid or flexible articles, microcellular or otherwise, and (if required) to include fillers or reinforcing sections. Thermosets other than polyurethane can be used—like epoxides and polyesters (with the latter, the process is known also as 'resin transfer moulding').

Subject to suitable conditions, products with excellent surface finish can be made.

Blow Moulding

Blow moulding was adapted to making containers from plastics from traditional methods of blowing glass bottles. Essentially, the thermoplastic is heated and formed into a 'parison', which then is inflated by compressed air and brought into contact with the walls of the mould cavity. The method is used commercially for bottles and other containers in poly(vinyl chloride), high density polyethylene, and other thermoplastics, in two main forms as follows.

(i) Extrusion Blow Moulding

In extrusion blow moulding extruders are used to make required tubular lengths of melt, each of which then is sealed by closure round it of the mould and blown to the shape required. In some processes the parisons are made in a stage separate from the moulding; in others, bottles are formed, filled, and packed in a continuous operation (filling with a cold liquid product being in effect the final step in cooling the mouldings).

Good surface finish is necessary (especially for transparent bottles) and usually can be achieved. The technique is applied also to making large containers and barrels, in polymer of very high molecular weight—when there can be difficulty in obtaining good appearance.

(ii) Injection Blow Moulding

In this method a mandrel is placed in the mould of an injection moulding machine, the mould heated, and thermoplastic forced in to flow round the mandrel and form a tube with a closed end; mandrel and plastic then are removed and passed while still hot to the blowing mould, where air introduced through the mandrel blows the material to the shape required.

Since two sets of moulds are required the technique is more costly than extrusion blow moulding but can offer advantages when special shapes or a particularly high degree of uniformity are required. It is useful also when there must be no risk of contamination of the containers—as in products for use in medicine. Articles of high quality, including visual appeal, can be obtained.

Blow mouldings often are made in polyolefins and when they are required to be printed it usually is necessary to prepare the surfaces by some suitable treatment, such as flaming or corona discharge (see Chapter 13).

Most thermoplastics can be permeated by gases and vapours and this has been a particular concern with blown bottles, where there has for many years been interest in the possibility of packing carbonated drinks like beer, lemonade, and water in containers that would be less brittle than glass. From time to time various methods have been proposed and developed with a view to reducing permeability, with differing degrees of success. Those of particular interest currently include:

(i) Fluorination

With this approach the inside of a blown container is exposed to fluorine in such a way that hydrogen atoms on the polyethylene molecule may be replaced by fluorine. Thus polar groups are introduced into the thermoplastic and make it less attractive to non-polar contents such as motor fuel (it is understood that the procedure is applied sometimes to blow mouldings used as fuel tanks).

(ii) Co-Extrusion Blow Moulding

Co-extrusion is the process in which two or more streams of melt are combined in the die to make an extrusion of two or more layers of plastics; often, one of the layers is a barrier based (say) on poly(vinyl alcohol).

Co-extrusion blow moulding combines this with a blowing stage to give mouldings comprising more than one layer of material, usually in order to improve barrier properties. In some instance, to ensure the special degrees of protection and storage life necessary for the contents, the blown bottles may have as many as five layers.

(iii) Biaxial Orientation

If desired it is relatively easy to arrange for biaxial orientation to occur in injection blow moulding: the mandrel is made from telescopic sections which are extended before air is introduced; its extension causes orientation in one direction and the subsequent blowing stage orientation in the other. Stretching and orientation in this way can give strong products with improved resistance to gas diffusion. When a material such as poly(ethylene terephthalate) is employed the bottles resulting can be light in weight yet suitable for packing carbonated drinks like lemonade: containers with high surface gloss can be obtained.

(iv) Plasma Polymerization

In this context 'plasma' is a gas containing free electrons, ions, and neutral particles and may be created by applying external excitation to the gas. When used to polymerize ethylene a plasma can give coatings of higher molecular weight, cross-linked and without pores—and provide relatively impermeable barrier layers on blow mouldings.

Further Methods of Improving Properties

The great majority of blow mouldings are containers for products sold to the public and because of this are required not only to be serviceable but also attractive and capable of carrying appropriate identification and instructions for use.

Improvements of clarity in transparent materials can be obtained in a variety of ways. If unplasticized poly(vinyl chloride) includes an impact modifier with refractive index matching that of the polymer it is possible to use thinner sections and so increase clarity, rates of output, and gloss. With polyolefins, similar results may be achieved by including a nucleating agent to accelerate crystallization; even polypropylene, which normally is translucent, thus can be made in an almost clear form.

An an alternative to printing the bottles or to affixing labels subsequently it is possible to apply the label at the moulding stage. Paper or thermoplastic film labels are printed in advance and are placed in the moulds before parisons are collected, being held in position by vacuum; adhesion of the label takes place when the blown material fills the mould.

A similar technique may be used to make injection mouldings with elaborate patterns or decorations: paper or film may be printed flat with the design required then folded, say, in a box-shape and placed in the mould—where it is held in place electrostatically or by vacuum applied through pins; moulding then takes place on or through this substrate. Such an approach has been used to mould boxes for biscuits, showing pictures of the biscuits on the box. It eliminates difficulties associated with printing shaped objects.

Compression Moulding

Compression moulding was the earliest form of high-pressure moulding used in the industry and still is normally limited to thermosetting materials (plastics or rubbers). When automated it is capable of giving attractive items at high rates of output. However, for thermoplastics (unless, as with vinyl long-playing records, special equipment is used) the cycle times are too long in comparison with other methods.

Briefly, a measured amount of the thermosetting compound is placed in the cavity of a heated mould which is attached to the platens of an hydraulic press. The temperature of the mould may be (as an instance) 150 °C. When the press is closed and heat and pressure applied the material flows and fills the cavity; excess (known as 'flash') escapes from the mould. Heat applied causes cross-linking and hardening of the material; the moulding sets in the shape of the cavity and after a pre-determined time (perhaps three minutes) can be removed. Pre-heating the material (to, say, 70 °C) can be employed as a method of shortening cycle times.

When a moulding is removed from the press it still has attached to it a thin web of flash which must be removed by buffing or sanding before the item can be used. Although flash is created the amount of waste generated in the process is relatively low, bearing in mind that unlike injection moulding there is no feed system to be removed.

The method is appropriate for large and small mouldings where very close tolerances and delicate inserts are not required (inserts would be disturbed when the material started to flow). It is used for moulding materials such as aminoplastics, to make bottle caps, electrical and light fittings, and tableware.

If the plastic is in the form of fine powder and is heated before being placed in the mould compression techniques can produce at fast rates thermoset items with high gloss (essentially, this is because the powder is not required to flow very far and is in contact continuously with a hot polished-metal surface). The hardness and surface finish obtained in moulding can be modified in a variety of ways—as for example by coating in the mould with low-shrink resins (see 'Polyester Moulding Compounds' below). Labels may be applied at this stage, as is thought to have been done first with gramophone records, where two small labels were placed centrally one on each side of the pre-heated material. In an extension of this principle complex patterns (like 'willow pattern') can be moulded into or upon tableware made from melamine–formaldehyde.

Mouldings in melamine–formaldehyde are bright, attractive, resistant to scratching and to staining, and can be free from odour—very suitable for applications such as tableware and kitchen utensils. Decorations like emblems, patterns, or pictures can be added by introducing into moulds containing minimally cured mouldings, before completing the process, printed α-cellulose paper impregnated with uncured melamine–formaldehyde resin. In this way, a single moulding tool will make objects with a wide variety of decorative designs. Such work is done most easily on moulds that are flat or have smooth profiles but grades of paper are available for sharp and rectangular edges; since the patterns are moulded-in and protected by a layer of cured resin the surfaces are durable and may be washed repeatedly without untoward effects.

Melamine–formaldehyde resins are available in a range of colours and complementary or contrasting colours can be used in two-tone mouldings made with double-punch tools (as examples, cups and mugs that are white inside but a different colour outside).

In an extension of moulding-in printed decorations, using polyester resins and lay-up by hand, designs can be transferred from the printed substrate and sealed into the surfaces of articles—the feedstock originally carrying the effect being removed afterwards.

Polyester Moulding Compounds

Polyester compounds are based on unsaturated polyester resins with a reinforcement (frequently glass fibre) and additives such as catalyst, filler, release agent, thermoplastic material, and colour; there are many different types—one example known as 'dough moulding compound'. Although thermosetting the effect of including a small percentage of thermoplastic is to reduce shrinkage (hence descriptive terms like 'low profile' or 'low shrink resins' may be used for them); a small inclusion of polystyrene helps to keep the hot material in close contact with a highly polished mould and so obtain a high-gloss finish—important especially for markets such as domestic appliances and motor cars.

Improvements in both surface finish and hardness also can be obtained by coating articles in the mould, using unsaturated polyester resins re-inforced heavily with glass beads. In an example of this the coating is applied to the hot mould by robot electrostatic spraying and on contact with it flows and fuses; moulding follows, when the polyester bonds with the coating. In this way, very hard scratch-resistant coatings (suitable for applications such as kitchen sinks) can be obtained.

Compression Moulding of Thermoplastics

As noted above, when thermosets are moulded by compression techniques it is necessary to set the shapes of the mouldings by applying further heat after the compression stage. However, in uses where thermoplastics are moulded the reverse is true—in order to set the material to shape the mould must be cooled before ejection. Unless the equipment used was designed for the purpose the cycles of heating and cooling can take a considerable time, and even when rapid cycling is achieved they are likely to be inefficient so far as the use of energy is concerned.

It seems likely therefore that compression moulding of thermoplastics will continue to be confined to articles with which special advantages can be obtained. An instance is the preparation of specimens for testing: the compression approach gives reduced anisotropy compared with injection moulding, and hence specimens with more uniform properties. Since compression also will give particularly good reproduction in a material of the heated surface of the mould it was the method of choice for gramophone records, where the quality of sound obtained depended on accurate replication.

However, composite materials based on thermoplastics helped to create new interest in compression moulding; it was found that the fibre re-inforcements kept their length better than in the injection process, making it easier to produce mouldings with desirable properties like strength and rigidity. This advantage could well lead to further applications of compression moulding to thermoplastic compositions in future.

Control of Quality

Mouldings are made for a vast range of different purposes, in some of which (at one extreme) very tight control of quality is essential; in others (at the other extreme) appearance really may be the only thing that matters. The extent of supervision for quality control therefore can vary accordingly—from testing exhaustively to satisfy limits stated in a written specification to (say) a quick visual comparison of a selection of mouldings with colour standards. A programme for selecting representative mouldings and inspecting and testing them to a degree appropriate for the requirements should be settled in advance with the purchaser.

Preparatory to moulding it is always desirable to inspect the materials to ensure that features such as colour and particle size are as required. It may be necessary to dry a material before use, and forms of contamination other than moisture should be eliminated too. If the material itself is unsatisfactory in some way it is very likely that the mouldings will be.

Very often (since a melt process is to be used) a material for injection moulding or extrusion will be subjected in advance to one or more tests of flow at different temperatures, using relevant equipment and methods from British and international standards or other specifications.

In semi-automatic moulding the operator may inspect each unit for features including colour, dispersion and uniformity of colour, standard and uniformity of gloss or mattness, and the presence or otherwise of defects like 'silver streaks' or stringing. However, it is preferable if assessments can be made in numerical terms, expressed in units that may be compared quantitatively and subjected easily to statistical treatment and analysis. It is often convenient in the course of production to check the weights of mouldings, either as representative samples or automatically in every case. It may be possible to measure and to check dimensions (length, depth, *etc.*) but for obvious reasons the rapid measurement of a feature like wall-thickness can be difficult, especially with intricate shapes.

Developments in micro-processing have made it much easier to monitor and to control conditions in the course of manufacture, and to prepare and maintain written records. Injection moulding may be supervised with the help of melt thermocouples fitted in the nozzle, pressure transducers in the hydraulic line and in the mould, and a transducer to monitor linear movement of the screw. Information from these sensors may be displayed and depicted in relation to the control limits (upper and lower) and the operator thus made aware immediately of variations that could be significant. Conditions can be corrected more rapidly and if need be any mouldings that might not be satisfactory diverted for examination. Should the number of suspect products exceed a limit set in advance, a line can be shut down automatically.

Where robots are used for the removal of mouldings a measuring system can be used to assess the components: measurements of different dimensions can be taken by means of a video camera with appropriate programmes, and surface appearance judged (say) by gloss.

Information generated through such a system may be used in the first instance to approve or reject a moulding, then transmitted to records kept numerically or as charts. (For purposes of reference in future such records must be identifiable through catalogue or part numbers, together with batch details in some form.) While the moulding is conveyed to storage or assembly, automatic methods can be employed if required to assess every tooth of a moulded gear wheel—applying criteria such as size, shape, and appearance—and if any aspect be unsatisfactory can divert it for re-assessment or re-working.

References

1. W. V. Titow, 'PVC Plastics', Elsevier Science Publishers, London, 1990.
2. A. Whelan and D. Dunning, 'The Dynisco Extrusion Processor's Handbook', published by the authors for Dynisco, Inc., 1st Edn., 1988.
3. A. Whelan and J. A. Brydson, 'The Kayeness Practical Rheology Handbook', published by the authors for Kayeness, Inc., 1st Edn., 1991.
4. A. Whelan and J. P. Goff, 'The Injection Moulding of Engineering Thermoplastics', Van Nostrand Reinhold, New York, 1991.

CHAPTER 9

Extruded Surfaces

A. WHELAN

Introduction

It is possible, using ram extruders, to process thermosets by extrusion but as with injection moulding the overwhelming bulk of the materials fabricated in this way are thermoplastics. Both moulding and extrusion are melt processes, the essential difference between them being that the former yields separate items while extrusion is continuous—the resulting profiles having to be separated into films, sheets, bags, and so forth, in subsequent operations.

The plastic is softened by heat in the cylinder or 'barrel' of the extruder, and most commonly is conveyed by means of a rotating screw along the length of the barrel until it is shaped by being forced through a die. A finish may be imparted to the extrusion by the design and conditions at the die, or this may be modified in later working by orientation, applying polishing or embossing rolls, and so forth.

The following are typical products of extrusion:

(i) feedstock for other processes: melt-mixing by means of a compounding extruder gives good dispersion in a formulation and the continuous extruded rods produced at the die can be cut to make regular pellets for moulding

(ii) film (flat, or layflat—folded tubular forms which often are shaped, cut, and sealed to make bags of various sizes and capacities)

(iii) sheet (flat or corrugated)

(iv) pipe (in a range of diameters and wall thicknesses)

(v) tubing

(vi) wire and cable insulated with plastic, for conveying electricity and communications

(vii) profiles (frames for windows and doors, curtain tracks, troughs, guttering, *etc.*)

TABLE XII *Qualitative expression of effects on high density polyethylene of changes in melt flow rate, density, and distribution of molecular weight*

Property	*Effect of increasing*		
	Melt flow rate	*Density*	*Breadth of distribution*
Translucence		diminishes	
Surface gloss			diminishes
Softening temperature		increases	
Melt strength	diminishes		increases
Melt elasticity	diminishes		increases
Melt fracture	diminishes		diminishes
Pseudoplasticity			increases
Tensile strength			
at yield		increases	
at break	diminishes	increases	
Elongation at break	diminishes	diminishes	
Impact strength	diminishes	slightly diminishes	diminishes
Modulus		increases	

(viii) filaments and fibres—for making brushes, rope, twine, woven as awnings and other heavy-duty fabrics, and for fibre optics
(ix) nets—for packing (particularly fruit and vegetables) and protecting bushes and plants in gardening and horticulture
(x) plastic coatings on substrates such as paper, metal, or other plastics.

The first section of the previous chapter—'Material Used for Moulding and Extrusion'—summarized the main types of plastic employed, and subsequent sections of that chapter included relevant information about additives of various kinds. It is not appropriate here to consider the extrusion of the various materials in detail but some of the most important points linked with the finishes obtained are noted as follows:

(i) Polyethylene

Like many other plastics, polyethylene is a family of materials differing in attributes like molecular weight, distribution of molecular weight, and the degree of branching of the molecular chains. In processing, such factors affect the ease (or difficulty) with which the melt flows, crystallization, and behaviour on cooling. The material is available commercially with densities ranging from 0.910 g cm^{-3} ('low') to 0.959 g cm^{-3} ('high'): Table XII shows how changes in the melt flow rate and distribution of molecular weight may affect the properties of high density polyethylene and Table XIII indicates differences in properties between polyethylene of different types.

(ii) Poly(vinyl chloride)

PVC plastics are manufactured in an even wider variety of forms than polyethylene; polymerization may be as an emulsion or suspension, the results in

TABLE XIII *Qualitative comparison of properties of polyethylene on different densities*

Property	*Polyethylene*		
	Low	*Medium*	*High*
Transparency	highest	higher	
Surface gloss	highest	higher	
Softening temperature		higher	highest
Melt strength		higher	highest
Moulding cycle times		higher	highest
Resistance to shrinking and deformation	highest	higher	
Resistance to environmental stress cracking	highest	higher	
Resistance to absorption of oil and grease		higher	highest
Resistance to permeation by gases and liquids		higher	highest
Tensile strength at break	highest	higher	
Elongation at break	highest	higher	
Impact strength	highest	higher	
Strength at low temperature	highest	higher	
Modulus	highest	higher	
Long-term load bearing		higher	highest

either case being material with a syndiotactic structure—that is, with chlorine atoms alternating on either side of the main chain. The chlorine atoms generate strong attractions between the chains, so the material is harder and stiffer than polyethylene—though this can be altered by adding plasticizers. The monomer is harmful and care must be taken to ensure that its residues in the polymer are below the minimum level stipulated.

Additives are helpful, even essential, in processing many thermoplastics but they are particularly important for satisfactory use of PVC. Some of the additives necessary in unplasticized formulations are:

(*a*) heat stabilizers, which often are based on mixtures of calcium and zinc salts, or (especially for transparent or translucent formulations) on tin complexes. Low concentrations of epoxidized soya bean oil are included in some systems. It is important to ensure not only that a formulation contains sufficient heat stabilizer for the original process but also enough to protect the material if need be during re-grinding and re-processing.

(*b*) lubricants, both internal and external, are necessary. Lubricants are made specifically for particular combinations of applications and equipment, in amounts normally below one part per hundred of resin (so as not to affect unduly the quality of the material). Calcium stearate is a typical internal lubricant, while synthetic waxes and fatty acid esters are common external types.

(*c*) processing aids, perhaps based on acrylic polymers.
(*d*) impact modifiers—based on plastics such as acrylonitrile–butadiene–styrene, or methacrylate–butadiene–styrene—which may be used at high levels (about 12 parts per hundred of resin) to give good impact strength in non-oriented products.
(*e*) fillers, such as calcium carbonate, China clay, inorganic fibres, and titanium dioxide.
(*f*) pigments; as the material is naturally translucent a wide range of colour is possible—including translucent as well as opaque colours.

In preparation for extrusion the composition is melt-mixed, perhaps in a batch mixer or a continuous machine such as a twin-screw extruder; in this stage it is necessary to work the material just sufficiently to ensure that it is gelled fully (fused) but not to the extent that degradation has started. Degradation can be caused by overheating or by excessive shear and to avoid it it is essential that the correct temperatures be employed at each stage. A characteristic first indication of decomposition is a change in colour; once started it can spread rapidly. One of the products is hydrogen chloride and this will catalyse further degradation—besides attacking metals like steel, causing corrosion, pitting of surfaces, removal of protective layers, and rusting of mild steel: in forms such as gas, acid liquid, or vapour with water, hydrogen chloride is harmful to human beings.

Unplasticized poly(vinyl chloride) compositions can be extruded in transparent, translucent, or opaque forms—to give (as examples), film, flat or corrugated sheet, and profiles such as gutters, downpipes, and troughs. For uses such as pipework the extruded products may be fixed or joined with fittings injection moulded from the same or similar material.

It was discovered in the 'twenties by Dr Waldo Semon when attempts were being made to dissolve PVC that it could be plasticized to give elastomer-like compounds, and since then an extensive variety of plasticizers has been suggested and many introduced commercially. The addition of plasticizers provides materials with a considerable range of flexibility and softness: those used most commonly today are esters of C_8 to C_{10} alcohols, such as phthalates, phosphates, and sebacates—particularly dioctyl phthalate (known also as 2-ethylhexyl phthalate), di-iso-octyl phthalate, and dialphanyl phthalate.

The conditions of processing and finishing can have influence but in essence the properties of a plasticized composition are linked with those of the original polymer and the types and amounts of plasticizer and other additives present. As a general comment, plasticized compositions are more flexible and less resilient than unplasticized PVC, less resistant to attack by other substances, to permeation, and to elevated temperatures: under some circumstances of use the plasticizer can be extracted, leading to the gradual embrittlement of an item.

Plasticized PVC can be extruded in various forms—including layflat film for bags and other containers with gusseted walls, flexible pipework, hoses, and continuous seals in many different configurations.

(iii) Polystyrene

Polystyrene may be extruded to produce film and other transparent articles with high gloss finish but in commercial practice the formulations toughened with rubber are used to a greater extent (extruded transparent film is toughened by stretching and orientation while it is still hot from the die).

The amounts of rubber present vary with grades but about 9% would be typical. The rubber particles should be distributed evenly throughout the matrix and while they have the effect of improving the impact strengths of products there is also a certain loss of hardness, stiffness, and resistance to heat. The rubber can be attacked or permeated more easily than polystyrene, so grades toughened in this way generally are not so good in these respects: polystyrenes, whether toughened or not, tend to degrade on exposure to light; with transparent fittings, yellowing can become apparent fairly soon after installation.

Toughened polystyrene and other related plastics like acrylonitrile–butadiene–styrene can be extruded to produce opaque, coloured sheet for thermoforming, and as profiles and pipe. Techniques are available by which good surface finish may be obtained, but under typical conditions the extruded surfaces will be matt or 'satin'.

Preparation of Material

Many of the comments in the previous chapter about the selection of grade, additives and mixing before moulding apply equally in preparation for extrusion. It is important of course that the material should be appropriate for the purpose, uniform, dry, and free from contamination. It should be tested for flow and while many tests have been devised for this it is convenient to classify them as either for 'low' or 'high' rates of shear. The main terms used in such testing ('viscosity', 'shear rate', 'shear strain', *etc.*) are defined in words and expressed as formulae in ISO 472, and it is not necessary to repeat them here. Viscosity may be regarded as the resistance to flow or the internal friction in a polymer melt and often will be measured by means of a capillary rheometer, in which shear flow occurs: with flow of this type—one of the most important with polymer melts—when shearing force is applied one layer of melt flows over another in a sense that could be described as the relationship between two variables—shear rate and shear stress.[1] In the capillary rheometer the relationship between the measurements is true only if certain assumptions are made, the most important of which are:

(i) the flow is isothermal
(ii) there is no slip on the die wall
(iii) the melt cannot be compressed
(iv) the pattern of flow remains the same throughout
(v) there is negligible dissipation of energy at the die entrance and as a consequence of factors such as uncoiling of chains.

Flow curves may be obtained by plotting shear stress against shear rate and apparent viscosity (the uncorrected viscosity) can be calculated by dividing the shear stress at the wall by the apparent rate of shear.

Melt flow rate or melt flow index may be obtained by applying a specified weight to force a sample of the plastic, heated to a specified temperature, through a hole of stated diameter: the amount extruded in this way in ten minutes is the parameter required. The actual conditions of test may be varied according to the material and the requirements of processing but in general the rate of shear is far lower than is experienced in reality in most melt processes (in other words, this is a 'low shear rate' test).

Checks on raw material sometimes are made by using a small single-screw extruder as a rheometer, or a test machine designed for the purpose—such as a Brabender 'Plastograph' (a device for measuring torque, which may be attached for the purpose to a small extruder).

In such tests the extruder is fitted with a rod die and specified cylinder temperatures set so that the appropriate heating of the material occurs: extrusion behaviour is measured over a range of screw speeds and graphs plotted (against screw speed) of output, melt temperature, and die swell. Any alteration in speed will alter conditions in the cylinder and the temperature of the melt can be changed significantly by a change in speed; in practice, a range of screw speeds is employed, so rates of shear can be quite high (this is a 'high shear rate' test).

Establishing Conditions

Quite commonly, after softening a material by heating to a known temperature, flow testing is carried out by forcing the melt by ram at known speed through a specified die (for example, a die with a ratio of length to diameter 20:1. The pressure opposing flow is measured and recorded: the ram speed then is changed progressively, and the forces necessary to maintain flow under different conditions are measured, recorded, and compared.

If the dimensions of the cylinder and the speed of the ram are known it is possible to calculate the volumetric flow through the die; the shear stress and shear rate may be estimated, flow curves constructed, and the material characterized by shear flow over a range of temperature.

The work of a laboratory capillary rheometer—a 'high shear rate' rheometer—can be performed by a screw injection-moulding machine provided it is equipped with a suitable (nozzle) pressure transducer and the injection speed can be set and held at a specified value. The cylinder is charged with material with the screw rotating slowly and low back pressure, the charge being allowed to remain for, say, two minutes: such conditions will promote uniformity of temperature and give residence times similar to those in a laboratory capillary rheometer. The melt then is purged into the air and the pressure, speed, and temperature are recorded (if the melt temperature cannot be taken directly that of the purged material may be measured via a probe). In further stages the conditions are changed and readings taken again.

Information gained by methods such as these is of value in establishing conditions both for injection moulding and extrusion. It can help in ensuring, as examples, that the diameter of a runner system or the size of a die is suitable for a good flow of material, satisfactory cycling, and obtaining the finish required.

Single-Screw Extrusion

Figure 65 shows the essential features of a single-screw extruder; the material passes from a hopper, is heated electrically, forced by the rotating screw along the length of the cylinder, and out through the die. An extruded shape may follow the conformation of the die or this in combination with some method of distortion after extrusion (like blowing film into a bubble), then is fixed by cooling.

The drive unit usually is based on an electric motor and turns the screw at a suitable speed set in advance; the electric elements in the cylinder are connected to temperature controllers that maintain the levels required. While screw speed and temperature are of critical importance the ability to extrude a given material continually to a satisfactory standard depends also on more fundamental considerations—like the suitability in design and construction of an entire line, including the operation of the die and of any haul-off or other technique applied after material passes the die.

The essential features of an extruder screw are illustrated in Figure 66. Since the screw is required to rotate within the barrel and to push forward the material its overall diameter is slightly less than the barrel's internal diameter (an extruder with cylinder diameter of 100 mm might have a screw diameter of 99.8 mm). The length of the screw may be divided into regions or zones—typically the feed zone, compression (plasticating) zone, and the metering or pumping zone. The feed section commences behind the hopper and is of constant depth; from the end of the feed section to the start of metering the root diameter is increased gradually, and in the metering zone it becomes constant again. The feed section might represent half the overall length, compression 30% and metering 20%. The ratio of the volume of the flight in the feed zone to that in the metering zone gives the compression ratio: the lower volume at that stage is to compensate for the reduction in volume of the material as it melts and fuses.

With the conventional three-zone screw the compression ratio will serve to generate a considerable amount of heat, which may well be undesirable when (as in extrusion blow moulding) consistent high rates of output are required. To obviate this, machines are available with screws that have zero compression (that is, the depth of the flight is the same along the entire length): such screws prevent excessive temperatures but have little mixing effect; however, mixing can be improved by adding at the tip suitable fluted sections.

The rate at which the screw turns controls how much of the resin or composition is pumped forwards, and this in turn controls also or affects the mixing, the temperature of the melt, and the extent to which temperature varies.

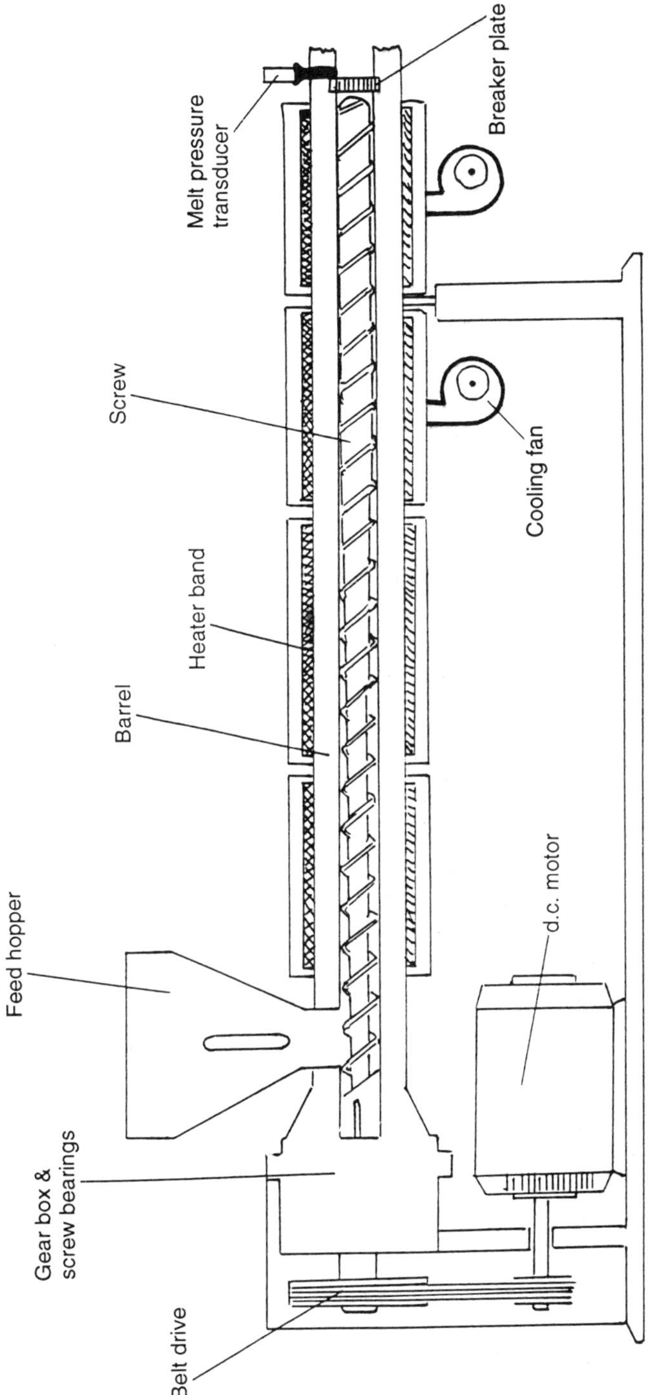

Figure 65 *Essential features for a single-screw extruder*

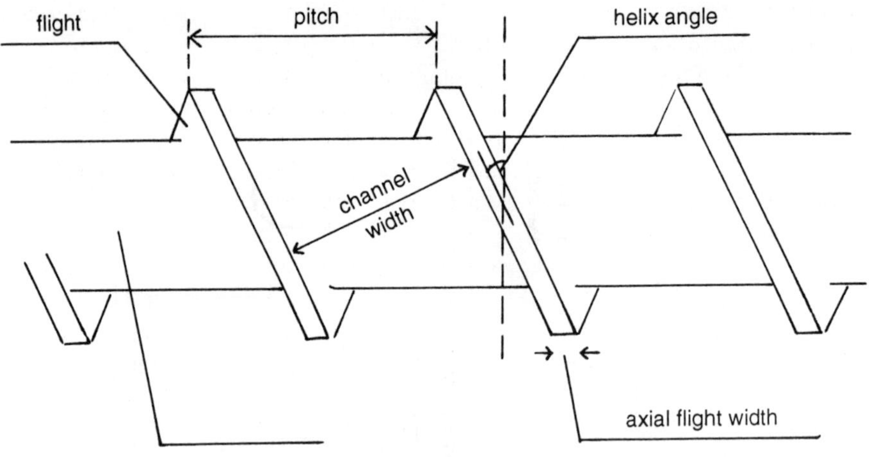

Figure 66 *Essential features of an extruder screw*

It must be possible therefore to set accurately the speed of the screw, speed must be capable of being read and recorded accurately, and it must be possible to hold it to a set value. Therefore, the machine must have a display of screw speed, and the drive system must be powerful enough to keep it constant.

Extruder screws and the range of conditions available on a particular machine may be versatile enough for several materials and grades but if it is desired, say, to extrude polyamide 6.6 or an unplasticized poly(vinyl chloride) more consistent and satisfactory results can be obtained by fitting a 'dedicated' screw designed specially for the material concerned. For polyamide 6.6, for example, the unit should be capable of generating higher temperatures than are needed for an amorphous thermoplastic.

For a variety of reasons the mixing in the extruder may not be uniform and the melt inhomogeneous as a result. With a view to preventing this the root of the screw may be fitted with pins that protrude into the material to help break-up laminar flow; or, more usually, the tip of the screw may be modified with mixing elements or sections (in effect the screw is lengthened, so that a section of rings, cams or kneading discs can be fitted to it).

Mixing can be improved also with barrier-design screws, which hold back unmelted material. In a conventional single-start screw the flight often is filled with a mixture of solid and melt, the solid floating in the melt so that it is hard for the screw to grip it: barrier-design screws have two flights, separated by the flight land, and as resin melts it is transferred from one to the other. The result is improved output at a given speed, and a lowering of melt temperatures.

Twin-Screw Extruders

For certain uses, twin-screws are popular. The screws may be designed to rotate in the same direction ('co-rotation') or in opposite directions (counter- or contra-rotation), while the flights of the screws may or may not intermesh (though intermeshing or partly intermeshing types usually are preferred).

Such machines are particularly suitable when there is a requirement for compounding as well as extrusion—as with unplasticized PVC. If compounding and extrusion of the formulation can be carried out in series, so that it is not necessary to heat it twice, the quantity of heat stabilizer added can be reduced. For reasons such as this the manufacture of unplasticized PVC pipe has become a typical use for twin-screw units.

In a conventional single-screw machine the hopper is filled and the screw takes from it at each rotation a maximum amount of material: this is in contrast with a twin-screw extruder where the action has the effect of drawing material down and forcing it forward much more positively. In the latter, flood feeding could result in the generation of very high forces, and even ultimately in the failure of the thrust bearings of the machine (so that large thrust bearings can be incorporated, the screws in some machines are tapered so that they are larger at the rear). For reasons such as these the twin-screw unit often is 'starved-fed' and the through-put of material is independent of speed. With increasing output, residence time is diminished and thus the shear and the requirement for energy are decreased. The rate of output may be virtually independent of the size of the die. Because twin-screw machines mix so effectively, greater volumes can be mixed, melted, and conveyed within shorter cylinders than are necessary with other types.

Twin-screw machines in which the screws are counter-rotating have particularly positive feed and conveying characteristics; residence times and temperatures should be more even throughout, so configurations of this kind are satisfactory for unplasticized PVC. The main disadvantages are: the possibility of entrapment of air in the material, the generation of high pressures, comparatively low maximum speeds, and low rates of output.

Co-rotating machines are used for compounding, the flow between channels giving better mixing. In such machines the screws wipe each other clean and, especially with a material not very sensitive to heat or shear like polyethylene, high speeds and outputs are possible. Wear on the screws and barrel also may be lower with similar outputs. On the other hand, the rate of production depends upon the pressure obtained at the die-head and at high pressures, because in this type of extruder the clearances between the flanks of the screws usually are greater, the differences in residence time of the materials become wider; since much higher rates of shear are possible there is a chance that some decomposition of heat-sensitive materials may take place.

Extruders may be classified conveniently using a three-figure system—for example, 1–60–24—where the first number states how many screws the machine has, the second gives the diameter of the screw in millimetres, and the third the effective length of the screw as a multiple of the diameter. In a 1–60–24 there would be a single screw of diameter 60 mm and length 24 diameters.

The efficiency of compounding extruders may be enhanced in a number of ways, and twin-screw units are designed also for special purposes. Mixing elements in a variety of forms (reversed flights, kneading discs, pins, *etc.*) may be incorporated part-way along the screws, and in some types can be changed easily as required: material is softened in an extrusion section, passed to a mixing zone, then to another extrusion section, and this sequence may be repeated (as it is in machines designed to eliminate volatiles, or in which chemical reactions are required to take place).

Special Configurations

(i) Cascade Extruders

The simplest type of cascade extruder consists of two screws, one mounted above the other, the output from the first feeding the second. Interconnection is by a passage that can be vented to allow volatiles to escape. The screws are not in direct contact and can be driven at different speeds, the output from each of them being set at an optimum with the help of valves.

(ii) Elastic-Melt Extruder

The elastic-melt extruder makes use of the Weissenberg or 'rod climbing' effect—which is observed when an elastic fluid is sheared or rotated inside a container by a rod. Because of viscoelasticity, the fluid climbs the rod.

The machine consists essentially of two heated discs, one of which rotates against the other. The thermoplastic material is introduced into the gap at the edge, becomes a polymer melt and moves towards the centre; the reduction in volume taking place on melting is compensated by a narrowing of the gap or nip between the discs (typically, one of the discs is convex) and the melt is forced through an aperture at the centre from which it is fed to the die. To improve uniformity of the melt a short screw (of low ratio of length to diameter) is placed between discs and the die. Since the discs can be arranged vertically, such machines can be very compact.

(iii) Gear Pump Extruder

In an extruder of this type a system of intermeshing gears is used to force the plastic towards and through the die. A gear pump can be fitted to a single-screw extruder in order to obtain steadier pressure and smoother output, or it may be used to extrude fibres from polymers which at processing temperatures are liquids of low viscosity.

Elastic Effects in Polymer Melts

Polymers are long-chain molecules and become distorted when subjected to shear during processing. Shearing itself tends to straighten the molecules but if still molten they will coil again when it ceases. When the cooling after shear is rapid, the re-coiling may not be completed. These changes can give rise to several effects which, because they are related to uncoiling and re-coiling in vulcanized rubber, often are referred to as 'elastic effects'. The most important of them are 'die swell', 'melt fracture', 'sharkskin', 'frozen-in' orientation, and 'draw-down', with related phenomena. Each is taken in turn.

(i) Die Swell

When polymer melt emerges from a die the extrudate swells so that its cross-section, immediately on leaving the die, is greater than that of the orifice. This happens because as the melt is sheared through the die the molecules become extended, with the greatest orientation near the die wall: on emergence they tend to coil—and thus to contract in the direction of flow and to expand in directions perpendicular to the flow. It has been shown by experimental work that:

(1) die swell increases with increase in the rate of extrusion or, more specifically, with rate of shear—up to a critical shear rate (see 'Melt Fracture' below)
(2) at given rates of extrusion or shear it decreases with increase in temperature
(3) at a fixed shear stress, it is affected little by temperature
(4) at a fixed shear rate, it decreases with increased length of die parallel
(5) through a slit die, swell is somewhat greater than through a capillary die, and it increases more rapidly with increasing shear rate
(6) die swell increases with increase in the ratio of diameter of reservoir to diameter of capillary (though it is affected little at ratios above 10:1).

The problem of correcting die swell can be approached in a variety of ways. With solid extrudates it is a common practice to draw down and so stretch the material, so that it can just pass a sizing die. (In this technique, it is not necessary even that the dimensions of the orifice be exact.) It should be remembered however that the drawing will result in orientation, increasing the strength in the direction of flow but decreasing it transverse to the flow—which may be desirable in the extruded shape, or may not.

When the extrusion is a profile with sections of different thickness the shear rates and die swell will be higher in the thinner sections than elsewhere—so drawing again may be of limited value. The thinner sections may have a shorter die parallel (to ensure that the rate of extrusion is constant over the entire cross-section), and this will alter die swell even further.

Usually when making pipe and tubing the extrudate is inflated to the dimensions of a sizing die: in such instances it must be taken that on emerging from the die the swell will be in line with the rate of shear, that the thicknesses of sections will be reduced in proportion to the amount of inflation, and with ratio of sizing die to the external diameter of the die orifice.

(ii) Melt Fracture

When extrusion is carried out at high speeds, distortion of extrudates may occur: it may be attributable to melt fracture (known also as 'elastic turbulence', and as 'bambooing'), or to sharkskin. These phenomena are not understood fully but appear to have different origins.

Melt fracture occurs when the rate of shear exceeds a critical value for the melt concerned at a particular temperature (that is, the 'critical shear rate'). There is a corresponding critical shear stress and the relevant point on the flow curve (or the shear rate–shear stress diagram) is known as the 'critical' point. It is believed that it is reached in the die entry region (that is, where material is being funnelled from the die reservoir into the capillary of a capillary rheometer)—which, in an extruder, corresponds with the point at which melt moves into the die parallel portion of the die. Some further complicating effects may occur at the wall of the die.

The form taken by the distortion varies between types of polymer but generally it is helical. With polyolefins a feature resembling the thread of a screw may appear, and with polystyrene the extrudate may form a spiral; with other melts, ripples or repetitive kinks like bamboo may be seen. For all melts, at rates well above the critical point, the helical nature becomes obscured by severe distortion which looks quite random.

The problem is most likely to occur when extrudates of small diameter are being made at high speed—as for example in wire covering.

Melt fracture has been studied in some detail, experiments having shown:

(1) the critical shear rate increases with temperature
(2) for materials of high molecular weight, melt fracture will begin at lower shear stresses than are possible for lower molecular weights
(3) polymers of similar melt viscosities (even if they differ in degrees of branching) tend to have similar critical points
(4) the quality of an extrudate may be improved markedly if the die entry is tapered; by this means, extrudates undistorted externally may be obtained at rates well above the former critical points (though there is evidence of some internal distortion still): it is possible too that tapering the die parallel may have the effect of raising the critical point
(5) there are indications that changes in the design ratio of the die parallel can increase the critical rate of shear.

Provided factors such as these are appreciated and taken into account, melt fracture should not present undue difficulties in practice.

(iii) Sharkskin

Sharkskin has been studied less widely than melt fracture but in fact may be a more common problem in actual operations. The distortion takes the form of transverse ridges (as distinct from the helical shapes in melt fracture) and is thought to be a consequence of the melt tearing as it emerges from the die. An explanation, briefly, is that melt close to the inside wall of the die is moving very slowly (not at all in the case of the layer immediately in contact with the wall) and as melt emerges and moves from the face at constant speed the outer layers may be stretched and tear.

Experimental studies have indicated that the critical rate of shear for onset of sharkskin is inversely proportional to the radius of the die—in other words, it is much lower with dies of larger diameter: this means in turn that with small dies (such as are used in typical laboratory rheometers) melt fracture may occur at rates below those giving rise to sharkskin, but with industrial dies the reverse can be the case.

Because of the relationship between rate of shear and radius of die it can be shown that sharkskin will occur, regardless of the die, above a critical linear rate of extrusion—that is to say, for a certain hypothetical melt it might happen at a rate of extrusion of 0.0125 m s^{-1} irrespective of the size of die.

It appears worst when a melt is partly elastic and of a consistency akin to that of a friable cheese. Sometimes the condition can be ameliorated by reducing melt temperature, so that as it emerges from the die a melt is more strongly elastic. In an alternative approach, improvements have been obtained by heating the exit of the die—so that the surface layers of the melt are more fluid and tearing is less likely.

Within a single polymer type, the occurrence of sharkskin may be associated with the distribution of molecular weight: the tendency to show the effect seems to be much reduced if a thermoplastic is of a broad range of molecular weight.

There may be enormous variation in the severity of sharkskin. At one extreme, the distance between ridge and adjacent trough may be equivalent to one-third of the cross-section of an extrudate; at the other, the fault may be barely detectable by the naked eye. However, even in the second case it will be visible as a matt finish, and the unevenness of surface can be felt by applying light finger pressure.

(iv) Frozen-In Orientation

When molten and not being subjected to external stresses the molecules of a typical plastic show a tendency to coil (they seem to seek a random coil configuration). Under external stresses—such as in extrusion, moulding, or other shaping processes—there is a propensity to change from this random state to one of orientation. In most kinds of thermal processing it is desirable to set the material as soon as possible after shaping (for example, in order to achieve rapid production cycles) and in extrusion this may be done by cooling the profile in a water bath shortly after it leaves the die. In such circumstances the molecules may not have time enough to coil completely before, in effect, being frozen—so giving rise to frozen-in orientation.

Because of such orientation the products may be anisotropic in behaviour, with mechanical properties differing if measured in different directions (tensile strength will be higher in the direction of orientation, and impact strength also will be affected—fracture can take place more easily parallel to the direction of orientation).

Orientation may be obtained deliberately by stretching the melt just before it is frozen: mono-axial orientation is important for the production of fibres and filaments, and bi-axial orientation (that is, stretching both lengthwise and transversely to an equal extent) is essential for the qualities required of some types of film. Bi-axial orientation may be desirable also in bottles, other hollow containers, and pipe, when it is required to improve hoop strength and resistance to fracture.

When orientation is introduced deliberately it usually is controlled so that its effects are known in advance. However, frozen-in orientation can be most undesirable when not created under controlled conditions and its effects in consequence are uncertain. This is most likely to happen when a melt is subjected to high stresses combined with reduced intervals between shearing and setting: such conditions would arise when melt temperatures were low, and low temperatures were applied after shaping (as an example, by means of a cooling bath).

(v) Draw Down

In many processes based on extrusion the material is subjected to further manipulation after leaving the die—as examples, by stretching or casting on chill rolls in the manufacture of film. In all such cases it is essential that an extrudate withstand the forces applied to it and not tear—in other words, while there should be some strength and elasticity the main requirement is that the molecules of which it is comprised can flow relative to each other (in this sense its viscous behaviour is the most important feature).

Two examples may be used to illustrate the complexity of problems of this kind. When film is made by extrusion followed by casting on chill rolls there can be a tendency for the extruded web to shrink inwards towards the centre of the rolls—the phenomenon known as 'neck-in'. The edge of film concerned becomes thicker than the rest. It has been found that more elastic melts, capable of keeping a tension in the direction of extrusion, are less liable to exhibit this fault.

In extrusion blow moulding a parison may sag under its own weight after leaving the die, with unwanted and perhaps random thinning as a result. In part this behaviour may be attributable to an elastic effect (to uncoiling of the chains) but also to viscous flow as the molecules move in relation to each other. It is reasonable to believe that the elastic element will become more important with:

(1) increases in molecular weight (and hence in viscosity)
(2) decreases in melt temperature (and hence increasing viscosity), and
(3) increases in length of parison per unit of weight.

(The last is because an elastic deformation under standard load depends on the length of the part being stretched but—so long as the weight of the parison remains constant—the viscous flow does not depend on its length.)

Extrusion of Film and Sheet

Plastics film and sheet can be made in a number of different ways (including casting film from solvent or plastisols, pressing, and skiving from material in block form) but in commerce the most important processes for making these intermediates from thermoplastics are calendering and extrusion.

Calendering and calendered finishes are considered in detail in another chapter of this book but it may be helpful here to include a brief review of the process and a comparison of its main features with extrusion.

The calender has a long history in paper-making and was used first for polymeric compounds by manufacturers of rubber: many rubber factories have calenders that are capable of making sheet suitable (as examples) for fabrication into products as diverse as boots or ducting, but in the world of polymers some of the most elaborate and impressive units are those used to manufacture PVC film and sheet at high speeds (for example, 150 m min^{-1}) with extraordinary accuracy (within tolerances of plus or minus $2.5\ \mu$m).

For plasticized film the first stage is blending polymer and plasticizer in a ribbon blender, followed by melt-mixing in an internal mixer (for example, a Banbury mixer). Other ingredients are added at the melt-mixing stage—heat stabilizer, pigment, lubricant, *etc*. The melt may be discharged for storage to a two-roll mill, then passed to a strainer extruder—from which the output in strip form travels (*via* a detector for any metallic contamination) to the four-roll calender. Adjustments to the rolls to control thickness may be made while the calender is running, and the film may be trimmed while hot, embossed if required, cooled, measured, and wound up in rolls of appropriate length.

The finish on the surfaces of the calender rolls is very important: they may be highly polished or matt, and these surfaces affect behaviour during processing as well as the appearance of film or sheet produced. Unless it is grossly over-lubricated the hot material will adhere to the hotter of two similar rolls forming a nip, or to the roll running faster—except that irrespective of differences in roll speed or temperature there is a preference to adhere to a matt rather than a polished surface.

Any compound for calendering must be formulated so that its elasticity is not too high—otherwise the amount of swelling that occurs as the material emerges from between rolls could be excessive (see 'die swell' above). The compound must have reasonable melt strength and must not adhere too strongly to the rolls, properties that are necessary for successful handling—for example, to ensure peeling from the rolls smoothly and consistently.

If a material is to be stretched a high melt strength will be a requirement; it must be capable of forming film of the thickness wanted while being run at speed and without necessitating changes in operating arrangements or causing variations in properties. If, say, a compound is not stabilized adequately, or heat not removed quickly enough, a fast-running calender could generate shear heat sufficient to cause thermal degradation.

Melt viscosity over a convenient range of temperature should not depend markedly on the temperature, and it should be possible to control and adjust the viscosity if need be (for example, high viscosity would be required to make thick sheet from a rubber compound).

Poly(vinyl chloride) may be calendered in plasticized or unplasticized form and thin film can be produced by calendering followed by stretching (with stretch ratios of up to 8:1). Such products have a very wide range of uses.

Unlike an extruder, the calender is not required to de-aerate, homogenize, and fuse the composition—it simply accepts hot fused PVC which it forms through contact with the rolls. On the four-roll configuration, with three nips, a surface is formed at the feed nip and then is re-worked twice; the core material is not re-worked substantially. Cooling of surface and core at different rates can give rise to stresses and lead (for example) to wrinkling of the product. Stresses can be created also if the temperature is not uniform across the bank, with the prospect of their being released and associated problems later in service.

For long runs and high outputs calendering probably is the most economical method for making thermoplastic film or sheet (in the case of conventional rubbers, as distinct from thermoplastic elastomers, it often is the only process available for making wide sheets). Speaking broadly, extrusion is the preferred method for manufacture of film and sheet in materials other than PVC.

References

1. A. Whelan and J. A. Brydson, 'The Kayeness Practical Rheology Handbook', published by the authors for Kayeness, Inc., 1st Edn., 1991.
2. A. Whelan and D. Dunning, 'The Dynisco Extrusion Processor's Handbook', published by the authors for Dynisco, Inc., 1st Edn., 1988.

CHAPTER 10

Electroplating and Electroless Plating on Plastics

A. C. HART

Introduction

Over the past forty years in many applications plastics have been replacing metals, one of the main reasons for this being the comparative ease with which—at relatively low temperatures—the newer materials can be formed into intricate shapes. Often, components of complex geometry can be made in one operation, whereas several separate stages would be required for similar shapes in metal.

Usually a plastic component will be lighter in weight—which in the aerospace industry especially always has been an important consideration. In recent years weight savings have assumed greater importance in other markets—for example, motor vehicles, where a key factor in achieving greater fuel economy has been the reduction in weights overall brought about by changes from metals to plastics.

Plastics offer other advantages, including better resistance to corrosion than metals like aluminium and steel used for general purposes.

On the other hand, many metals have characteristic bright, reflective surfaces that may be enhanced by polishing and surface treatments such as electroplating. The natural appearance of plastics is quite acceptable for many applications but sometimes there is need to make components with reflective surfaces similar to those in metals; this may be no more than a question of appearance—the designer or manufacturer feeling that an item will look better—while in other circumstances a metallic finish may be functional. One instance of the second would be fittings for bathrooms, where a bright surface not only is attractive but also resistant to wear and easy to keep clean.

For reasons such as this various methods have been devised to provide metallic coatings on polymeric substrates. It can be done with metallic paints and by metallizing (as described in other chapters of this book) but it is fair to write that electroplating can give the most durable of these finishes and may be preferred in uses like external trim of motor vehicles—when essential requirements include good resistance to abrasion and to corrosion.

Commercial processes for the electroplating of plastics were developed during the 'sixties, particular interest being shown by the motor industry, and the technology soon came to be used in other fields.

In theory, any metallic finish that can be produced by electro-deposition can be applied to plastics; however, in practice the range of systems used is quite limited, nickel–chromium being of greatest importance.

From time to time important new uses for plated plastics have emerged and among these one of significance at present is the shielding of electronic equipment against interference (that is, electro-magnetic and at radio frequency). A European Directive on this (89/336/EEC) came into effect on 1.1.92, and in USA legislation was implemented by the Federal Communications Commission. The Directive requires that all electronics equipment offered for sale in the Community must comply with standards for shielding and this means in turn that plastic covers and housings for electronic devices must have over one surface an electrically conductive film or layer. A metallic film of high quality can be applied by means of technology similar to that developed for decorative electroplating.

Typical Process Sequences

The central problem to be overcome when attempting to electroplate plastics is that generally they are good electrical insulators—they lack any intrinsic electrical conductivity. In contrast, metallic components can be cleaned and pre-treated if necessary, immersed in the plating solution, and when they are connected to a d.c. supply the coating will be produced.

With polymeric materials it is essential first to provide on the surface a conductive film as a basis for electro-deposition; there must be good adhesion of this film to the substrate, and the quality of the finished work will depend on the method used to apply it and the standard attained.

Various methods are available for preparing the conductive layer. In the early stages of development the surface of the plastic might be roughened by mechanical means or treated with solvents before being coated with paints containing graphite or silver: silver reduced chemically also was used. However, such methods really were not suited to production on a large scale and there was difficulty in maintaining consistent quality. In the 'sixties the technology was improved by introduction of the chromic acid process for etching the surface of the plastic; this gave genuine adhesion between polymer and metallic coating and made the surface hydrophilic and therefore suitable for processing with aqueous solutions.

At the same time, rapid advances were made in technology for electroless (or 'autocatalytic') deposition of nickel and copper—which made possible application of conductive layers efficiently on a large scale. The techniques employed today are based on these principles, although many variations have emerged depending upon the supplier of the plant and the particular plastic to be coated.

Quite recently an entirely new process has become available in which specific areas of plastic components are coated selectively, a primer being applied usually by spraying. This gives a layer that can be metallized subsequently by conventional electroless deposition and so provide the adhering surface necessary for further processing.

Initial Metallization

(i) Conventional Immersion Processes

In these processes the plastic component is treated in an etching solution to render the surface hydrophilic and to promote adhesion between it and the metallic film; the surface then is activated so that later it can catalyse electroless deposition of thin layers of copper or nickel. The thin films so produced are the conductive surfaces on which further metal may be deposited by conventional electroplating.

The technology was advanced in the 'sixties by the etching of components made from acrylonitrile–butadiene–styrene with solutions based on chromic acid. These solutions are strongly acidic and highly oxidizing and attack the plastic in a controllable manner. The structure of items made from ABS is not homogeneous and the effect is to give on the surface numerous microscopical pits which promote a mechanical key between substrate and the metallic film. Figure 67 is a photomicrograph of such a surface showing the selective removal of butadiene. The mechanism of adhesion obtained in this way remains a matter

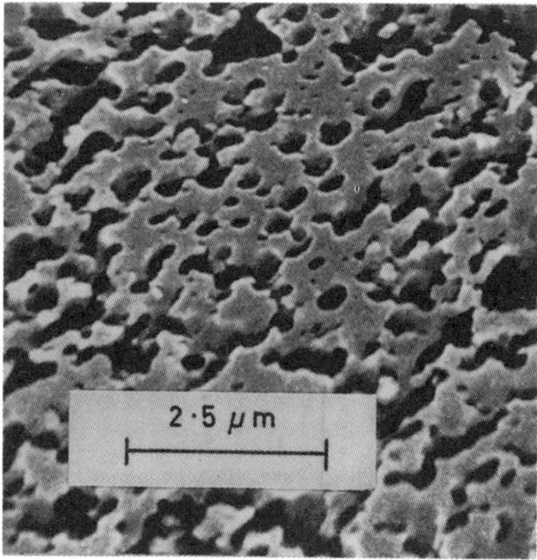

Figure 67 *Photo-micrograph of surface of ABS after etching with chromic acid* (Courtesy of J. K. Dennis and T. E. Such)

of debate, molecular attraction having been suggested to be a factor in addition to the physical links.

Etching solutions based on chromic acid are available in a number of proprietary and patented forms but the following would be indicative for an aqueous type:

(1) 430 g chromium trioxide (CrO_3) per litre of water

(2) 180 ml concentrated sulphuric acid (specific gravity 1.84) per litre.

(Note: It is preferable to use proprietary products and attempts to prepare solutions are not recommended; in all stages of storage, transport, handling, and use appropriate safety precautions should be taken.)

The plastic components may be immersed in etchant for periods between three and six minutes, at operating temperatures of 60 to 65 °C; both too much and too little etching have adverse effects, so the conditions must be controlled closely.

Different grades of ABS may not behave in the same way: the amount and distribution of butadiene in the matrix are of critical importance and because of this special grades have been developed for electroplating. However, even when electroplating grades are used the conditions of moulding can influence results.

Chromic acid etching was developed originally for ABS but has been modified since to cover a wider range of materials. Polycarbonate is more difficult to etch in this way and mouldings in the material may be treated first with solutions containing organic solvents—which cause swelling and micro-cracking of the surface, so rendering it more sensitive in the etching stage.

The solutions accumulate gradually dissolved material from the plastics and from time to time have to be discarded and replaced; disposal presents environmental difficulties and for this reason efforts have been made to find alkaline rather than acidic liquids which are capable of etching satisfactorily but do not contain chromium. At the time of writing, unfortunately, development has not reached the stage at which the process based on chromic acid can be superseded for the entire range of substrates.

After etching the components are passed through a neutralizer to remove all chromium residues preparatory to further processing.

At this point the surface is hydrophilic but still not capable of promoting deposition either of copper or nickel; in order to create sites at which these metals will form it is necessary to activate the surface by depositing nuclei of palladium. Originally this was done in two steps, immersion of the components in a solution containing stannous chloride and then transferring them to one based on palladium chloride—from which the metal was deposited. Now however these are combined in a single immersion in a colloidal solution of palladium chloride containing also tin salts in both stannous and stannic states. A typical solution of this type would comprise:

(1) 0.007 g palladium chloride (as $PdCl_2$) per litre

(2) 35 g stannous chloride (as $SnCl_2$) per litre

(3) 4 g stannic chloride (as $SnCl_4$) per litre

(4) 500 ml hydrochloric acid (specific gravity 1.6) per litre.

Normally, immersion for three to six minutes at a temperature in the range 25 to 30 °C would be sufficient to activate the surface.

It is thought that the palladium is adsorbed on the surface of the plastic in the form of minute discrete particles, which are protected and partly inactivated by the presence of tin; the latter is removed by immersion in a solution of accelerator (such as ammonium bifluoride), which takes away the tin and gives full effect to the palladium.

The process has been improved over a period so that now very small particles of palladium may be deposited very evenly, and this has been demonstrated to give improved adhesion of the layers of coating to the substrate.

Development efforts have been directed also towards better activation with plastics other than ABS—some of which can be quite difficult to process satisfactorily. In such instances, ionic surfactants are used to wet the surfaces and make it possible to produce an electric charge; this increases significantly the effectiveness of the activation.

After activation, an initial conductive metallic film—on which layers of metal will be accumulated later—can be applied. Metals that may be derived from their salts by chemical reduction at normal pressures and modest temperatures are suitable; in practice three have been employed—copper, nickel, and silver.

The last (silver) was used at one time but is comparatively difficult to process, and more costly, so now finds little application. Solutions for reducing copper and nickel have been made quite sophisticated and both are used in electroplating plastics—the preference for industrial work probably being with nickel.

Electroless copper solutions contain copper salts and a reducing agent, such as formaldehyde; the preparations for industrial use contain also stabilizers like 2-mercaptobenzothiazole, to prevent decomposition other than on activated areas of the workpiece, and accelerators like ethylenediamine-tetra-acetic acid—which increase the rate at which metal is deposited. Formulating such solutions requires achieving a balance between stability on the one hand and speedy deposition on the other.

The copper solutions give satisfactory rates of deposition at slightly above room temperature (with plastics especially operating conditions can be a factor; if too high a temperature is necessary, there might be distortion of some items).

In the nickel processes the reducing agent used most commonly is sodium hypophosphite, which gives a deposit comprising a nickel–phosphorus alloy rather than pure nickel; the accelerators commonly are fluorides, and thiourea is a typical stabilizer.

The solutions for nickel may be classified in two groups—those working, respectively, under acidic conditions or in mildly alkaline environment (at pH of about 9). Since acidic solutions normally are used at around 90 °C they are largely unsuitable for plastics; the alkaline ones can be formulated for temperatures as low as 25 °C and so are preferred. With these the rate of deposition is lower but the initial stage requires only a thin film of nickel so lower speed need not be a serious disadvantage.

There was for some years debate over the advantages and disadvantages of electroless copper or nickel for plastics in decorative applications. It was said that since copper was more ductile it would give the better adhesion; however, more recent improvements in etching and activation have tended to nullify any such benefit. Nickel offers reliability and advantages in cost and now is the method of choice for many purposes; for a nickel solution of alkaline type a useful life in a bath of up to six months is not uncommon.

When metallic coatings are applied for radio-frequency or electromagnetic shielding electroless copper always is used (its superior electrical conductivity gives excellent performance) but in order to protect the copper against abrasion and corrosion it is necessary to apply a top layer of nickel.

(ii) Selective Metallization

In both shielding and decorative work demand has been growing for processes that will coat only selected parts of a component. For shielding usually it is only necessary to coat one surface (the inside) and any additional coating just adds to cost. Also, when an outer surface is coated there may be difficulty over the external colour finish—problems, say, over adhesion of paint to metallic film. Un-metallized plastics are comparatively easy to paint and if the outer surface is 'natural' it either may be left like that or painted without trouble. For decorative applications selective coating can be helpful for visual effects—as an example, by contrast between bright nickel–chromium and matt black surfaces.

There are two main methods of approach to selective coating: in the first, the required areas of the components are protected by suitable material that absorbs chromium from the etch stage and prevents activation; more recently, primers have been formulated that may be applied to areas it is required to metallize and after a curing at modest temperatures direct plating with electroless copper or nickel is possible. The second technique dispenses with both the etching and activation stages; it does not give such good bonds as conventional chromic acid etching, *etc.*, but the adhesion is sufficient to meet specifications for many purposes.

(iii) Plastics to be Electroplated Directly

From time to time attempts have been made to make plastics conductive enough to be plated directly, by immersion in the solutions in the same way as with metals. The method usually was to incorporate a conductive material, such as carbon black.

One proprietary compound in particular was offered containing agents to provide good adhesion and to promote growth of the deposit on the surface. However, notwithstanding such advantages, for other reasons it was not successful commercially; there was difficulty in moulding a substance filled so heavily with results consistent enough to ensure consistent behaviour at later stages.

At the time of writing development is proceeding on polymers that will be conductive inherently, without the use of additives. However, it may be doubted whether such materials can be made cheaply enough and with the properties required to replace those employed at present.

Finishing Subsequently

(i) Electroplating

Once the initial conductive metallic film has been prepared as described virtually any coating suitable for electrodeposition can be applied to it. In practice the most usual system is nickel–chromium ('chrome plating'), which—provided application is as it should be—gives excellent resistance to abrasion, to corrosion, and generally is very durable. For certain types of work, coatings of precious metal (especially gold) are popular.

With plastic components nickel–chromium finishes always are applied over a thick undercoat of ductile copper, which can compensate for the differences in thermal expansion and contraction between the plastic and metallic finish. Such a layer, immediately on top of the electroless initial film, is regarded as essential if the standard of adhesion required is to be reached.

Generally the layer of copper is deposited from a bright acid copper solution containing approximately 250 g l^{-1} cupric sulphate and 100 ml l^{-1} sulphuric acid, with proprietary additives. This supplies what is needed and smooths surface imperfections to give good appearance, without mechanical polishing, ready for the next stages. It overcomes also the effects of etching and other pre-treatments, which may leave rather a dull surface finish. To comply with current specifications the layer of copper should be some fifteen to twenty microns thick.

In the early stages with plastics it was thought—bearing in mind that the substrates did not corrode like metals—that quite thin metallic layers would be enough. However, in practice and since the copper layer is essential, the nickel–chromium must be thick enough to protect it: if it is not adequate in this respect, corrosion of the copper can cause complete exfoliation of the coating.

The deposit of nickel on top of the copper provides the essential resistance of the system to corrosion. Generally it is deposited from a solution based on the familiar Watts formulation—300 g l^{-1} nickel sulphate, 30 g l^{-1} nickel chloride, and 40 g l^{-1} boric acid—and as in the copper processes the solutions contain additives to help give a bright deposit with good 'levelling' characteristics.

A single layer of bright nickel of thickness 5 to 10 μm will supply good durability in less corrosive environments but for more exacting conditions thicknesses of 15 to 20 μm are necessary. An example of the latter is exterior trim for motor vehicles, where specifications demand a double-layer system of nickel.

The double-layer system was developed with a view to obtaining the maximum resistance to corrosion in bright nickel electroplated coatings. Usually the lower layer consists of 'semi-bright' nickel with good levelling qualities and must not contain any sulphur. The upper layer is a conventional bright nickel deposit and normally does contain a very small amount of sulphur, which is derived from the additives in the solution and contributes to the bright appearance. The upper layer is less resistant to corrosion than the sulphur-free layer and hence corrosion of the nickel tends first to spread laterally in this layer, with the semi-bright nickel below continuing to protect the substrate. The use of a double-layer system like this has been shown to improve dramatically the resistance to corrosion that can be achieved with any given thickness of coating, both for metallic and plastics substrates.

Frequently these systems are described as 'chrome plating' but it is nickel rather than chromium that gives the primary resistance to attack. When it was introduced in the 'twenties the original purpose of the layer of chromium was to prevent tarnishing of the nickel, but it has been found since that the physical nature of this layer can influence significantly the rate at which the underlying nickel deteriorates.

Typical coatings of chromium applied over nickel are about 0.3 μm thick and are under high stress—which gives rise to cracking. Cracks in the chromium are initiated by corrosion of the underlying nickel and (ultimately) of the substrate; however, development in the 'sixties indicated that an increase in the extent of cracking would spread corrosion more widely over the surface of the nickel and so reduce the speed of penetration—a conclusion that led in turn to chromium solutions giving 'micro-cracked' deposits and to overall enhanced resistance to corrosion.

Similar results may be obtained by another technique in which, before application of the chromium, a thin layer of nickel containing small inert particles is deposited on top of the bright nickel. The inert particles equate in the chromium to a very large number of tiny pores—hence, this type of coating has become known as 'microporous chromium'.

Usually the chromium layer is applied using a bath of hexavalent chromium, containing typically 250 to 300 $g\,l^{-1}$ of chromium trioxide (CrO_3) and small quantities of sulphuric acid. The solutions used to deposit 'micro-cracked' chromium contain also additives to this effect.

The combination of a 'double-layer' deposit of nickel and either micro-cracked or micro-porous chromium provides the very best system of protection available by these means. Plastics components intended for particularly demanding applications (including exterior trim on the most expensive cars), are finished to such specifications.

(ii) Other Electroplated Finishes

For decorative purposes especially precious metal finishes frequently are applied to plastics—for example, on buttons. In one process, a layer of copper is deposited on the initial conductive film, then electroless nickel, and lastly gold is deposited by immersion. Precious metals, particularly gold, may be used as an alternative to chromium on top of the conventional thick copper and electroplated nickel system.

Other applications require an electroplated brass finish, which generally is superimposed over a thick copper and nickel system—thus ensuring good adhesion and resistance to corrosion. It always is necessary to apply over the brass a coating of lacquer to prevent tarnishing.

For uses in electronics, copper alone may be applied—in order to enhance the thickness of metal and its capacity to carry current.

(iii) Electroless Processes

In shielding for radio-frequency and electromagnetic interference an electroplated coating may not be necessary. Highly effective performance can be obtained with deposits of thickness as low as 2 to 3 μm and these can be applied economically by electroless processes, which give very uniform thicknesses of deposits even in recessed areas.

The most popular method involves deposition initially of a layer of electroless copper up to 2 μm thick. The excellent conductivity of the copper ensures that this gives good results for attenuation (that is, for shielding). The copper layer is followed by electroless nickel up to 1 μm thick, which protects against abrasion and corrosion. Obviously it is important that the nickel not be porous and with this in view an activation stage may be interposed, with chemical deposition of palladium on the surface of the copper in preparation for the nickel.

Grades of Plastics for Electroplating

The technology for deposition from aqueous systems of metal on plastics was developed originally for acrylonitrile–butadiene–styrene (ABS), and it remains still the plastic used most widely in the field—particularly for the more traditional uses where the coating is essentially decorative. However, in the past decade requirements have changed and in response to this the range of plastics available for these processes has been extended. Some comments about the materials that are most important at present are given below.

(i) Acrylonitrile–Butadiene–Styrene

ABS plastics are available readily and offer excellent mechanical and physical characteristics; they can be moulded easily, give a surface finish of high quality, and the shrinkage of material in the mould is comparatively low. For plating, the last feature in particular is important—since high shrinkage can lead to imperfections which, with a bright finish, can be all-too-evident and serious enough to require rejection of the items concerned.

Long experience of the use of ABS means that a body of knowledge is available to help ensure good results and the rapid correction of any defects. Special grades are offered for electroplating and with these, using the conditions recommended, mouldings should be obtained with the distribution of butadiene at the surface necessary for good adhesion of the metallic layers.

Electroless and electrolytic coating of ABS are still among the simplest processes and other plastics that can be coated from water-based systems almost always require additional stages of processing.

(ii) Polypropylene

The properties and moulding qualities of polypropylene are rather different from those of ABS. Systems used for coating it are analogous to those for ABS

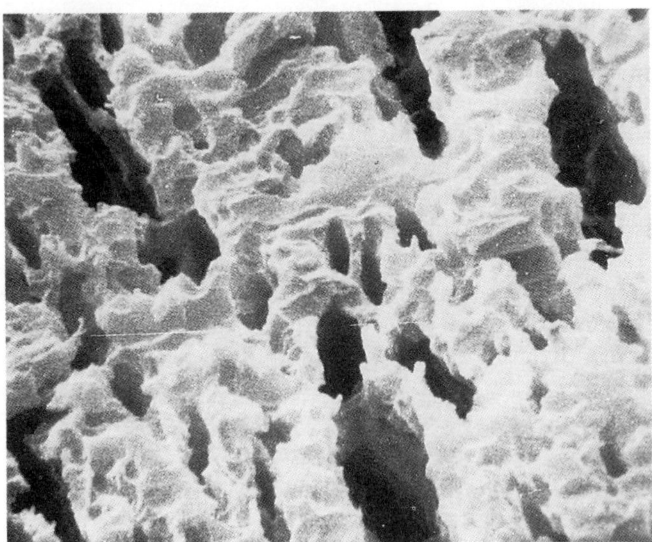

Figure 68 *Photo-micrograph of surface of polypropylene after etching with chromic acid* (Courtesy of MacDermid G.B. Limited)

and have been available for some years. Etching of the surfaces can be by means of solutions based on chromic acid and these are effective because the moulding includes areas both crystalline and amorphous in nature (the etchant attacks the amorphous areas more readily—with results as depicted in Figure 68).

Grades of polypropylene were developed for these applications and since the plastic was less expensive than ABS and good bonds with the coating were possible they had considerable popularity. The main disadvantages were that the mechanical properties were less satisfactory than for ABS and especially that there was rather more shrinkage after moulding, with associated problems of visible surface defects. In general the plated surfaces available with polypropylene were never quite so good as with ABS, and the difficulties mentioned have resulted in some loss of favour for the material—especially during a period in which standards of quality have been rising.

(iii) Polycarbonate

Polycarbonate combines the advantages of good mechanical properties with satisfactory moulding and dimensional stability—attracting increasing interest in consequence for moulding covers or housings for computers and other electronic equipment. Such uses meant in turn that it was essential to develop satisfactory methods of plating the material.

In recent years, work to this end has been extensive. Polycarbonate generally is more resistant than ABS to etching solutions and in preparation for plating it almost always is necessary, before etching, to expose the items to attack by organic solvents—creating swelling and cracking at the surface. This preliminary is followed by chromic acid etching, and subsequent processing is on lines similar to those for ABS.

In view of its application in the field of electronics there is particular interest in the coating of polypropylene mouldings against radio-frequency and electromagnetic interference: the successful development of aqueous processes for polycarbonate provides a viable alternative to other methods for shielding the housings of electronic products—such as by paints containing nickel.

(iv) Polyamide

When attempting to plate nylon the central problem is its resistance to etching. With a view to overcoming this, special grades have been evolved containing filler which is removed in the etching stage to give the surface necessary for adhesion. With such grades, the stages of processing afterwards are similar to those for ABS.

In the United States, nylon mouldings are electroplated in quite large quantities.

(v) Blends of ABS and Polycarbonate

Over recent years the range of materials available has been extended by the addition of blends intended to combine the advantages of more than one plastic. Among these, blends of ABS and polycarbonate are of special interest; they are available in various ratios—those with relatively high polycarbonate contents offer mechanical properties closer to those for polycarbonate alone and are preferred for applications such as shielding; in decorative work, blends with a higher proportion of ABS (around 55%) are preferred. Those with more ABS offer enhanced adhesion and resistance of the coated product to thermal cycling, particularly to the effects of higher temperatures.

(vi) Poly(butylene terephthalate)

Another comparatively recent development is poly(butylene terephthalate) heavily filled for electroplating applications. For a plastic it has unusually high density and so gives components much heavier than would be made with conventional unfilled polymers. In some instances this is regarded as desirable—the heavier mouldings are closer in weight and in 'feel' to components cast or machined from metal. Since they are more familiar they are perceived to be of better quality—a concept apparently quite important for uses such as bathroom fittings and plumbers' requisites.

Subject to suitable modifications to the process, the filled poly(butylene terephthalate) can be plated successfully.

(vii) Others

From time to time processes have been evolved for coating other materials, including polysulphone and poly(phenylene oxide): however, at present there seems to be little commercial interest in electroplating newer plastics.

Figure 69 *ABS radiator grille for 1971 model Ford Mercury, electroplated in nickel–chromium*
(Courtesy of MacDermid G.B. Limited)

Applications

(i) Decorative

Some indications were given earlier in this chapter of the main advantages to be gained—in comparison with metallic casting, machining, and so forth—by moulding articles in plastic and finishing by electroplating. In general, items made from metal required more extensive preparation for plating and were heavier and more costly in terms of labour and other respects. An important impetus for change came from the motor industry, which was using already a variety of plated metal components, and in the USA an early success for plastics was the radiator grille, which in the 'seventies and 'eighties quite often stretched across the entire front of a vehicle. (Figure 69 shows a typical example from this period, on a 1971 Ford Mercury.)

Mouldings like these were very large (weighing sometimes as much as 5 kg) and were made in very large numbers, so that an important sector of the industry came to be occupied with one type of component. Previously for such work die-cast zinc alloy was employed, and with plastic it was not necessary to carry out the polishing essential to obtaining the necessary standard of finish in zinc.

Radiator grilles like these were never so important in Europe but several models here have used rather smaller plastic grilles.

Other plated products popular in the USA include exterior mirrors, horns, and lamp units for freight tractor units, and decorative accessories for 'custom-

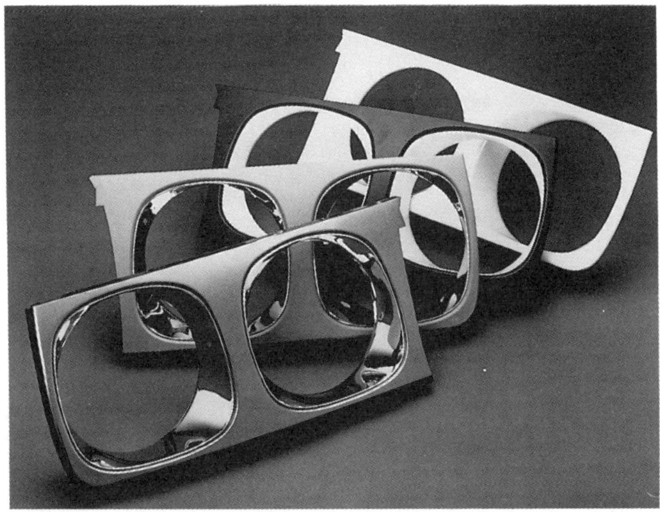

Figure 70 *Electroplated ABS surrounds for motor headlamps, showing a combination of body-paint colour and nickel–chromium*
(Courtesy of Borough Limited)

ized' vans and cars, for cross-country and beach transport, cycles, and motor cycles.

In Europe, typical external bright trim included bezels for front and rear lights, which commonly were ABS with coating systems of copper, nickel, and chromium. Figure 70 shows a component for lamps for the Jaguar car made in Britain—an attractive combination of bright nickel and chromium with the body-work colour.

Further examples include: housings for wing mirrors, for signal lamps, bezels for door handles, and for car radios. Badges and name plates almost always are electroplated plastics, sometimes (for special decorative purposes) combining plating with coloured paints.

The electroplating of reflective surfaces for exterior mirrors is an important and growing application. Moulding is to a high standard of precision and the electroplated surface creates a mirror with excellent reflection. It will resist impact much better than glass so a longer life in service can be expected—hence, such mirrors are fitted extensively on commercial vehicles, 'buses, and lorries. ABS is moulded and electroplated also into small mirrors that can be fixed by adhesive on vehicle wing mirrors to cover 'blind spots'.

Outside the motor industry another important field is fittings for kitchens, bathrooms, lavatories, and washrooms. Traditionally such items would be from brass or zinc alloys, finished by electroplating with nickel–chromium to provide resistance to cleaning and frequent polishing, and advantages were to be gained by moulding in plastics and electroplating. Examples are illustrated in Figure 71 and include: holders for toilet rolls, nuts, racks (for tooth-brushes), 'roses', shower-heads, sprays, stoppers, taps, and towel rings.

Figure 71 *Plastic bathroom fittings electroplated in nickel–chromium* (Courtesy of Borough Limited)

In this sector of the market nickel–chromium systems are used most widely but other finishes have become fashionable, including gold for luxury bathrooms. Generally in this instance a thin coating of gold is provided over thicker layers of copper and nickel (similar to those used beneath chromium). An important feature recently was the introduction of the special filled poly(butylene terephthalate) mentioned in an earlier section, which conveys a sense of weight and quality well-suited to products with more costly finish.

Equipment for the catering trade, particularly for bars, also offers opportunities for electroplated plastics: items such as display heads for beer pumps may be made by combining electroplating with painted surfaces; drip trays for glasses are moulded from plastics and finished with nickel–chromium to enhance their resistance to abrasion and for convenience in cleaning.

Fancy closures for bottles for perfumes or other luxury items can be moulded from plastics and coated with precious metals like gold; containers of this kind are required to be especially attractive in appearance, and to remain so over the whole life of the contents.

Another important application is the decoration of buttons; the annual production in Britain of plated buttons is estimated at between 100 and 150 millions. Usually the processes employ barrels rather than mounting the buttons on jigs, with application of nickel–chromium systems; however, finishes such as brass, bronze, and gold also are popular.

(ii) Shielding

Recently the shielding of electronic equipment against radio-frequency and electromagnetic interference has attracted considerable interest.

Unwanted signals may be generated by a variety of sources, including natural ones like lightning, but in practice the most common origin is other electronic equipment. Interference may be encountered over a very wide range of frequencies, from 10 kHz to 100 GHz. The effect may be trivial and unimportant (as, for example, the brief distortion of a television picture) but it could be critical—as with a computer system controlling an aircraft in flight. While not so threatening to life, the malfunction or failure of commercial computers also can be very troublesome and expensive to put right.

Interference has become more serious both because of the increasing numbers of electronic systems for many different purposes, and because the plastic housings that often are used are not inherently conductive. In the 'eighties in USA legislation was introduced and implemented by the Federal Communications Commission, and Standards have been issued by the Association of German Electrotechnical Engineers. A European Directive (89/336/EEC) was brought into effect in January 1992; briefly, it requires that all electronic equipment offered for sale in the Community complies with Standards for shielding against interference of this nature.

Various approaches are available (including application of conductive paint films containing nickel, and vacuum metallizing) but the techniques developed originally for decorative metallic coatings have been found, with some modifications, to be highly suitable.

Electroless deposition offers the advantages of a coherent metallic film of uniform thickness, even in recessed areas of the plastic moulding, and excellent results can be obtained with very thin coatings. Attenuation of 30 decibels is regarded as the minimum necessary for effective shielding and is taken to imply that 99.9% of radiation will be eliminated; a figure of 50 decibels suggests the elimination of 99.999% and would be regarded as 'average to good'. Performance of this order is achieved with copper coatings of thickness 1.5 to 2.0 μm, with nickel over-layers of 0.3 to 0.5 μm. The effectiveness of the shielding is attributable largely to copper but the nickel protects the copper layer against corrosion and physical damage.

In some applications (such as military equipment) a very high standard of effectiveness is required—and it can be attained by electroplating over the electroless copper to make thicker deposits. When components will be subjected to handling it may be desirable to enhance the durability of shielding by adding electroplated nickel–chromium finishes.

As described earlier, selective techniques are available and electroless deposits are so effective it is not necessary to coat both the inside and outside of housings. This in turn opens the opportunity to apply decorative or matching coloured paint finishes to the external surfaces of housings, as illustrated in Figure 72.

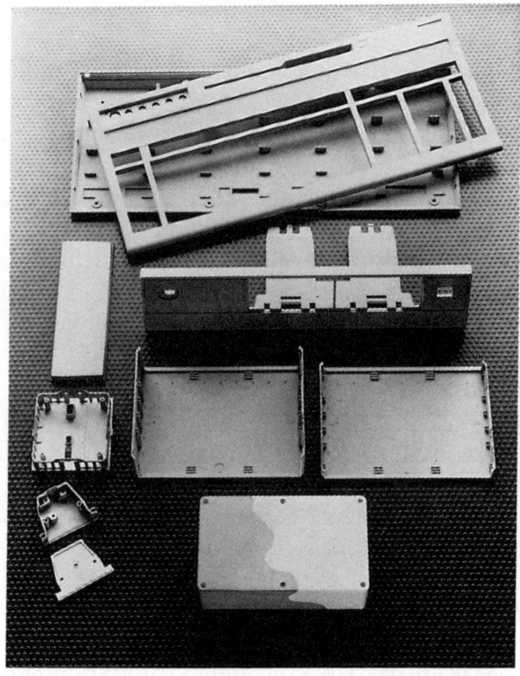

Figure 72 *Plastics components with selective copper–nickel electroless coatings for shielding against radio-frequency and electro-magnetic interference*
(Courtesy of Shipley Europe Limited)

(iii) Printed Circuits

Printed circuit boards based on plastics are well-established and electroplating has an important place in the technology. In general, circuits are obtained by etching selectively copper foil laminated to an epoxy–glass fibre composition; however, it is necessary to make electrical connections through holes drilled in the boards and this is done by electroplating on the material exposed by the drilling.

The epoxy composite is etched and activated by processes derived from those used in decorative work but modified quite specifically for the application (which is known as 'through hole plating'). In the early stages the geometry was found to present difficulties and it was necessary to evolve special solutions for depositing copper into the holes.

Frequently, printed circuit boards are finished by deposition of an alloy of lead and tin, which facilitates soldering.

For Further Reading

ASTM B 604: 1980, 'Specification for decorative electroplated coatings of copper–nickel–chromium on plastics'.

British Standard 4601: 1970, 'Electroplated coatings of nickel plus chromium on plastics materials'.

J. K. Dennis and T. E. Such, 'Nickel and Chromium Plating', Butterworths, London, 1986.

R. W. Furness, 'The Practice of Plating Plastics', Robert Draper, Teddington, 1968.

International Standard 4525: 1985, 'Metallic coatings—electroplated coatings of nickel plus chromium on plastics materials'.

G. Muller and D. W. Baudrand, 'Plating ABS Plastics', Robert Draper, Teddington, 1970.

R. Weiner (ed.), 'Electroplating of plastics', Finishing Publications, Stevenage, 1977.

Acknowledgment. The author wishes to acknowledge the assistance and support of the Nickel Development Institute in the preparation of this contribution.

CHAPTER 11

Vacuum Metallizing

R. R. READ

Introduction

The deposition under vacuum of a thin metallic film is not a new technique and its application to plastics dates back now some years (the author's own earliest involvement was in the early 'fifties, with a view to the replacement of silver electroplating with aluminium reflective surfaces in motor vehicle headlamps).

The main difficulties experienced in practice in production on a large scale are the successful deposition of even films, and especially their 'keying' to the substrates. The surface of a film will replicate that of the substrate, so the latter should not be uneven but—on the other hand—an intractable material with a smooth surface may prove hard to metallize successfully.

Flat plastic films and simple mouldings afford the best opportunities for good results in repetitive work; in the early stages transparent materials such as acrylics were used, but now coatings may be applied by this method to a somewhat wider range of materials and products. Table XIV is a selected list of plastics being coated currently by vacuum deposition of thin films.

In practice, the metal used most frequently for the decorative coating of plastics is aluminium, but other metals, alloys, and compounds can be applied also.

Basic Technique

The metallic films may be deposited on the upper surface of the plastic (this is known as 'first surface metallization') or on the under-side ('second surface metallization'). In the first instance it usually is necessary to protect the metal layer by means of a surface lacquer; in second surface metallizing the integrity of the film is protected on one side by the plastic, which normally in such cases

TABLE XIV *Selected list of plastics coated currently with vacuum-deposited films*

Acrylic
Acrylonitrile–butadiene–styrene (ABS)
Dough moulding compound
Polyamide (PA)
Poly(butylene terephthalate) (PBT)
Polycarbonate (PC)
Polyesters
Polyethersulphone (PES)
Polyethylene (PE)
Poly(ethylene terephthalate) (PET)
Poly(phenylene oxide) (PPO)
Polypropylene (PP)
Polysulphone
Poly(vinyl chloride) (PVC)

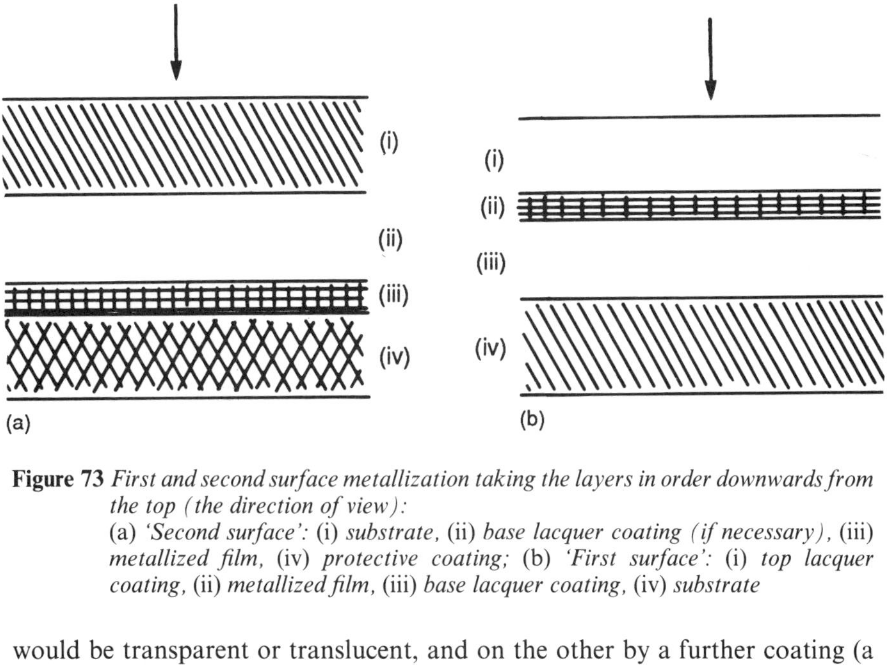

Figure 73 *First and second surface metallization taking the layers in order downwards from the top (the direction of view):* (a) *'Second surface':* (i) *substrate,* (ii) *base lacquer coating (if necessary),* (iii) *metallized film,* (iv) *protective coating;* (b) *'First surface':* (i) *top lacquer coating,* (ii) *metallized film,* (iii) *base lacquer coating,* (iv) *substrate*

would be transparent or translucent, and on the other by a further coating (a 'back-coat') of paint. To assist with keying to the substrate (or 'super-strate' in the case of second-surface metallizing) a base coating of lacquer may be applied between the metal and the plastic. Figure 73 illustrates diagrammatically first surface and second surface metallizing.

Composition and Application of Lacquers

The type of plastic to be metallized will be a factor in determining the composition of the lacquers. Several proprietary formulations are available for preparing surfaces and for the protection of metallic layers, and while details of their composition are specific to the manufacturers they may be of the following general nature.

(i) Base Coating

(*a*) A modified nitrocellulose resin in a fast-drying solvent system of ketones, alcohols, glycol ethers, and aromatic hydrocarbons (such a composition may be dried in air)

(*b*) An amino-modified alkyd resin in aromatic hydrocarbons, alcohols, and ester solvents

(*c*) A modified alkyd resin in aromatic and aliphatic hydrocarbons and glycol ether

(*d*) Epoxyacrylate, unsaturated acrylate, and acrylate ester resins in a solvent system of methyl isobutyl ketone and isopropyl alcohol, with photo-curing agents which, on being exposed to ultra-violet radiation, will initate curing (such a base coat requires ultra-violet light for curing).

(ii) Top Protective Layer

An epoxy-modified acrylic resin in n-butanol and xylol, sometimes with the addition of 2-ethoxyethanol.

Lacquers for base coating may have non-volatile contents ranging between 30% and 50% w/w, while those for top protection are typically from 20% to 30% w/w.

(iii) Dipping

At first sight the application of lacquer by dipping components into a suitable tank or reservoir might seem the simplest method but in fact it is rather difficult to control and has serious limitations. Only very simple components can be prepared satisfactorily in this way; more complicated shapes with recesses and reverse angles often are not coated effectively (since their configurations can lead to the entrapment of air) or may show variation in thickness of coating. For satisfactory results, even with simple components, the viscosity of the lacquer in the reservoir must be kept constant by continuous monitoring and adjustment, and the entry and removal of the components must be very smooth, well-controlled, and uniform.

(iv) Flow or Rain Coating

In this method the lacquer is fed from the tank in a continuous flow by means of a weir, passing over the components, and any excess allowed to drain away. Alternatively, the components may be mounted on jigs and lacquer allowed to fall on them (*i.e.*, to rain) from perforations in a container above. In the latter process, to ensure even distribution of the lacquer, the jigs are rotated during the coating and subsequently to assist the removal of surplus material and to help achieve uniform thickness. The speed at which jigs are rotated is specific to the design and configuration of the component concerned.

Again it is essential that the viscosity of the lacquer be monitored continuously and kept constant; it also should be filtered before being allowed to pass from the reservoir, so that all foreign matter is removed.

(v) Spraying

A wide choice of equipment is available for spraying and the lacquers may be applied by this means. The simplest approach is manual operation, where material either retained in a cup fitted to a hand-held gun or (more usually) fed from a pressure pot is forced by compressed air through a suitable orifice and so impinges on the component or components concerned. The pattern of spraying can be adjusted in various ways (such as by altering the settings, or by changing the viscosity of the lacquer). In suitable circumstances (such as for spraying large numbers of similar products) fully automatic equipment is available (and normally would be more economical for work of this kind).

(vi) Curing

Whatever method may be used for curing the lacquers it is of course most important that this be complete—especially in the case of base coating. If cure is not complete there will be in the next stage in the metallizing chamber a high volume of fumes and gases which will have to be removed before the necessary level of vacuum can be reached. Volatiles trapped in the lacquer can migrate in the course of evaporation, permitting penetration by the metal and giving rise to white or irridescent discolorations. Similarly, the top protective layer must be cured so that the component will withstand testing to specification and the actual conditions of service.

Cleaning of Components

Steady production of components to the standard required depends also on good practice at an earlier stage of preparation. In a number of ways in the course of ordinary factory operations the pieces may become contaminated—such as by release agents from moulding, by swarf, or by other particles (which often are drawn to and kept on the surfaces through the attraction of a static electrical charge) arising from de-gating, drilling, trimming, and so forth. When moulding components for metallizing the use of release agents should be discouraged, and under no circumstances should those of the silicone type be applied. (Such release agents are carried easily through the air, and because of this it is best not to employ them too on equipment in the same shop moulding plastics items for other purposes.)

In units carrying out both the preparation of components and the metallizing it is possible to have controls adequate to ensure that the components are clean, but when items to be metallized are prepared elsewhere there can be great difficulty at times in keeping dust and dirt arising from packaging and transport to the minimum necessary to give good results consistently.

It may be possible in operations where the base lacquer is applied by dipping or flow methods to make use of the solvent in the lacquer as a kind of washing or cleaning agent (fine organic particles may be dissolved, while the pieces or mouldings as a whole are not affected), but really this is suitable only with instances of minor surface contamination of the parts and as a practice is to be discouraged because of the danger of accumulation in the lacquer of foreign bodies and other material which in turn would spoil the appearance of the coating.

Generally it is preferable to select one of a variety of other methods of cleaning, involving either water-based materials or organic solvents. The former have advantages like the comparative ease of disposal of the washing agent after use, and as a counter to static electricity, but many plastics will absorb water and after washing with aqueous preparations it is necessary before beginning the next stage to dry the pieces thoroughly by some appropriate means.

On the other hand, while cleaning with organic solvents which have low boiling points and good affinity for greases and oils will be effective and leave the parts dry, care must be taken to select a solvent or solvent system which does not attack the components and which will be satisfactory on grounds of health, safety, and environmental considerations. Concentrations of solvents in the atmosphere at the place of work must not exceed specified limits and there should be an effective system for recovering and re-cycling the solvent or solvents—in order to keep costs down and to prevent the release of vapours outside.

The cleaning may be carried out by solvent vapours, by immersion in the liquid phase—and by both methods in combination. Immersion also should overcome static attraction, but if this phase is not used a gun emitting deionized air under low pressure may be employed to clean individual parts and to take away debris. Agitation of components by ultrasonic means can be helpful in removing contamination from blind holes and recesses.

In instances in which the plastics components are not lacquered the cleaning can be carried out in the vacuum chamber, by adding a plasma-based chemical vacuum deposition station and thus eliminating cleaning as a separate stage in the process. By this means, excellent adhesion of the metallic films can be obtained.

Evaporation under Vacuum

The metal used most commonly is aluminium and this can give a decorative finish to items such as components for motor vehicles, containers for cosmetics, jewellery, and so on. An overlay of tinted lacquer sometimes is applied to enhance the versatility of this form of coating.

In the early days comparatively simple vacuum evaporative equipment was

TABLE XV *Selected applications for metallized articles, with processes used**

Application	*Process*		
	Evaporation	*Sputtering*	*Ion Plating*
Automotive badges, instrument panels, lamps, trim, *etc.*, for motor vehicles	√	√	
Domestic appliances badges, control panels, trim, *etc.*	√	√	
Closures for packs for cosmetics, toiletries, *etc.*	√		
Conductive coatings transparent		√	
Costume jewellery	√		
Filters	√		√
Mirrors	√		√
Neutron optics	√	√	
Optics spectacles, sun glasses	√	√	
Packaging	√	√	
Radiography	√	√	
Shielding against electro-magnetic and radio-frequency energy	√	√	√

*Indicated thus: √

used, but today there are more sophisticated systems—including sputtering and ion plating. Table XV shows a selection of components of different types, together with the process employed in each instance.

The process is carried out in an air-tight chamber—originally a simple bell jar and now normally in commercial production purpose-built plant some metres in diameter and length. The chamber is fitted with the necessary pumping system, work holder, source holder, power supply, control equipment, and gauges. It is possible for an experienced operator to control the process with the precision necessary but for routine production to a specified standard automatic control is regarded as essential.

In the first stage, the chamber is evacuated to a high vacuum and the source of material to be deposited heated to its melting point. Additional heat is applied so that the vapour pressure exceeds the pressure within the chamber and the vaporized molecules of material travel in straight lines from the source—to be deposited on the work pieces, the walls of the chamber, *etc.* Subject to correct adjustment of pressure the vapour molecules do not collide with any molecules remaining from the air and take the 'mean free path' (the average distance a molecule will travel between two successive collisions)—which for a given vapour is inversely proportional to the vapour pressure.

Bearing in mind that the deposition of the metal thus will be on a 'line of sight' basis it becomes obvious that in this elementary form the technique would be suitable only for a limited range of simple articles. (The number of pieces that could be coated in such a static arrangement would be restricted, and the more complicated shapes and forms would not be metallized well enough.) In order to overcome this, arrangements may be made for the rotation of the articles in the course of the deposition cycle, so that the area of material within the line of sight is increased, and more work pieces can be accommodated by the provision of facilities for planetary rotation of the holders.

A typical chamber would be of cylindrical design, manufactured from stainless steel with a high surface finish. At one end, a hinged or sliding door is fitted, in which is a glass inspection panel so that the work may be viewed during the cycle. It is possible to use the chamber set up either vertically or horizontally, although in the main modern units are of horizontal design and so can be mounted at floor level—whereas vertical arrangement may require the evacuation of a pit to take the pumping system. The latter may be connected to the sealed end of the chamber or to the perimeter wall; usually the route is selected with a view to keeping inter-connecting pipework to a minimum. There also will be a suitable low-tension electrical supply, to provide current to the source holder.

Evacuation of the chamber is achieved by means of a series of pumps: initially a rotary piston pump takes the pressure down from atmospheric to around 30 Torr, then a blower of the Rootes-type reduces it further to about 0.1 Torr; lastly, an oil-diffusion pump brings the chamber down to the condition required for the evaporation cycle—in the region of 0.0002 Torr.

The holder for the work piece or pieces consists essentially of a metal rack or jig, on which the articles can be held by means of clips or other such accessories: it may be used stationary, or in more elaborate forms with rotary or planetary movement; in such cases the drive is by means of an electric motor mounted externally on the chamber, with the shaft led into the chamber through a vacuum-tight seal. Planetary movement of the work holder in particular requires further description: typically, the articles are clipped to individual cages (called 'planets'), which are attached to the main frame of the holder. The frame is driven and by means of gearing and inter-connecting drive chains the planets in turn can be made to rotate in a selected ratio to their speed of rotation (the usual ranges of ratios are from 5:1 to 10:1).

The source material may be in forms such as wire, rod, or powder; it may be held in a separate assembly fitted inside the chamber or as an integral part of the work holder (although insulated from it electrically). Filaments such as tungsten helical wires or molybdenum boats (described also as 'crucibles'), holding the source material, are attached to the source holder, or holders (single-, double-, and triple-filament holders may be used, situated at different positions in the chamber, and evaporated either concurrently or consecutively as desired). If the source material is contained in boats it should be remembered that as this is a 'line of sight' technique coating of the articles will take place only in the segments of the chamber above the positions of the boats.

The necessary start and stop buttons, timers, and gauges are mounted on a control panel, where often there is also a mimic system showing the current stage in the cycle. It is normal for the information on the control panel to be visible but it is common practice also—and highly desirable—to have a lockable transparent cover for it, with immediate access only to the main control button: this ensures that the parameters of the cycle, once set, are not changed in any way without authority.

Many different materials of construction are used in plants for vacuum coating; it is considered that the most suitable is stainless steel but whatever is chosen it is important to ensure that it is homogeneous in structure, free from pores, and to a high surface finish. Since trapped air has to be removed, the use of material which absorbs or can retain air will lead to longer production cycles. In appropriate cases, consideration should be given to the sealing of surfaces and perhaps the application of other finishing treatments, such as electroplating or stove enamelling.

An important aspect of the unit that sometimes is overlooked is the means of venting the chamber to atmosphere at the end of each cycle. So that the metallized articles are not disturbed, air must be re-admitted at a controlled rate, and it must be dry and clean. An air intake located close to a chimney or exhaust outlet from other processes is not likely to be a suitable source.

Sputtering

The theoretical basis for sputtering has been known for a long time but more recently commercial methods have been developed along lines similar to those in vacuum evaporation just described. Essentially, a discharge of argon gas plasma is established between an anode and cathode electrode; in this instance, the source material is the cathode, and the work piece the anode. Gas ions charged positively are attracted to the cathode, where they collide with it and remove atoms of the source material—which in turn travel to the anode and form a coating of sputtered material on the work piece.

In general, the speed at which transfer and deposition take place is low, but it may be improved by magnetron sputtering, of which the types available include planar, closed field, hollow cathode, and post cathode—all giving coatings with good geometrical array.

By this means, a wide range of metals, alloys, and compounds may be

TABLE XVI *Comparison of processes*

Parameter	*Process*		
	Evaporation	*Sputtering*	*Ion Plating*
Use of inert gas	none	argon	argon
Pressure (Torr)	< 0.0005	< 0.005	< 0.0005
Throwing power	poor	good	excellent
Rate of deposition (μm min^{-1})	high > 1	low < 1	high > 1
Heating of component	low	high	high
Uniformity of coating (tolerance, %)	very good ± 1	good ± 10	good ± 10
Adhesion of coating	good	very good	excellent

deposited in the form of thin films: continuous production is available, by virtue of plant fitted with vacuum locks in series.

Ion Plating

In practice, ion plating is used mainly to deposit aluminium on non-plastics, but it is included here for the sake of completeness.

The technique may be said to combine the advantages of vacuum evaporation and sputtering, so that excellent qualities of adhesion are obtained without a limitation of maximum thickness of the coating—while at the same time the rate of deposition can be comparatively high. Many metals, alloys, and compounds may be deposited, on both metallic and non-metallic articles. However, its use at present is mainly for functional and protective applications, particularly where high resistance to corrosion is required. Thus, as examples, aluminium may be deposited on various types of steel and on titanium for uses in the aerospace and defence industries—and can be regarded as a less hazardous replacement for cadmium electroplating.

A plant for this purpose comprises a vacuum chamber with ancillary equipment. The articles to be plated are mounted on a water-cooled cathode and, following evacuation of the chamber and filling with argon to a pressure of 0.002 Torr, direct current from a supply in the range 2000 to 5000 V is applied—to give a glow discharge in the area of the cathode. The article or articles mounted there thus are cleaned by ionic bombardment. Next, the evaporation source is excited, and the atoms of source material are deposited on the work pieces.

A selection of different methods of evaporation is available—including electrical resistance heating, electron beam, induction heating, and sputtering.

Table XVI provides a comparison of the three processes (vacuum evapo-

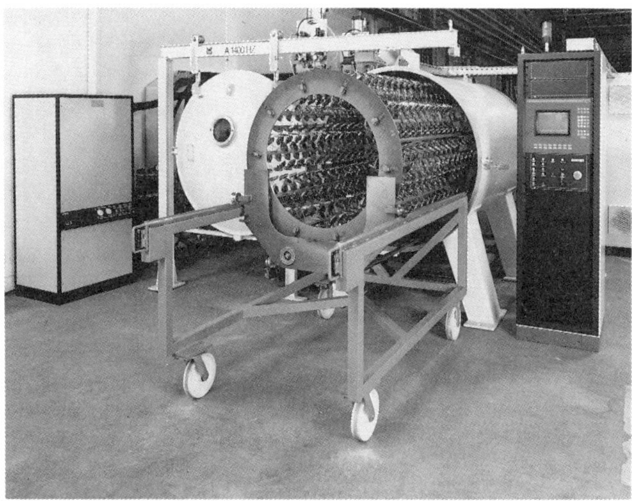

Figure 74 *A typical plant for coating small components in batches, using a 'Planetary' work holder*
(Photo by courtesy of Leybold Limited)

ration, ion plating, and sputtering); Figure 74 shows a typical batch coating plant for small components.

Conditions for Satisfactory Operation

Reference was made earlier to the importance for good results of cleanliness of the work pieces and of factors such as the maintenance of standard operating conditions. Consistent satisfactory production depends to a large measure upon the circumstances in which a coating unit is run. Ideally, 'clean room' techniques, as in the electronics industry, should be applied but for a variety of reasons, including costs, this is not always possible in practice.

The metallizing unit should be situated as far as possible from other processes, especially if these generate dust or fumes: it should be in its own enclosure—which should be constructed from materials that do not give off or retain dust, and preferably fitted with inter-locked double doors. The floor of the enclosure should be level and smooth, and treated or covered with hard-wearing, non-slip material. The temperature and humidity within the enclosure should be controlled. Often the units are built into enclosures with the pumping system outside. Control panels also may be mounted outside, as in Figure 75.

Only those employees working in the unit should be admitted to it (and it should not be on the route to the canteen, or—say—open to groups of visitors). Operators should be issued with (and required to wear) lint-free clothing and head covering: while they are in the area, overshoes should be worn to prevent

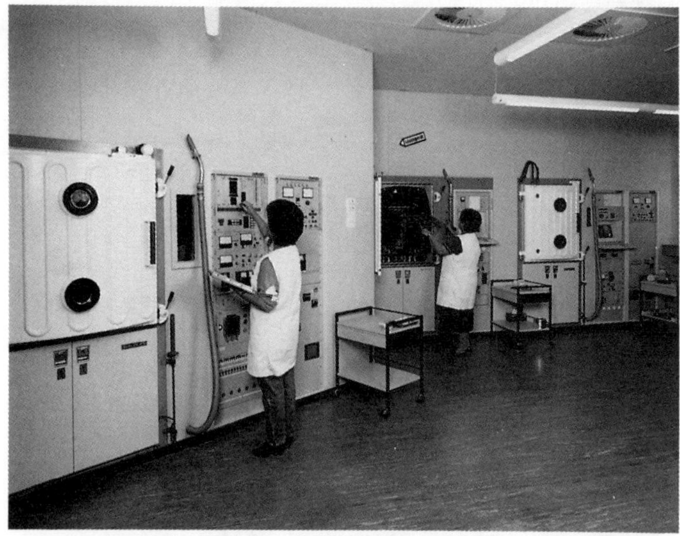

Figure 75 *An example of a good installation, with only the chamber doors and the fronts of control panels in the room*
(Photo by courtesy of Balzer High Vacuum Limited)

dust and debris from outside being brought in and deposited from ordinary footwear.

The wearing of gloves while handling the articles to be coated, and when removing them, must be mandatory.

It will be apparent from the description of the process that the units are likely to become increasingly contaminated as time goes on (in addition to being deposited on the work pieces, the source materials will coat the walls of the chamber, jigs, and other fixtures). With a view to maintaining regular cycles and conditions, such deposited materials must be removed at frequent intervals—a task that can be facilitated by means such as removable liners around the perimeter of the chamber, and by having several interchangeable sets of jigs and fittings. Items like these can be removed for cleaning and replaced with other equipment cleaned earlier (the replacements must be free from any residues, including cleaning preparations, and dry). After cleaning, a test evacuation should be carried out. In general, when not in use, the unit should be kept under a measure of vacuum.

Maintaining Records

Units may be fitted with pen-recording chart systems or more sophisticated electronic means of references so that any problems, including the gradual increase of cycle times, can be monitored and identified easily. Efficient operation depends on the keeping of records, which can indicate, as examples, whether cycle times are lengthening because of normal use, because of leakage from seals, because of greater emission of vapour from articles being metallized (arising either from deficiencies in moulding, or from incomplete curing of lacquers), from reduced efficiency in the pumping system, and so forth.

Where units are run on a daily basis (that is, not on a shift system), a check of the facilities each morning before starting production is to be recommended.

Plastics for Metallizing

The metallizing of plastics offers new scope for design engineers, extending the opportunity to make improvements and to replace components made previously in other materials (like glass, or ferrous or non-ferrous metals). On the other hand, there are also certain new problems, which must be considered and overcome if work of satisfactory quality is to be achieved.

It will be apparent that the articles to be metallized should themselves be consistent in composition, homogeneous, and not retaining volatile materials, either in the substrates or the lacquers applied. Sometimes, with plastics in particular, difficulties of this nature can be serious—even insurmountable. Typical plastics, as good electrical insulators, have also the disadvantage of retaining static charges; they attract dust in this way and require special care in cleaning and preparation (to remove dust and to discharge the electrical forces).

It is essential to devote particular care to the quality and properties of the surfaces: many plastics may be metallized direct but the results in practice generally are unsatisfactory except for acrylic and polycarbonate materials. With films especially, any surface contours and defects will be reproduced faithfully in the reflective coating applied. It is for such reasons that the surfaces usually are prepared by the application of a lacquer base coat as described—even, in some instances, by primer or pre-base coat lacquers. These help to ensure satisfactory adhesion, fill small defects or voids in the surface of the underlying film or moulding, and provide the smooth surface essential for an effective metallic layer.

The main qualities required of lacquers for base coating are:

(i) compatibility with and good adhesion to the plastic concerned
(ii) hardness, but not such as to detract from good adhesion of the metal
(iii) low content of plasticizer or other volatiles
(iv) resistance to water and to other solvents
(v) suitable optical properties
(vi) ease of application, by methods such as those described previously.

Usually when plastics are metallized it is necessary to safeguard the integrity and durability of the metallic layer by applying also a top lacquer. Aside from features such as those noted above, the top lacquer must be capable of protecting the metal (that is, it must be resistant to the conditions experienced in service) and must be stabilized against degradation caused by exposure to ultra-violet light.

Each layer of lacquer must be cured individually, and cured fully in order to obtain good results. The curing may be by means of conventional ovens, or, with specialized lacquer systems, by means such as exposure to infra-red or ultra-violet light.

If not removed, volatiles from the lacquers can contaminate the metallizing system and spoil the work: unfortunately too, plastics (unlike glass and metals) will desorb gases and vapours very slowly. In most instances the commercial vacuum plants are equipped fairly generously with pumping capacity, but it would be uneconomical to fit pumps with the theoretical capacity necessary to overcome this desorption and an alternative approach is needed. Typically, a 'cryocoil' is fitted, in such a way that vapour from the interior of the chamber condenses on its cold surface. In some units, liquid nitrogen was used as the cooling medium, passing through a wound copper tube which was situated in the port connecting the vacuum chamber to the pumping system. In the course of the work, water vapour condensed on the unit and was retained as ice; at the end of the cycle, the unit would be defrosted. More recently, proprietary refrigerants have been used in closed-loop systems (so that there is no discharge to atmosphere of the substances concerned).

It might appear from the foregoing that the metallizing of plastics can be a somewhat complicated procedure, requiring particularly careful preparation and handling. On the other hand, the difficulties mentioned are well understood and in large operations especially it is possible to make suitable arrangements (such as by using the same jigs and work holders for the lacquering, metallizing, and top coating stages) and so to meet the problems without there being too much complexity in handling.

Special Applications and Techniques

So far in this chapter attention has been given in particular to the metallizing of mouldings; however, a most important application in recent years has been the coating of films for a variety of purposes. Metallized film has been shown to be an economical and satisfactory replacement for foil in packaging, labelling, hot-stamping, and metallic filaments. Aside from its bright and attractive appearance, such film is a satisfactory barrier to gases that could cause spoilage of food, and its electrical properties are useful for electromagnetic shielding and the packing of sensitive electrical components.

Film is metallized continuously from the roll in fully automatic units. The facilities for de-coiling and for re-coiling are built within the vacuum chamber, in such a way that the material can be coated on one or both sides. A typical plant is capable of coating film up to a width of 2.5 m, in rolls with diameter up to 1 m, with a coating speed to 10 m s^{-1}. Evaporation may be by means of conventional resistance heating of a wire source fed continuously, by electron beam, or by sputtering. Besides aluminium, materials that can be deposited include copper, nickel, nickel–chromium, silver, and dielectrics such as aluminium, silicon, and magnesium oxides. A manufacturing unit for work of this type is shown in Figure 76.

Reference was made earlier to the use of a top coating of lacquer to protect the metallic layer. It was found possible in some instances to eliminate the top

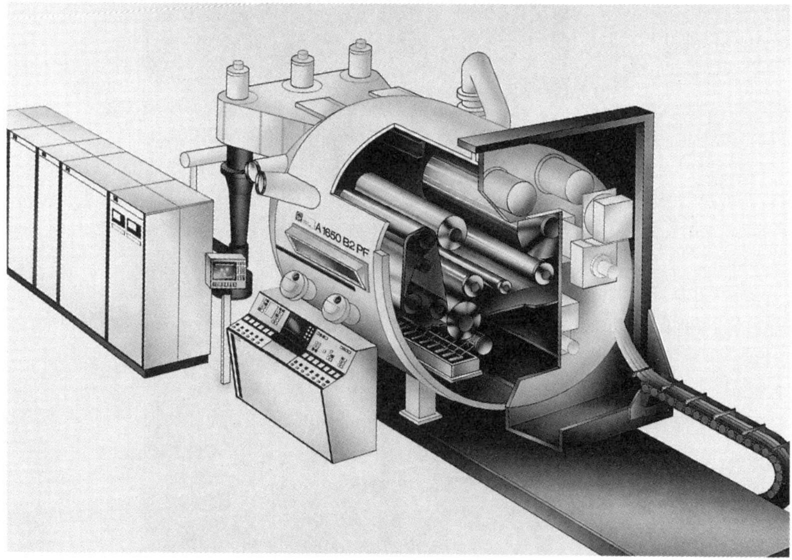

Figure 76 *Schematic diagram for a vacuum web coater for metallizing plastic film continuously*
(Photo by courtesy of Leybold Limited)

lacquer by depositing a transparent protective film polymerized from a direct current plasma. However, the method required the cleaning after every production cycle of the coating source, and could not be applied to continuous coating. More recently, a proprietary process has been developed in which a plasma polymerization source excited by microwaves is used and this both gives surface film of good quality and can be integrated in systems for continuous coating. The resulting material is an amorphous, hydrogenated, silicon oxycarbide film, which adheres well to aluminium, resists attack by many solvents, and meets specifications for resistance to heat, corrosion, salt spray, and so forth. Examples of applications for this technique include mirrors, reflectors for automotive lighting, and reflector arrays for lighting in buildings.

A further special application is the deposition of anti-reflective coatings on the lenses of spectacles. Such coatings are better than earlier methods of controlling the amounts of light passing through the lenses, and so help to reduce or to eliminate some problems associated with wearing spectacles. The coating used is made up from one or more layers in order to give the refractive index and optical characteristics required. They comprise metallic oxides or fluorides. For sun-glasses, the salt is not oxidized completely, and so will absorb the visible spectrum without altering colours. Apart from filtering out ultraviolet light and reducing reflection, such coatings provide a harder surface than the underlying substrate, and enhance resistance to abrasion and corrosion.

The plants used for this purpose often are referred to as 'Box coaters' and are illustrated in Figure 77. Such units will take batches of from 100 to 200 lenses

Figure 77 *'Box coater' in use for coating spectacle lenses*
(Photo by courtesy of Balzer High Vacuum Limited)

up to 80 mm in diameter; the lenses may be turned automatically during the process so that both sides are coated. Evaporation of the source is by an electron beam system; the optical characteristics of the coatings are controlled by a high-resolution quartz crystal thickness monitor.

Reference was made earlier to in-line continuously operating coating plants to meet requirements for mass production. Such plants consist of a linear array of modules interconnected by vacuum-tight doors, seals, and valves, and each module has its own pump system and ancillary equipment. The articles to be coated are arranged on suitable fixtures and transported through the plant, being rotated as necessary in the various stages of the sequence (a typical sequence is shown in schematic form in Figure 78). The coating material may be metals, alloys, or metal oxides and these are deposited using the sputtering technique, with plasma-enhanced chemical vapour deposition as the means of pre-treatment and top protection. Since only the first and final modules are vented to atmosphere the total time of processing is kept to a minimum. Obviously, such complex plant requires a comparatively high initial investment of capital and should be considered only when establishing the production of very large numbers of similar components using minimum labour but with the requirement for consistently high standards of quality and productivity. Figure 79 illustrates a plant of this type in production.

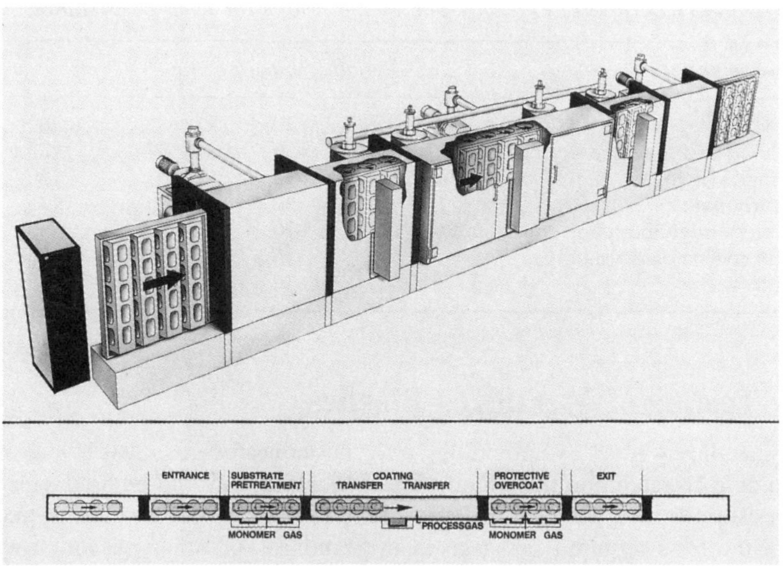

Figure 78 *Schematic diagram for a continuous in-line coating plant*
(Photo by courtesy of Leybold Limited)

Figure 79 *A continuous in-line vacuum coating unit in production*
(Photo by courtesy of Leybold Limited)

TABLE XVII *Periods of exposure and temperatures for testing resistance*

Type of material	*Period of exposure*/min	*Temperature*/°C
Acrylic	30	85
Acrylonitrile–butadiene–styrene	30	85
Polyamide (nylon)	30	150
Polycarbonate	30	135
Polyester dough moulding compound	10	200
Polyphenylene oxide (modified)	30	105
Polypropylene	30	135

The Testing of Metallized Plastics

The metallized films or mouldings may be subjected to a variety of tests depending largely upon the circumstances in which they will be used—including resistant to wear, to heat, to exposure to environmental conditions, and so forth, the tests required being given in Standards or other relevant specifications. However, a selection of tests to satisfy basic requirements is summarized below.

(i) Adhesion

The following is a simple practical test which may be applied to the adhesion of the lacquer base coat, the metallic layer, or to the top coat. It should be used finally at least 72 hours after completion of the processing.

Representative samples of the lacquered, metallized, or coated materials are taken and by means of a sharp knife a cross-hatched pattern of squares of dimensions 1 mm by 1 mm is cut into the surface. A strip of transparent adhesive tape of good quality is pressed firmly over the test area, then pulled away in a single continuous movement. Finally the test area is examined and whether squares have been removed (and in what number) noted.

(ii) Hardness

The hardness of the metallizing may be tested by means of various instruments available commercially, the essential principle of which, in all cases, is to apply consistent forces.

A simpler method is to inscribe the surface using lead pencils in a range of hardness from 2H to 8H, and, if an impression is obtained, to note the type of pencil used.

(iii) Resistance to Heat

Representative samples are placed in an oven and subjected to dry heat for stipulated periods of time (the temperatures employed and the periods of

TABLE XVIII *Cycles of temperature and humidity*

Period of time/h	*Temperature*/°C	*Conditions*
24	40	Relative humidity: 95% to 100%
24	80	In an air-circulating oven
48	40	Relative humidity: 95% to 100%
48	80	In an air-circulating oven
24	– 40	In a refrigeration cabinet

exposure to heat differ with the plastics being tested and their composition but indicative figures are given in Table XVII).

(iv) Cycles of Temperature and Humidity

Representative test pieces may be subjected to a series of cycles of exposure to differing conditions of temperature and humidity—the cycles and conditions being decided by the circumstances likely to be encountered in practice in the transport and use of the articles concerned. Meanwhile, an appropriate number of control specimens taken from the same production batches are kept under ambient or standard atmospheric conditions.

An example of the sequence of cycles in a typical test of this nature is given in Table XVIII.

After exposure to such a sequence, the specimens should be tested for adhesion of the metal coating by means of the cross-hatch test described in 'Adhesion' above. In a comparison with the unexposed control samples, there should be no deterioration in the adhesion of the coating, reflectivity, or the condition of the surface.

Further Opportunities

Perhaps in this chapter an unusually large amount of attention has been given to the special requirements of metallizing techniques and to problems that may be encountered. On the other hand, metallizing is well established, with routine production on a large scale in many different units. It extends further the scope of applications for plastics, and as the control of the qualities of films and mouldings made for metallizing continues to be improved (and tightened) so should it be easier to apply the technique successfully and to extend its versatility. In some uses, the surfaces of substrates have been enhanced already to the extent that a base coat of lacquer is necesssary no longer—while for a limited range of applications the top lacquer has been eliminated. For instances in which base coat lacquer is a requirement still, research in Europe is under evaluation currently for the development of water-based products that, when proved, should be more acceptable with regard to health, safety, and effect on the environment at large. The combination of plastic with a thin coating of metal often is economical, and the use in a large way of some of the more toxic metals especially (like cadmium, chromium, and nickel) may well be controlled even more strictly in the future on grounds of health and safety.

Some Terms Used in Lacquering and Metallizing

Accelerator. An additive employed in lacquers to increase the speed at which curing will take place; generally not part of the reaction nor consumed in it, but acting in a way similar to that of a catalyst.

[ISO 472 gives cross-references to 'activator' and 'catalyst' and the following slightly more general definition:

> *accelerator*: *promoter*: A substance used in small proportion to increase the reaction rate of a chemical system (reactants plus other additives).]

Aurora Borealis. The effect of a rainbow created in thin films by the deposition of inorganic compounds under high vacuum.

Back Coat. A coating applied as a protection over second-surface metallized films.

Back Streaming. A diffusion of vaporized oil entering the vacuum chamber.

Bloom. 'Bloom' is defined in ISO 472 as:

> A visible exudation or efflorecence on the surface of a plastic

—and it is noted that it may be caused by substances such as lubricants or plasticizers. In metallizing, the material migrating to the surface usually is white or off-white in colour.

Cracked Oil. Oil from a diffusion pump exposed while hot to the atmosphere, and thus being oxidized or burnt.

Electromagnetic Interference ('EMI'). Impaired performance of electronic equipment consequent upon electromagnetic forces from sources either associated with the equipment or from elsewhere.

Gettering. The combination of metals in volatile and chemically active form with residual gases in the chamber, the compounds resulting being deposited on the walls of the chamber.

Irridescence. An optical effect akin to a milky rainbow, associated usually with a defect such as movement of the base coat.

Newton Fringe. A rainbow effect in films that have been metallized under high vacuum, caused usually by inadequate thickness of the top lacquer.

Out-Gassing. The release from either the plastic or the lacquer base coat during evacuation of the chamber of gases and vapours.

Radio-Frequency Interference ('RFI'). Impairment of performance consequent upon radio-frequency energy from sources elsewhere.

Rainbow, 'Rainbowing'. The effect of a visible colour spectrum, usually

indicating a defect of some kind in the lacquering or metallizing (see: '*Aurora borealis*', 'irridescence' and 'Newton fringe', above).
Top Coat. A coating, of lacquer or of a transparent material deposited under vacuum, applied to prevent deterioration of a decorative coating and to improve its resistance to being handled.

For Further Reading

L. G. Carpenter, 'Vacuum Technology', Adam Hilger, Bristol, 1983 (ISBN 0-85274-481-1).
N. S. Harris, 'Modern Vacuum Practise', McGraw-Hill, London, 1989 (ISBN 0-07-707099-2).
L. Holland, W. Steckelmacher, and J. Yarwood, 'Vacuum Manual', E. & F. N. Spon, London, 1974 (ISBN 0-419-10740-1).
J. F. O'Hanlan, 'A User's Guide to Vacuum Technology', John Wiley & Son, New York, 1980 (ISBN 0-471-01624-1).
A. Roth, 'Vacuum Technology', North Holland Publishing, Amsterdam, 1982 (ISBN 0-444-86027-4).

Acknowledgment. The author acknowledges the assistance of Balzers and of Leybold AG with photographs.

CHAPTER 12

Painting Plastics

T. A. WILDE

Introduction

Why paint plastics? Bearing in mind the range of pigments available, and the surface graining techniques that can be obtained when moulding, why should a painted finish be required? There are various reasons, which can include in any particular case:

(i) when an assembly comprises several components, or for matching parts made from other materials paint is applied in order to give a unified result (even when an assembly is of parts made from plastics that have identical pigmentation and surface finish the pieces often will look different until they are painted)

(ii) to extend the range of visual appeal beyond what is available from pigmentation and moulding techniques; with paints a range of colours can be achieved even wider than from the pigmentation of plastics

(iii) in addition, the levels of gloss obtainable when moulding may not be consistent—or maybe the mouldings are too glossy or not glossy enough

(iv) variations may be visible on moulded surfaces for other reasons, such as differences in the rates of flow of material in different parts of a mould

(v) on occasion, painted finishes may be used to alter perceptions of the substrate—such as when plastics are used in place of leather, metal, or wood; they help to combine the properties and ease of fabrication of plastics with the appeal of the traditional and familiar

(vi) to give a product more than one colour, or to apply designs or lettering—improving its appearance and imparting information without a need for separate stages in fabrication, engraving, or machining in some other way.

Examples of plastics painted for reasons such as these may be found in the automotive industry but are by no means confined to it; the painting of plastics is applied in fields ranging from cartons and cups to toys and television sets—indeed in many areas, some quite unexpected.

Which Material?

Since painting is most likely to be for one of the reasons indicated above, it follows that the plastic will be selected for a particular purpose: not because of appearance, but because of its inherent properties—chemical, electrical, mechanical, or thermal. The materials encountered most frequently are acrylonitrile–butadiene–styrene, polyamides, polycarbonate, polyethylene, and polypropylene. ABS and polycarbonate may be selected for reasons such as strength and stability, polyamides when resistance to higher temperatures is important, and the polyolefins are being used increasingly because of ease of moulding and their comparatively low cost. They also can be re-worked fairly readily (which is important in an environmentally conscious world) although painting does tend to make re-processing more difficult.

Since the painting aspect is not the prime reason for the selection it follows that the type of paint and the method of application must be suited to the plastic; further, the properties that make a material useful in engineering often also make it harder to paint satisfactorily. As an indication, generally it is easier to paint ABS and polycarbonate than nylon, and easier to paint nylon than polyethylene or polypropylene. Other plastics that can give problems in painting include acetal (polyformaldehyde) and poly(phenylene oxide).

Some Typical Problems

Problems of painting which seem specific to plastics may be summarized as follows.

(i) Adhesion

With conventional paint systems rather poor adhesion is obtained with acetal and the polyolefin plastics; adhesion with polyamide normally is good at first but under adverse conditions can deteriorate. In general, ABS and polycarbonate do not give comparable difficulties.

(ii) Compatibility

If a paint contains aggressive solvents there may be attack on the plastics, including the accentuation of flow lines from moulding, with these showing through a painted surface. Materials like ABS and polycarbonate are susceptible to attack in this way, while polypropylene, especially, resists solvents—with the consequence that obtaining a good 'key' to the surface of this material can be troublesome.

(iii) Static Electric Charges

Reference has been made elsewhere to the propensity of plastics, because of static electric charges, to attract particles from the air; this factor also must be overcome if good surface appearance is to be obtained consistently. Any dust contamination present immediately prior to painting or landing on wet paint will create unsightly surface irregularities.

(iv) Thermal Properties

When thermoplastics, as mouldings or in other forms, are exposed to temperatures in the range 80 to 130 °C there may be some release of strain, softening, and distortion—which makes it impossible to employ with them conventional painting methods that require higher temperatures for curing.

Some of the methods used with plastics to improve adhesion and to apply paint films are described in the sections following.

Reasons for Poor Adhesion

Problems of adhesion can arise with most substances that are painted—not only plastics—especially if the surfaces to which paint is applied are not clean. Hence, for the best results, the plastic articles always should be free from loose particles and from other surface contamination (such as moisture, or oils). However, the plastics that are difficult to paint exhibit also one or both of the following characteristics.

(i) Chemical Resistance

There may be resistance to the formation of chemical bonds between the paint and the plastic; further, or alternatively, resistance to solvents in the paint may impair or prevent its diffusion into the surface of the substrate and the formation of a physical 'key'.

(ii) Low Surface Tension

Low surface tension (poor 'wettability') may be a factor, with apparent repulsion of the paint and prevention of its rapid, even spread over the whole surface; test inks of different values are available so that wettability can be assessed in advance with samples of the items concerned; for satisfactory use with paints based on organic solvents, surface values of greater than 50 mN m^{-1} are desirable, and for water-based systems values greater than 70 mN m^{-1}.

Methods for Improving Adhesion

Polypropylene exhibits all the characteristics noted above—it is chemically inert, resistant to solvents, and has low surface tension—yet is being used in many applications that require painting. Steps which have been taken towards improving adhesion of paint to polypropylene include.

(i) Additional Solvent Treatment

Preparatory to painting, the polypropylene components are subjected to trichloroethylene or 1,1,1-trichloroethane which (especially if fillers or other materials are present in the composition) will remove grease and permeate the surface. However, it should be remembered that for environmental and health reasons the use of solvents like these is regulated strictly.

(ii) Priming

The surfaces of the items may be primed by applying adhesion promoters, which are chlorinated polyolefin resins dissolved in xylene—again, however, subject to careful control as in (i).

(iii) Other Surface Treatments

The structure of the surface may be modified in other ways to enhance its reactivity and wettability. Frequently, the effect is partial oxidation, with the formation of C—OH and C=O groups; with polypropylene, especially, rather powerful methods of treatment may be necessary, and sometimes they are employed in combination with priming. Some typical surface treatments are as follows.

Etching with Acids. Acid etching would be applied usually on a small scale, with limited amounts of acid on restricted areas.

Corona Discharge. In this approach the surface is exposed to a high-voltage electrical discharge; the technique is used most frequently for treating film in preparation for printing and is described more fully in the chapter following. Since the gap between the electrode and the surface of the plastic is required to be precise it is not applicable to mouldings other than those with very simple shapes.

Flame Treatment. In this the surfaces are flamed, usually by application of a gas burner for a brief period. Factors such as the temperature of the flame (the ratio of gas to air in the fuel), distance of flame from the surface, and speeds of travel of flame and objects, all are critical—and small variations in any of these factors can lead to unsatisfactory results (that is, under-treatment, or over-treatment). Because of this, standardization and consistent results are best achieved through programmed control with robots. Even so it is difficult to treat more complex shapes satisfactorily, and normally such items would be primed after flame treatment—that is, two methods of promoting adhesion would be used in combination.

Ultra-Violet Irradiation. In this method the items first are coated with a specific adhesion promoter, then passed under sources of ultra-violet light of specific wavelength to activate the adhesion promoter. The result usually is a good chemical bond with strong adhesion, but still with the disadvantage in respect of environmental control that the promoters require.

Plasma. For plasma treatment the mouldings are arranged in a suitable chamber which then is evacuated to the pressure appropriate for the type of plasma being used. A gas or gas mixture is introduced (typically, a mixture of oxygen and air), and plasma formed by energizing the gas for a short time.

Evenness of treatment depends on factors such as the designs of the chamber and of the electrode, the levels of energy and pressure, and the period of time of exposure; once established at satisfactory values, these parameters can be controlled fairly easily. The surface chemistry is similar to that with simple flame treatment but the use of a gaseous reactant means that even treatment of quite complex shapes is practicable. Besides even and repeatable results the method offers the additional advantage that adhesion promoters are not necessary—with associated savings in costs and enhanced environmental implications.

Polymer manufacturers offer 'paintable' grades of polypropylene or copolymer blends with talcum or other mineral fillers; with such grades the tolerance of over-treatment seems to be increased—which can be advantageous, especially with flame treatment.

As with other methods, the effects of plasma treatment can be measured by assessing wettability and the extent to which it has been enhanced—but it should be kept in mind that over-treated components also will show good wetting characteristics. The surface chemistry will differ, with reduced numbers of C=O groups.

At the time of writing, three basic forms of plasma treatment are in use, the essential differences between them being the levels of vacuum and the energy frequencies (basically, the lower the energy frequency, the greater the vacuum, or the reduction in pressure, that is necessary).

For plasma induced by a *high-frequency microwave* system, extremes of vacuum are not needed, but particular care in the design of the system is essential in ensuring even treatment. Since microwave energy is employed, the use of steel for chamber and jigs is precluded.

Plasma created by *radio wave at megahertz frequency* does not require extremes of vacuum, and steel may be used in making chamber and jigs. To a certain extent the treatment remains specific to the shape, but with good electrode design it can be very effective.

Plasmas generated by *low-frequency direct current* require less-specific electrode design and therefore are more flexible for even treatment of various shapes. However, low pressures are involved and these may restrict the shape of the chamber and call for large pumping capacity and/or long periods of evacuation.

In summary, effective surface treatments can be obtained in a variety of ways and in selecting a process for a particular purpose account should be taken, *inter alia*, of the sizes, configuration and numbers of components to be handled, and the associated costs of materials, operation, and satisfying the relevant legislation (particularly over health, safety, and environmental questions).

Selection of Paint

General factors that influence the choice of paint include the following.

(i) Appearance Required

It may be that a component must appear to be the same colour as, or match, one or more other items—in which case assessments of the precise colour and the appropriate level of gloss will be important.

(ii) Chemical and Physical Properties

Properties determining whether a paint will be suitable for the purpose include: its flexibility (while in service, will it suffer abrasion or knocks?); whether it will be used inside or out-of-doors, and if the latter will it resist humidity, strong sunlight, and other ambient conditions; will it be expected to serve in an unusual environment (say, exposed to airborne gases, oil mists, or vapour)?

(iii) Compatibility

The paint must be compatible with the plastic, capable of adhering to it, and resistant to the range of temperatures to which it will be exposed in processing and in use afterwards.

(iv) Availability

In fact the ideal paint for a particular type of work may not be available, and that selected may represent a compromise between needs such as the use of suitable solvents, the temperatures of processing and curing, and the speed of evaporation ('drying times') necessary, especially if several coats are to be used.

There are stringent controls over emissions of particulate matter and solvent vapours—not only must a factory be of standard sufficient to meet the relevant exposure limits (which are controlled under the Health and Safety at Work Act, 1974, and subsequent regulations) but manufacturers have a legal duty to ensure that exposure and emissions are (in the official phrase) as low 'as is reasonably practicable'. Specific values as limits for emissions of particulates and vapours are laid down under the Environmental Protection Act, 1990. In effect this means that there must be continuing attention to the maintenance of operating standards, and to the improvement of formulations—and on the latter, paint technologists are following two main lines of advance:

Water-Based Paints. For some purposes, water-based formulations are being used already but further improvements will be needed before there are resilient, high-gloss finishes suitable for exacting work in, say, motor vehicles. So far, water-based paints have been restricted to inexacting finishes, and to primers or intermediate coats over which conventional paint films are applied.

Some key problems with paints of this type are:

- their indifferent wetting qualities, which must be overcome by plasma treatment of the plastics (by this means, as an example, water-based paints can be used with ABS)
- high boiling points—which means long cycle times (more time is necessary for evaporation of the solvent (water) and for stoving)
- since water-based paints quickly corrode conventional plant, stainless steel construction is necessary.

High-Solids Paints. Hitherto, for paints for spraying purposes, the mixed solids contents have been of the order of 5% to 15% by volume. However, with more modern spray equipment and other improvements in plant, the use of paint with solids contents in the range 30% to 40% is possible—which means in turn a diminished need for organic solvents.

Design of Products to be Painted

With most processing activities there are likely to be some rejected parts, particularly perhaps at the start and towards the end of each run. One of the many advantages of thermoplastics is that they can be re-processed with comparative ease—so that a process like injection moulding can be very economical. It should be remembered however that painting is an additional process and that after this stage it is more difficult to re-work the underlying material. For this reason it is more desirable still that the numbers of rejects be kept to the minimum, and with this in view it is well worthwhile to devote time and attention in advance to aspects in which problems can arise. Among these at the design stage would be the following.

(i) Mould Lines

The surfaces to be painted should be affected as little as possible by features like mould split lines and the locations of gates, cores, ejectors, and so forth. Otherwise, if marks from the moulding are visible, considerable preparation of the surfaces (to remove irregularities and flash) may be necessary before satisfactory results are obtained.

(ii) Section Thickness

Attention at the design stage to differences in section thickness also can be beneficial. The position of features such as strengthening ribs can be important—they may give rise to 'sink' marks on the opposite surface, the unsightliness of which may be exaggerated by painting. Marks of this kind are especially noticeable when otherwise a surface is flat and featureless.

(iii) Recesses

For the best finish it is important so far as possible to apply paint evenly and a component with recesses, especially if they are deep and narrow, could well be troublesome in this respect. With configurations of this kind further application of paint, from different angles, may be needed to obtain the standard of finish specified; however, it is not always practicable to do this.

(iv) Surface Effects

The appearance of the surface of the component will be reflected in the appearance after painting—if, say, it is a grained pattern rather than smooth, a grainy effect will be obtained. To some extent this difficulty can be overcome with a thick coat of primer, but obviously it will be helpful from the point of view of quality if the mould surfaces and the natural finish of the material chosen are in keeping (rather than conflict) with the final finish required.

(v) Masking

At times designs are developed that require the 'masking' of parts of the surfaces of components, so that areas in different colours can be obtained. Plainly this approach requires more labour and can lead to increased numbers of rejected components—so it is worth considering at the design stage whether the appearance sought might not be obtained more easily and cheaply through an assembly of two or more separate pieces.

Design of Plant

Table XIX presents a summary of the essential factors to be considered in designing plant for the painting of plastics.

While a degree of versatility may be thought desirable the preference should be to design and build the unit specifically for a particular type of product and process. Greater versatility will lead to higher costs in terms of initial capital, and also probably in cost per unit made.

Preferably, the components to be painted should be moulded at the same site, and the moulding and painting stages integrated so far as possible. In-line and continuous processing become more relevant as the numbers of units to be handled are increased: maintaining the speed of output and avoiding delay at intermediate stages are significant considerations, but keeping a close relationship between the fabrication or moulding stages and the final painting is highly desirable in order to minimize the extent to which it is necessary to clean the pieces before painting. If mouldings can be taken automatically from clean tools and transported swiftly to an adjacent painting unit the expense of a preparatory cleaning stage can be avoided. In addition, pre-treatments are more effective with recently moulded articles and, as an example, plasma

TABLE XIX *Factors to be considered in plant design*

Factor		*Comment*
Type of work	(i)	the quantities of components to be produced (including expectations for the future)
	(ii)	configurations and dimensions of the components
Type of plastic	(i)	the nature of the plastic substrate
	(ii)	the forms in which it will be received and presented for painting
Special problems	(i)	associated with the specification
	(ii)	associated with the substrate chosen
	(iii)	in maintaining standards, testing, and delivery on schedule
Type of paint	(i)	whether solvent-based or aqueous
	(ii)	requirements for applications and curing
Type of treatment before painting	(i)	physical preparation of components
	(ii)	chemical treatment
	(iii)	other surface treatments
	(iv)	priming
Requirements of legislation	(i)	in respect of health and safety
	(ii)	environmental factors
	(iii)	other, including liability to the public
Requirements for treatment of effluents	(i)	to prevent plant becoming 'clogged' with paint
	(ii)	to enable collection of paint waste for disposal
	(iii)	to maintain air, soil and water quality
Requirements for:	(i)	handling and jigging of components
	(ii)	sub-assembly
	(iii)	inspection
	(iv)	packing, storage, and dispatch
	(v)	re-working
Degree of automation		
Costs overall		

treatment immediately after moulding will give fuller utilization of the moulding robot, operator, and the conveyor system.

Automated and semi-automated plants are served best by floor-mounted conveyor systems rather than overhead; the latter can carry dust into the painting area and it then can come down and contaminate products. The speed of the conveyor system will be governed by the speed at which paint can be applied (in other words, in this type of work the speed of the conveyor is not a central factor determining plant capacity).

The sizes of ovens and evaporating capacity will be related closely to the types of plastic and the paints being used, while the dimensions of painting chambers and arrangements of jigs must be planned in relation to sizes and configurations of pieces being painted.

Air fed to the plant should be at controlled temperature and filtered to prevent the ingress of dust. The feed should be balanced with the rates of exhaust, so that air to compensate, and dust, are not drawn in from other sources. If dust is present, plastic components will attract it; exposing them at the entrance to the unit to de-ionizing jets of air may be helpful in this respect, but further cleaning may be necessary.

The paint may be applied in a variety of ways, such as:

(i) by dipping
(ii) by spraying from hand-held guns
(iii) by spraying from fixed nozzles, or from jets reciprocating automatically
(iv) spraying by robot
(v) by electrostatic deposition.

Each approach offers both advantages and disadvantages. Dipping may not be suitable for certain configurations, and with this method it can be difficult to obtain even distribution. The human operator, with a hand-held gun, can see defects and respond almost immediately to control the level of spraying in any particular area—but consistent standards are more difficult to maintain (there is a possibility that quality will suffer if an operator is tired). Fixed nozzles and reciprocating units should be considered only for painting flat panels. For complex shapes, especially, it is desirable to consider spraying by robot—the reach can be greater and movements can be repeated in a more regular pattern than with a human operator; control of the transfer of paint also can be more precise.

In all such work the concentrations of volatile organic compounds should be kept within the limits prescribed (in Britain, in the relevant publications of the Health and Safety Executive—such as 'Guidance Note EH 40 Occupational Exposure Limits 1993'—and also the Environmental Protection Act, 1990), but the rapid movement of air through a painting area can remove paint before even it reaches the pieces; for this reason, air movement at speeds in the range 0.3 to 0.5 m s^{-1} is specified. This is relevant because the limits for particulates and volatile organic compounds are specified in the Environmental Protection Act as dilution ratios and therefore a greater dilution should be sought by increasing the sizes of booths and ovens rather than by increasing air speed. Airborne concentrations of volatile organic compounds *etc.* can be brought down by improving the efficiency of transfer of paint—by using high-volume, low-pressure spray guns, robot controls, and so forth. Transfer efficiency will be influenced also by the configuration of the product and will be better with parts with a high ratio of surface to edge than for small parts with holes, recesses, or slots in them which have greater proportions of over-spray. These efficiencies can vary with the configuration between extremes ranging from 70% to 30%—and this must be kept in mind in any monitoring for compliance with the regulations. In this regard, another significant consideration is the different speeds at which different solvent systems evaporate: fast-evaporating solvents will give a large amount of solvent vapour in the booth, whereas those evaporating more slowly may be lost after the booth, or in the oven.

It has been suggested already that if it can be justified by the volume and type of work, an automated or semi-automated plant is to be preferred; it will be capable of being controlled more easily, safer for the operators, and can be kept clean with less difficulty. With other units, access should be limited so far as practicable—so that operators are not exposed more than is necessary, and with a view to maintaining high standards of cleanliness. Appropriate protective clothing (as lint-free as possible) should always be worn, and a regular cleaning programme should be an integral part of routine working. The design and details of arrangements should include the provision of space sufficient for satisfactory inspection of the articles, for mounting and de-mounting, masking, and other work necessary—so that operations of this nature may be carried on without damage to the components or to paint films.

In Conclusion

As with many other branches of technology, the painting of plastics is a field in which changes and improvements still are being brought about: there can be little doubt that as time goes on the application will increase in importance, with paint systems and methods of application becoming more sophisticated and effective.

CHAPTER 13

Surface Treatments for Plastic Films and Containers

P. B. SHERMAN and M. P. GARRARD

Introduction

With some plastics, particularly the polyolefins, full exploitation of the markets for films and containers required the development of special techniques to prepare the surfaces for decorating and printing.

With regenerated cellulose films, polystyrene, and the vinyls printing presented no serious difficulties, as the inks available at the time either adhered readily to these plastics or could be modified easily so that they did. However, with polyethylene and polypropylene this was not the case: none of the printing inks that was available in the early days of the development of these materials would key firmly to their surfaces and, in consequence, printed matter could be rubbed, flaked, or scraped off quite easily. This was highly unsatisfactory and it was necessary to research techniques to permit the printing, laminating, and coating of polyolefins with good surface adhesion at high rates of output.

The primary reason for poor results with polyethylene was the chemically inert nature of the material; the carbon–hydrogen bonding of the molecules offered no linkages for physico-chemical adhesion by molecules of ink. It was realized very soon that some modification of the surface would be necessary in order to permit adhesion by printing inks. The most obvious approach was oxidation of the surface by some means, and the breaking down of the non-polar carbon–hydrogen bonds to form amine or carbonyl groups which would give more polar and therefore more wettable surfaces.

The main purpose of this chapter is to describe methods of modifying polyolefin surfaces in such a way as to render them printable, and also to describe the tests that are available for determining the effects of treatment and hence the standards of ink adhesion that may be anticipated. (The early part of the chapter is based largely upon the review by Gray in the first edition of this book.) The treatment and printing of materials other than polyolefins are not discussed, since the problems are mainly those of ink formulation and other needs for surface modification.

In the early 'fifties several techniques were developed to give ink adhesion to low-density polyethylene, and even today these basic techniques are still the principal methods employed, whether the material be in the form of film, blow mouldings, or injection mouldings. These methods include chlorination, chemical oxidation (using strong oxidizing agents such as sulphuric acid, dichromates, *etc.*), hot air, gas-fed flames, or electrical discharge treatment.

When other polyolefins (such as high-density polyethylene, polypropylene, and the ethylene copolymers) were introduced into commercial use they also were found to be unreceptive to inks and adhesives, but it became apparent that many of the techniques that were in use for treating low-density polyethylene could be applied also to these newer materials.

A Survey of Methods of Treating Polyolefin Surfaces

(i) Chemical Methods

The chlorination process[1] was the first to be used in industry.[2] It consisted of exposing the surface of the polyethylene film to chlorine gas in the presence of light, excess chlorine and hydrogen gas being removed subsequently by passing the film through a chamber in which fresh air was circulated. Because of the hazardous nature of the gases involved and for other technical and economic reasons the process became of little commercial interest.

Berry, Rose, and Bruce[3] found that the surface of polyethylene film could be made more receptive to printing inks simply by exposing it to ultra-violet light.

Wolinski[4] proposed a method by which polyethylene film could be treated as it emerged hot from the extruder: in this process, the film passed after extrusion through a chamber in which it was exposed to the action of ozone and of an accelerator, preferably in the presence of ultra-violet light.

It was found that the application of certain strong oxidizing solutions would make polyolefins receptive to inks and adhesives, and Horton[5] suggested the use of a strong sulphuric acid–dichromate solution. Later, Ziccarelli[6] described a similar process in which a moulding could be treated by immersion in a sulphuric acid–potassium chromate solution.

Bruce[7] drew attention to certain drawbacks in this, since in some instances when washing off the acid sufficient heat would be generated to distort the moulding. With a view to avoiding the problems of using solutions containing sulphuric acid Bruce proposed as a substitute an aqueous solution of chromium trioxide. Smith[8] suggested mixing chromium trioxide with concentrated sulphuric acid and after heating, to dissolve the trioxide, adding the mixture to water. (Obviously, particular care would be necessary in the use of most if not all of these processes.)

Following any of the methods of acid etching, the treated mouldings were rinsed with water and allowed to dry before being printed or otherwise decorated. While such methods are not of general interest now they could be used on occasion for preparing polyolefins for vacuum metallizing or for other coating processes. The principal advantage of such techniques is that the treatment is homogeneous, almost irrespective of the geometry of the moulding. The main disadvantages are the hazardous nature of the chemicals employed and the needs for special vessels and rinsing facilities.

Ziccarelli[6] described a solvent method of treatment in which the moulding was immersed in a hot solvent—such as perchloroethylene, trichlorethylene, or toluene. While a chemical oxidation method alters the chemical nature of the surface of a component, solvent treatment alters it physically. Application of the solvent has the effect of swelling the surface layer, giving the component a roughened finish similar to that obtained by sand blasting. A surface of this nature provides a good key for a wide range of coatings.

The main disadvantages of the solvent method are that the treatment is effective only for about 48 hours, and also that it can be destroyed by friction on the surface. In addition, the immersion of the components in hot solvent may result in their deformation.

Sand blasting can be used to prepare polyolefins to receive certain coatings and Ziccarelli[6] discussed this, placing emphasis on the importance of selecting grit of the correct size. In general terms, the main disadvantages of sand blasting are associated with the extra time required for carrying out a separate operation, including the need for complete removal of blasting medium.

(ii) Flame or Hot-Gas Treatment

One of the most important methods of preparing the surfaces is flash heating by flame or hot gas. The first technique to be used commercially was Kreidl's heat differential process, and various patents are involved[9] (see Table XX).

In essence, the process involves the application of a flame or hot gas plasma to the outer surface of a polyethylene film while the film is being passed over a chilled metal roll. (The surface also may be heated by other means, such as infra-red radiation.) Because of the contact with the metal roll the greater part of the thickness of the web remains at a much lower temperature than the treated surface.

Kritchever[10] also described methods in which the surface of polyethylene was made receptive to printing inks and coatings by direct application of plasma flame.

More recently, advances in the technology of control systems, growing demand for better performance, and new uses, together have inspired a resurgence in the popularity of gas flame plasma treatment.

(iii) Electrical Methods

Various electrical methods of treatment were investigated and early work was reported by Rossman,[11] who described the use of:

TABLE XX *Patents for treatment processes*

Process	*Patent number*	*Described by*	*Current use*
Chlorination	US 2 502 841 Brit. 581 717	Henderson Myles and Whittaker	Not used but is part of Alkor-werke laminating process
Ultra-violet radiation	Brit. 723 631	Berry, Rose and Bruce	Not used
Ozone, with ultra-violet radiation	US 2 715 075	Wolinski	Not used
Sulphuric acid–dichromates–chromic acid	US 2 668 134 US 2 886 471	Horton Bruce	Decorating polyethylene or polypropylene mouldings (for example, shoe heels)
Heat treatment:			
Hot air	US 2 632 921 US 2 704 382 US 2 746 084 Brit. 704 665	Kreidl Kreidl Kreidl Kreidl	
Gas flame	US 2 648 097 US 2 683 894 Brit. 718 715	Kritchever Kritchever Kritchever	Used widely for treating bottles
Electrical discharge	US 3 018 189 Brit. 715 914 Brit. 722 875 Brit. 771 234 US 2 859 480 US 2 881 420	Traver Traver Traver Traver Berthold Berthold	Used widely for treating film

(i) a glow discharge method at reduced pressure
(ii) high voltage discharge treatment from a Tesla coil.

Today, corona discharge is the electrical method used most widely and is one of the most popular for preparing the surfaces of polyolefin films. It relies on establishing a corona discharge between an electrode and an earthed roller covered by a dielectric layer: the film is passed over the roller and the discharge renders the surface of the film receptive to inks. Sherman and co-authors[12] described the theoretical electrical relationships in the treatment equipment and by means of the Lissajous technique for measuring power demonstrated how much true power was dissipated in the material under treatment. Table XX lists some relevant patents.

A further electrical method, the use of which is restricted to film (and which indeed is little used at present) is the electro-contact treatment process proposed by Rothacker[13] and described by Smith.[8] In this, the film is passed over a metal roll and contacted by a number of free-rolling electrodes. A direct current potential difference is applied between the roll and the electrodes, and an a.c. voltage superimposed over the d.c. The flow of electrons thus induced brings about changes in the surface of the film and improved receptivity to ink.

Two other electrical methods described broadly as 'plasma treatment' can be employed for mouldings and other three-dimensional objects—the 'suppressed spark'[8] and the 'glow discharge'. In the former, the moulding to be treated is placed in a chamber the top and bottom of which are made from parallel metal plates and are separated by thick slabs of polyethylene. When a high voltage is applied across the electrodes (the metal plates) the polyethylene suppresses or extinguishes the spark, and presumably the surface of the moulding is modified by being bombarded by positive and negative ions and electrons. In the glow discharge process, moulded plastic bottles are placed on metal pegs in an evacuated chamber and potential difference applied between the pegs and another electrode also placed in the chamber. The resulting glow discharge modifies the surfaces of the blow mouldings.

Table XX provides a summary of the various methods but today the predominant technique for treating polyethylene film is electrical discharge, and for bottles and containers it is gas flaming. The other techniques are used only for specialized purposes. Both the main techniques will be considered in more detail later.

So far, emphasis has been placed on the treatment of polyolefins to improve printability, but the same treatments affect also other processes. As an example, while an untreated polyethylene surface will not retain an adhesive the treated and modified surfaces are receptive to adhesives. A field of increasing importance for packaging is the production of laminates of film with other materials and, as examples, suitably pre-treated polyethylene film can be bonded securely to substrates such as cellulose or polyester films and aluminium foil, using standard adhesives, whereas this cannot be achieved without pre-treatment.

Polyolefins can be coated with other materials, such as poly(vinylidene chloride), in order to reduce the permeability by gases of the walls of blow mouldings or film (thus extending the shelf-lives of products contained in them); however, barrier coatings of this nature cannot be applied satisfactorily to the untreated surfaces.

Again, many cartons used for the packing of food products comprise paper laminated with polyethylene, the bonds being by means of simple adhesives which are effective only if the plastic has been treated suitably in advance.

The two main fields of treatment will be considered now in more detail—first, polyethylene film, and second, polyethylene blow-moulded containers—two applications which account for a large proportion of the total usage of this material throughout the world.

The Treatment of Polyethylene Film

In most countries between 50% and 70% of the polyethylene film produced is sold in printed form, either 'on-the-reel' or as bags. The preparative treatment used most frequently in this type of work is electrical discharge and a varied range of units for the purpose is available commercially—differing in design but

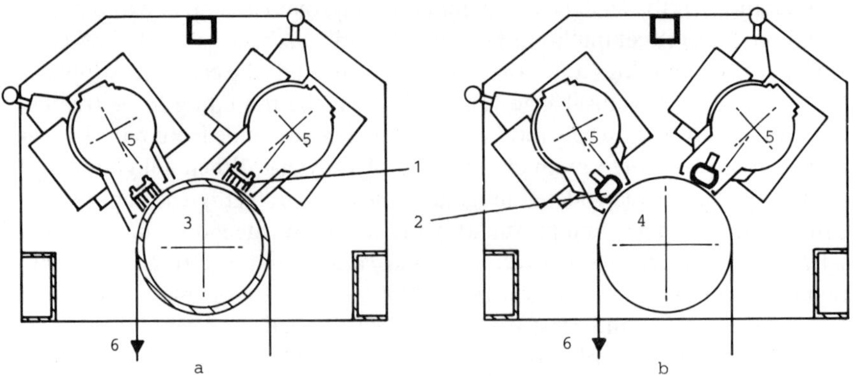

Figure 80 *Types of treater for applying corona discharge:* (a) *for non-conductive substrates, with an earthed base roll covered with a dielectric;* (b) *for material of all types, with a bare, earthed base roll and the dielectric on the discharge electrode The main components are:* (1) *Multi-fin electrode;* (2) *Electrode tube in ceramic or quartz;* (3) *Earthed base roll covered with dielectric;* (4) *Bare, earthed base roll;* (5) *Housing for the electrode assembly, enabling the extraction of ozone and cooling of the electrode;* (6) *Path of material being treated*

not in the basic principle. Some of the better-known systems in use in Britain at present are Ahlbrandt, Sherman, and Vetaphone.

Figure 80 provides a schematic diagram illustrating the principle. The equipment consists essentially of a high-frequency spark oscillator, operating at frequencies up to 40 kHz, with corrosion-proof electrodes in conjunction with a treater roll. Either the base roll or the discharge electrode is covered with a dielectric material that constitutes an electrical capacitance. Combined with the secondary induction of the oscillator circuit, these components form a tuned circuit. During each cycle, when the voltage reaches a peak, the air between the electrodes and the treater roll becomes conductive, breakdown in the form of a spark occurs, and there is a flow of ionizing current accompanied by the corona discharge.

The operating frequencies are determined by the capacitance of the electrode dielectric system; in general multiple electrodes require a lower frequency. If possible it is preferable to use more than one electrode—as this gives a more uniform discharge, more even treatment and a lower gap current—enabling the dissipation of more power and so lengthening the lives of electrodes and dielectric.

The efficacy of treatment is a function of the time of exposure to the corona, and because of this, especially with higher operating speeds, multiple electrodes are preferred. Higher speeds require also, of course, greater inputs of power. The main variables can be summarized:

(i) input of power: the higher this is, the greater the energy applied in treatment

(ii) linear speed of movement of the film: speeds of up to 750 m min^{-1} are feasible, but normal rates in practice are in the range 100 to 300 m min^{-1}

(iii) width of the electrode system: this is different for each width of film, the typical range being 1 to 9 m
(iv) number of electrodes
(v) air gap: normally, gaps of between 1 and 2 mm are used
(vi) oscillator frequency.

The dielectric layer has the effect of spreading the discharge evenly along the surface of the electrode and prevents its concentration at points at which electrode and treater roll might be close together. Materials commonly used as dielectric media include ceramics (and heat-resisting glass), glass-fibre, Hypalon*, and silicone rubber.

Under-Treatment and Over-Treatment

The only effect of under-treatment of the film is inadequate adhesion of ink—which of course is entirely unsatisfactory: the remedy is to increase the level of treatment.

The effects of over-treatment are more complex: even though the adhesion of ink may be excellent, over-treatment may alter the heat-sealing characteristics of film to the extent that weak seals are obtained; there is a tendency also for such film to adhere while on the reel, resulting in difficulties when unreeling.

Poor heat-sealing may be attributed to cross-linking caused by the corona discharge: as a result, the over-oxidized surfaces of the polyethylene do not flow and fuse readily.

Another common fault often aggravated by over-treatment is 'reverse-side' treatment, which occurs when there is a localized corona discharge on the reverse surface of the film. This might happen with tubular film when, during manufacture, air is trapped along folds and gussets at the edges (especially in thicker film): discharges can arise in the vicinity of the edge folds, as in Figure 81, and poor heat sealing results.

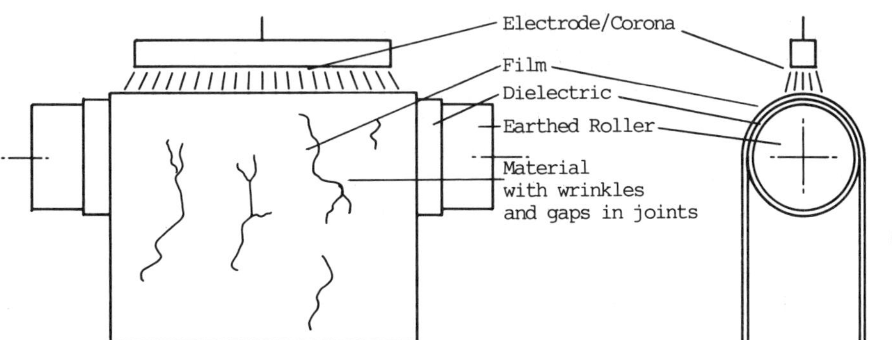

Figure 81 *Diagrammatic representation of the importance of correct flow of film for treatment, without wrinkles or entrapment of air through other causes*

* A trade mark of Du Pont for synthetic rubber.

Reverse-side treatment may be caused if the dielectric film is not wrapping the treater roll smoothly and becomes wrinkled; also at poor butt joints of the dielectric. In turn, unwanted treatment can cause the ink from the printed surface to adhere to the under-side when film is re-wound or collected in batches of bags, spoiling the work and rendering the printed material unsuitable for use. Such troubles are noticeable in particular with film of high surface gloss.

Problems of this nature may be reduced or eliminated by ensuring that during treatment the substrate film is flat against the base roll—which can be achieved by good control of the web, and by means of a 'nip' roll located immediately before the discharge electrode.

Methods of Assessing Treatment

Bearing in mind the difficulties that may arise it is important to have available reliable and easily understood tests for assessing the level of treatment. Since an effect of treatment is to increase the wettability of the surface the most simple test is to apply a liquid of low surface tension, in order to compare treated with untreated film. If the liquid spreads uniformly the surface may be judged suitable for printing; if, on the other hand, it coalesces into discrete droplets, the wettability is insufficient. There are available for such tests commercial surface tension fluids, which simply are applied to the polyolefin and their behaviour noted.

McLaughlin[14] described a technique in which a sample of treated film was held on a platform that could be rotated on a vertical axis. A drop of water from a special burette was placed on the surface and the platform then was inclined at a uniform rate until the drop just began to slide: the angle of inclination at which sliding began was a measure of the extent of treatment. Greater angles of inclination were required for treated than for untreated surfaces since (because of increased wettability) the drop of water had with the former a larger area of contact. Allan[15] gave another method in which a drop of water was applied to the treated film, illuminated, magnified (thirty times), and its image projected on a screen. The contact angle of the drop was measured by drawing tangents. Allan found that adhesion could be obtained when the level of treatment was sufficient to reduce the contact angle to 80°.

In a further method the wettability of a surface was determined by changing the composition of the test liquid (thus altering its surface tension), until uniform spreading across the film was achieved. Mixtures based on ASTM[16] standard concentrations of formamide and 2-ethoxyethanol were employed.

In carrying out all such tests it is essential to follow the same procedures precisely every time, maintaining comparable conditions, and with scrupulous cleanliness of samples, equipment, and solutions.

In practice the methods of test used most frequently are those that measure adhesion. Wechsberg and Webber[17] noted that some pressure-sensitive adhesive tapes adhered more strongly to treated rather than untreated film: when tape was pressed in the same manner on samples of film and peeled off under controlled conditions with a tensometer, the force required to separate the tape

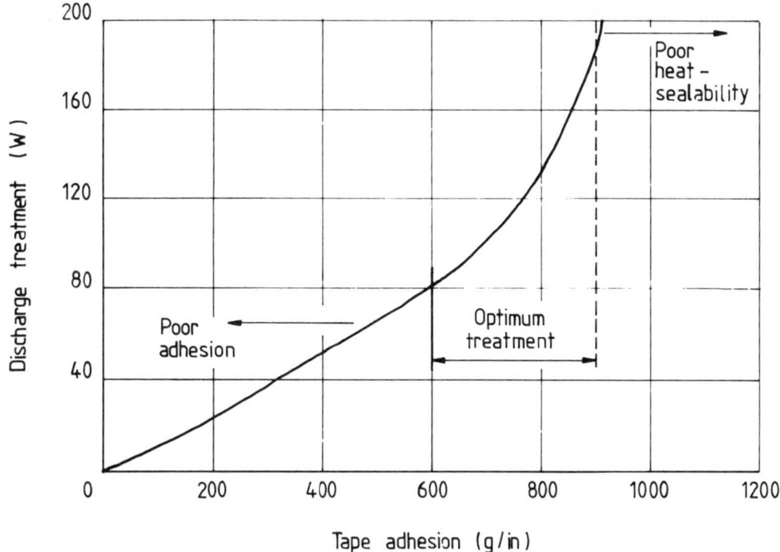

Figure 82 *Graph of results for tests of treatment, with adhesive tape as the indicator*

was a measure of the extent of treatment. Typical results of such tests range from about 200 g inch^{-1} (7.9 kg m^{-1}) for untreated surfaces to about 1100 g inch^{-1} (43.3 kg m^{-1}) for the treated. (Figure 82 presents in graph form the electrical energy applied in relation to the forces needed to separate the tape.) This is a useful procedure for control of quality, since the levels of treatment can be specified numerically in terms of a permissible range of peel strengths and there is a basis for rejecting film shown as 'under-' or 'over-' treated.

A simpler test that is used widely also involves adhesive tape. In this, a sample of the treated film is coated with a controlled thickness of a standard proofing ink and the coating allowed to dry completely. A specified pressure-sensitive adhesive tape then is applied under controlled conditions and stripped away: the percentage of the ink removed by the tape is assessed by eye. The test has the merit of being easy to perform and when experience has been acquired of assessing the extent of removal of ink it can be remarkably accurate.

Another test is based on the change in the strength of the heat seal between two samples of film as a function of the degree of treatment.[18] As treatment is increased, the strength of a heat seal between the two films diminishes. In the test, untreated film first is sealed under such conditions that the seal can just be pulled apart, and the same conditions then are used to seal treated film (treated side to treated side). The peel force required to break the seal provides a measure of the degree of treatment—highly treated films giving low peel strength.

Effects of Additives

Most polyethylene films contain additives with various purposes, some examples of which are:

(i) slip agents to reduce the tendency of inner surfaces of tubular film to cling together during manufacture ('blocking'), and generally to increase slip
(ii) anti-static agents to limit the propensity of film to acquire electrostatic charges while being handled on converting machinery, and to reduce the attraction of dust when on display in shops
(iii) anti-oxidants, to delay degradation of the polymer
(iv) colours
(v) fillers.

Some additives can interfere with the effects of others—like titanium dioxide diminishing the usefulness of slip agents—and many of them influence the efficiency of corona discharge treatment. The results vary with the additives concerned and with their concentration.[19] The presence of additives on the surface will screen the film against the effects of the discharge, while difficulties may be caused after treatment when additives migrate to the surface (it is intended that they should so do) and impair in consequence the keying of ink.

Problems such as these can be corrected to some extent by keeping to a minimum the periods of time between extrusion, treating, and printing; the longer the film is left between any of these stages the longer the period in which additives can migrate and begin to interfere.

Some additives are more soluble in printing inks than others: if an additive is highly soluble in the ink used it can have the effect of making the surface more accessible (as an example, an additive that is very soluble in isopropanol will have the least effect on treated film when printing with ink based on isopropanol).

It will be apparent therefore that the presence of one or more additives in a composition is not necessarily deleterious. (It depends on the additives, and their properties.) In a comparison, films containing different additives that showed markedly different responses to ink were wiped with a solvent and subsequently the adhesion of print was tested; the results obtained in these instances were approximately the same, suggesting that even interfering substances can be removed if need be. However, with some additives, should time be allowed to elapse between treatment and printing, it might be necessary to carry out some secondary preparation of the film.

Effects of Treatment on the Properties of Additives

(i) Slip Agents

It was mentioned earlier that the degree of treatment has the effect of increasing the tendency for blocking to occur: since this can hinder seriously the processing of film into bags and other products great care must be taken to avoid over-treatment.

Figure 83 illustrates the effect of treatment on the usefulness of the additives concerned. When coefficients of friction were measured on the two surfaces of polyethylene film containing slip agents throughout the results were consistently higher for the treated surface, regardless of the concentration of slip agent.

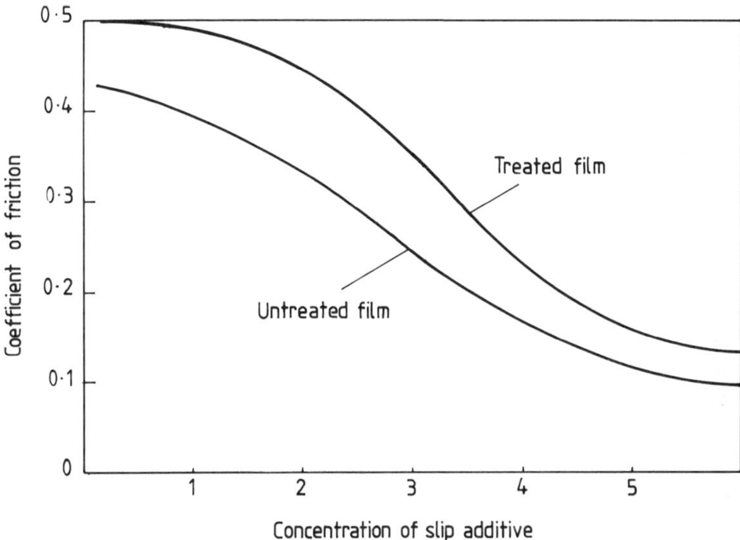

Figure 83 *Comparison of coefficients of friction of the surfaces of treated and untreated films at different concentrations of slip additive*

When such additives are at low levels the effect of treatment can be particularly noticeable.

(ii) Anti-Static Agents

These additives perform by migrating to the surface of the film, reducing the surface resistivity, and enabling a more rapid dissipation of static electrical charges. Many of them work best in conditions of high humidity, presumably through the absorption of moisture and a consequent increase in conductivity at the surface. The efficiency of such an additive may be measured by the rapidity with which static charges decay.

Langdon[19] reported measurements of electrostatic properties of film: a 5 kV d.c. voltage was applied and it was found that when film was treated by corona discharge the rate of decay of the applied static charge was increased considerably; as treatment was increased, so was the rate of decay.

The Treatment of Blow Mouldings

Various methods of treatment are possible with blow mouldings but gas flaming is the technique employed most usually, preparatory to printing or before the application of labels. Flaming has the advantage of versatility: it is possible to handle components of various shapes and sizes without a need for elaborate jets, and also it is economical. The electrical methods are more difficult to apply, mainly because of the need for electrode systems that could cope at speed with products in a range of configurations.

In a typical unit the mouldings are supported on jigs or mandrels and brought into contact briefly with an oxidizing flame in such a way that they rotate during passage and their entire surfaces (including necks and shoulders) are exposed to the flame. Alternatively, the arrangement of the burners may be such as to cover the entire surface. Exposures of one or two seconds usually are adequate. Burners are fed with controlled mixtures of compressed air and industrial gas (from town mains, tank, or bottle supplies), and setting up a unit must be done with care to ensure consistent running and uniformity of results. Current equipment may be fitted with programmable controls, which assist in maintaining standards of quality.

There has been discussion of the technology of flaming but the literature of flaming blow mouldings before printing is less extensive. McLaughlin[20] considered the thermodynamics and kinetics of combustion in flaming polyolefin components, concluding that to obtain the most effective treatment an oxidizing flame was necessary: it could be produced by exceeding slightly the ratio of air to gas that was needed to bring about complete combustion of all the gases involved. On the other hand, a large variation from the stoichiometric ratio could reduce the temperature of the flame, and hence the effectiveness of treatment. Ziccarelli[6] stated that a hot oxidizing flame was needed (1100 to 2700 °C), and described various burner nozzles.

The conclusions of a number of investigations were summarized by Buchel[21] and may be shown thus:

(1) treatment was effective only if in the presence of air and oxygen
(2) treatment in stages, using flames at lower temperatures for longer periods, of time, was effective
(3) it was possible to store treated bottles for considerable periods without affecting printing seriously; however, when the plastic contained additives such as anti-static agents there was after storage a marked deterioration in printability
(4) excessive treatment would interfere with adhesion—presumably because in such cases the layer of oxidized material on the surface was thick enough to impede the keying of ink
(5) running costs for gas flaming were small; an average consumption of natural gas for the purpose would be 15 to 30 litres per 100 square metres of surface.

The same author[22] reported that the treatment reduced sharply the permeability of the containers by gases, and that it increased slightly the tensile strengths of specimens of polyethylene.

Hurst and Schanzle[23] concluded that for the best results:

(1) the ratio of air to gas should exceed the stoichiometric ratio by about 5% to 10%
(2) the length of the inner cone of flame should be in the range 0.25 to 0.75 inch (6.35 to 19.05 mm)
(3) the distance from the tip of the inner cone to the surface of the moulding should be about 0.25 inch (6.35 mm), although results were obtained with distances up to 1.0 inch (25.4 mm)

(4) the surface should be in contact with the flame for about one tenth of a second.

Leeds[24] favoured the use of gases such as ethane, methane, and propane in pure form in order to ensure consistency, stressing the importance of using the correct ratio between air and gas (for propane, ratios preferred were from about 27:1 to 32:1).

In the patent literature, Kreidl[9] suggested a method in which bottles were passed through hot air at 400 °C, resulting in the outer surfaces being heated to about 150 °C while the temperature inside remained in the range 50 to 70 °C—the interiors of the bottles being cooled by a stream of cold air. In a later patent[9] he proposed disposing an annular gas burner concentrically with the bottle, so that the flame was directed towards the surface—the bottle being mounted on platform oscillating on a vertical axis. Kritchever[10] noted that bottles could be dropped or passed through an annular gas burner: Grow[25] postulated the use of high temperatures (flames at between 2300 and 2800 °C), these being formed by burning mixtures of butane, ethane or propane with oxygen.

To summarize, experience shows that the most important variables are:

(1) the ratio of air to gas fed to the burner or burners
(2) the period of time over which the surface of the container is exposed to the flame
(3) the distance between the flame and the surface of the container
(4) the nature of the plasma ionization mixture used.

At the outset of any treatment it is essential for good results to establish the best combination of these variables in relation to the products concerned: the most important single factor is the flame, and the ratio of air to gas must be adjusted to give a quiet blue flame with a well-defined pale blue cone. Just as with the electrical treatment of film, over- or under-treatment can give rise to difficulties—and for this reason too it is necessary to establish the parameters and to keep within them.

Assessing the Treatment of Mouldings

As with film, two basic approaches are available for the assessment of treatment of mouldings—tests for adhesion, and for wettability.

One of the simplest methods of assessing such treatment of a polyethylene moulding is to immerse it in clean, cold water, withdraw it and then to observe the behaviour of the water on the surface over a given period of time. If the moulding was treated adequately, so that the surface was fully wettable, the aqueous surface film will be unbroken. Comparisons may be made by recording the times taken for the films of water to break up on bottles having different levels of treatment.

A similar test involves immersing the moulding in an acid fuchsin stain solution, followed by rinsing under tap water, when the degree of uniformity in treatment can be established by a visual examination of the intensity of the stain and by comparison with a standard that has been treated to a known level.

In a further example, the surface may be painted with a solution of nitroethane and blue dye, when untreated areas show coalescent drops and the parts treated uniformly are dyed to the same degree.

Additives in Mouldings

The presence of additives, particularly anti-static additives, can affect the treatment and because of this it is recommended that bottles should be flamed immediately after manufacture, before an additive has had time to consolidate on the surface. Once a bottle has been flamed printing can take place, even some considerable time afterwards. However, if printing is carried out regularly after long periods of storage some care may be necessary over the selection of anti-static agent, since some can give rise to problems at the printing stage.

With certain of these additives the anti-static effect may be enhanced by the flaming, possibly as a result of the improvement in wettability of the surface.

The Treatment of Injection Mouldings

Polyolefin injection mouldings are printed less frequently than films and bottles but requirements for treatment for this purpose do occur, and also for pre-treatment for other purposes (like metallizing or being painted so as to match other components). Several different methods are available and may be used in practice, including:

(i) gas flaming
(ii) application of primers
(iii) dipping
(iv) vapour etching.

Gas flaming may be applied in much the same way as with blown bottles (and subject to similar constraints), while the use of primers is dealt with in Chapter 11 of this book, 'Vacuum Metallizing'. Some further details of dipping and vapour etching are given below:

(i) Dipping

A solution may be prepared by adding concentrated sulphuric acid to water in the ratio 1:40, heating the diluted acid to 60 to 65°C and adding crystals of potassium permanganate until saturation is reached. The moulded articles are immersed in such a liquid for four to five minutes then washed and dried before being metallized or painted.

Another suitable solution is made by dissolving potassium dichromate in water (5 g dichromate to 12.5 ml water) and adding concentrated sulphuric acid (87.5 ml). Immersion in this liquid would be for ten to fifteen minutes at 60°C, after which the mouldings would be washed and dried before passing to the next stage.

As a general comment, all such methods require particular care at all stages to ensure satisfactory working conditions and that the health and safety of employees and members of the public are safeguarded: as examples of what is meant by this—the plant used must be capable of withstanding the substances involved, and the operators required to wear appropriate protective clothing; suitable ventilation and vapour extraction should be available, with safety showers and appropriate disposal of waste.

(ii) Vapour Etching

The surfaces of polypropylene mouldings may be etched effectively by exposure to trichlorethylene vapour under controlled conditions. The temperature of the solvent should be kept below its boiling point and the period of exposure to the vapour determined according to the thickness and other dimensions of the mouldings (it may be in the range from a half second to a few minutes).

Surfaces etched in this way recover quickly and in order to ensure good adhesion the stages of metallizing or painting must be carried out immediately after treatment. Alternatively, the etched surface may be protected by applying a coating of resin immediately after the exposure.

Industrial degreasing tanks can be adapted for vapour etching, subject again to the maintenance of all appropriate safety measures. However, in view of the European programme to eliminate the use of chlorinated solvents within the coming few years the usefulness of the technique could be limited unless sufficiently effective and safe alternative etchants are found.

Some Theoretical Considerations

The methods of treatment have in common the objective of changing the nature of the surfaces so that wettability is achieved and the ability to absorb applied substances (such as inks) much enhanced.

Even so, detailed knowledge of the nature of the changes remains incomplete. So far as corona discharge is concerned, the effect is achieved by a combination of spark and high-frequency ion flow. High temperatures are generated and a considerable amount of heat is forced through the film, raising the temperature of the backing dielectric (which was of the order of 80 °C in some investigations with polypropylene). The profile of temperature through the film and the maximum surface temperature in the zone of the discharge are not known precisely (and would be difficult to measure) but it is likely that, at the surface, temperatures in the region of the melting point of the material are reached (possibly even higher at intervals of milliseconds during the discharges). The surface will be hottest in the areas hit directly by the sparks, so that under ambient conditions the activity of the film in those areas will be closer to that of the molten rather than the solid polymer.

If the discharge treatment is carried out in air the two main reaction products in the zone immediate to the surface will be:

(i) from the silent or glow-type discharge (the corona): ozone
(ii) from the sparks: nitrogen dioxide.

Both are strong oxidizing agents and nitrogen dioxide also can introduce nitro and nitrite groups into hydrocarbons. The extent to which these gases react with the surface depends upon the local concentrations and the temperatures involved.

In an investigation at laboratories of ICI the surface of treated film was scraped mechanically and the scrapings were analysed by infra-red and other spectroscopic techniques: this indicated that nitro groups, nitrate ester groups, and carbonyls were formed in the surface layer.

Toriyama and co-authors[26] exposed polyethylene film to corona discharge for relatively long periods of time then analysed the film directly by means of infra-red spectroscopy. In an air atmosphere they found the same structures but also carboxyl and ozonide groups. In a nitrogen atmosphere none of these was produced but ethylene double-bonds were detected, suggesting that hydrogen was abstracted and cross-linking took place. Aside from the chemical reaction at the surface, it seems likely that even when in air there was cross-linking. The theory was advanced[27] that this played a part in the improved adhesion of ink, since it improved the cohesive strength of the surface in comparison with unmodified polyethylene.[28]

References

1. W. F. Henderson, US Patent 2 502 841; British Patent 581 717.
2. A. Bosini, *Materie Plast.*, 1956, **22**, 9.
3. W. Berry, R. A. Rose and C. R. Bruce, British Patent 723 631.
4. L. E. Wolinski, US Patent 2 715 075; US Patent 2 715 076; US Patent 2 715 077.
5. P. V. Horton, US Patent 2 668 134.
6. J. J. Ziccarelli, *Mod. Plast.*, 1962, **40**(3), 126.
7. C. R. Bruce, US Patent 2 886 471.
8. E. A. Smith, *SPE J.*, 1962, **18**, 2, 157.
9. W. H. Kreidl, US Patent 2 632 921; US Patent 2 704 382; US Patent 2 746 084.
10. M. F. Kritchever, US Patent 2 648 097; US Patent 2 683 894.
11. K. Rossman, *J. Polym. Sci.*, 1956, **19**, 141.
12. P. B. Sherman, D. Clarke and J. Marriott, TAPPI Extrusion Coating Course, 1991.
13. F. N. Rothaker, US Patent 2 864 755.
14. T. F. McLaughlin, *Mod. Packag.*, 1960, **34**, 1, 153.
15. A. J. G. Allan, *J. Polym. Sci.*, 1959, **38**, 297.
16. ASTM 02578-67, American Society for Testing Materials (re-approved 1972).
17. H. E. Wechsberg and J. B. Webber, *Modern Plastics*, 1959, **36**(11), 100.
18. P. B. Sherman, Additive influence on corona treatment, TAPPI Film Extrusion, 1991.
19. S. J. Langdon, *Plastics (London)*, 1964, **29**(8), 43.
20. T. F. McLaughlin, *SPE J.*, 1964, **10**, 20th Antec, Session iv, Paper 3.
21. K. F. Buchel, *Br. Plastics*, 1964, **37**(3), 142.
22. K. F. Buchel, *Adhaesion*, 1966, **10**,. 506.
23. C. W. Hurst and R. E. Schanzle, *Mod. Packag.*, 1966, **40**(2), 163.
24. S. Leeds, *TAPPI*, 1961, **44**(4), 244.
25. H. J. Grow, US Patent 2 795 820.
26. Y. Toriyama, H. Okamoto and M. Knazanchi, *IEE Trans.*, 1967, **Ei-2**.
27. H. Schonhorn and R. H. Hansen, *J. Appl. Polymer. Sci.*, 1968, **12**, 1231.
28. L. K. Sharples, Conference on Printing and Decorating Plastics, The Plastics Institute, Bristol, 1968.

CHAPTER 14

A General Review of Printing Processes for Plastics

J. W. DAVISON

Introduction

It is an understatement to say that plastics are ubiquitous—in one form or another they are always close at hand throughout our lives. The extent of this close contact is growing all the time as the conventional materials of construction—wood, metal and glass—are replaced by new forms of plastic. For example expensive veneers of wood are being replaced with printed laminates, metal is being replaced by plastic in many parts of automobiles, and glass is being superseded as a material for the construction of containers. As well as the above-mentioned alternative outlets for plastics there are many areas where only the unique properties of plastics will meet a desired specification; as an example, many of the packaging materials used in the food industry. Such packaging techniques have allowed the safe storage of produce from one harvest to the next, thus providing a ready source of food throughout the year.

Plastics are defined as those materials that suffer irreversible deformation when subjected to a shear stress; that is, when subjected to pressure they will flow and take up a new shape. However, the term plastic is used in general to define those organic polymeric materials derived from synthetic monomers.

The range of such monomers is being increased all the time as is the number of permutations of co-monomers to give co-polymers. The conditions under which polymerization reactions take place also lead to families of polymers. With the introduction of laminates, composites, and alloys—physical mixtures of different polymers each member contributing a specific property to the assembly—there has been further proliferation. The nature and molecular weight of a polymer determine the method by which it may be formed into a useful product of commerce. A broad range of methods for forming the polymer into a particular shape exists: blow moulding, injection moulding, compression moulding, spinning into threads, casting into sheets, coating from solution, *etc.*

Shapes of plastic articles include sheets, blocks, hollow rectangles, webs, hollow cylinders and ovoids, complex three-dimensional shapes, rods, tubes, and woven threads. The plastics may be rigid or flexible, in the form of a woven thread, expanded sheet, a 'structural sandwich', a multiple dispersion of components, or a coating deposited from solution.

Once formed the plastic article may be polymerized further at a later stage, or cross-linked to meet a particular specification. Often plastics are modified by additives that serve their purpose only at the surface of an article—such as anti-static agents that help prolong the pristine condition of the item in dirty atmospheres.

From the above description of a plastic it can be seen that it may have a complex composition and a complex shape. The complexity of the shape determines what printing processes may be used to decorate the plastic and the composition of the plastic at its surface determines the type of ink with which it may be printed. There are now very few shapes and surfaces that cannot be printed.

As in other fields, technology changes and since the previous edition there have appeared a number of printing processes that at that time were either at an early stage of development, as in ink-jet printing, or were being modified significantly, as in the case of letterpress and cliche printing. There are now three ink-jet printing techniques in commercial operation, 'continuous', 'drop on demand', and 'bubble jet'. Since the advent of the wrap-round photopolymer plate and high-speed presses with anilox keyless ink distribution systems letterpress printing has found new markets. In fact, letterpress and flexography are beginning to look more and more as two variants of the same principle. Cliche printing has found many new markets since the advent of the silicone rubber pad, the photopolymer plate, the laser-ablated plate and the introduction of rotary as opposed to reciprocating mechanics. Cliche printing is known also as 'Pad Printing' (in Britain), 'Tampoprinting' (in Germany), and 'Autogravure' (in USA); however, a suggested generic name is 'tampography'.

All of the printing processes mentioned in the first edition (flexography, gravure, letterpress, silk-screen, *etc.*) have been improved with respect to quality, speed of production and reduction in cost per print as a result of advances in electronics, chemistry, and engineering. Microelectronics has allowed the development of automatic press control and platemaking to the point where the whole process from artwork to printing plate can be computer-controlled. Indeed systems now are available whereby artwork can be designed on a VDU screen and the printed image produced without any of the traditional stages of photography, platemaking, and complicated precisely engineered mechanical processes being involved—such as the four-colour process proof-printing system based on computer-generated artwork, digitized colour separation and computer-driven ink-jet printing. The possibility of your morning newspaper being delivered *via* the video printer beneath your television set has a high probability of coming true. The connection with plastics is that the 'paper' may be a polymer film derived from a formulated composition of polyethylene and polypropylene.

Decoration may not be the only reason for printing a plastic article. The packages for many products are required to carry regulatory information about their contents (this is so particularly in the food and pharmaceutical industries). For plastics components there has been a requirement, particularly from the aerospace industry, for such items to be marked indelibly so that their origin can be traced. This has led in turn to a number of highly specialized printing processes—such as laser-ablation printing, where the printed image is formed by the decomposition and charring of the organic plastic by a focused infra-red laser.

Stock control in many industries, particularly food retailing, is computerized and the printed bar code forms an essential element in the system.

Plastics That Can Be Printed

The broadest classification for plastics is the old 'thermoplastic' and 'thermosetting'. Examples of the former group are polyethylene, polystyrene, and poly(methyl methacrylate); examples of the latter are urea–formaldehyde condensation polymers, powder coatings based on polyesters, epoxy resins, and vulcanized synthetic elastomers.

Another broad classification is based upon the composition of the plastic, which gives four groups:

(i) plastics derived from a single monomer—that is, homopolymers such as polypropylene
(ii) plastics derived from a mixture of monomers—copolymers such as ABS (acrylonitrile–butadiene–styrene)
(iii) plastics filled with other materials—composites such as glass-filled nylon
(iv) plastics laminated together to give assemblies with specific properties—'boil-in-the-bag' food pouches made from polyethylene sealable by heat and heat-resisting poly(ethylene glycol terephthalate).

Plastics can be differentiated further by their degree of mechanical rigidity, giving two broad groups—the hard rigid plastics and the flexible (rubber-like) elastomers.

Yet another method of classification is by their method of polymerization—free radical, ionic, or condensation.

The method by which the plastic is brought to its final shape is also a means of differentiating the various types. Table XXI summarizes the main types of material for which printing is relevant at present.

Additives and Fillers

As noted earlier, a great variety of additives and fillers may be used in plastic compositions. A recent review report about fillers lists common minerals as

TABLE XXI *Common types of plastic that can be printed*

Type	*Unit of composition*
Acrylonitrile–butadiene–styrene	A blend of polymeric materials
Acrylics	Co-polymers of various methacrylates with methyl methacrylate
Ethylene–vinyl acetate	Co-polymer of ethylene and vinyl acetate
Polyacrylonitrile	$\left[-CH(CN)-CH_2- \right]_n$
Polycarbonate	$\left[-O-C_6H_4-C(CH_3)_2-C_6H_4-O-C(=O)- \right]_n$
Polyester	Prepared by condensation of ethylene glycol and terephthalic acid
Polyethylene available as : low density ('LDPE'), high density ('HDPE'), linear low density ('LLDPE'), *etc.*	$\left[-CH_2-CH_2- \right]_n$
Poly(methyl methacrylate)	$\left[-C(CH_3)(COOCH_3)-CH_2- \right]_n$ (printed as CH, with $O{=}C{-}OCH_3$ above and CH_3 below)
Poly(oxy methylene)	
Polypropylene	
Polystyrene	$\left[-CH(C_6H_5)-CH_2- \right]_n$
Polytetrafluoroethylene	$\left[-CF_2-CF_2- \right]_n$
Poly(vinyl acetate)	$\left[-CH(OCOCH_3)-CH_2- \right]_n$

Poly(vinyl alcohol)	$\left[\mathrm{CH(OH)-CH_2}\right]_n$
Poly(vinyl chloride)	$\left[\mathrm{CH(Cl)-CH_2}\right]_n$
Poly(vinyl fluoride)	$\left[\mathrm{CH(F)-CH_2}\right]_n$
Poly(vinylidene chloride)	$\left[\mathrm{C(Cl)_2-CH_2}\right]_n$
Styrene–acrylonitrile	
Styrene–butadiene	$\left[\mathrm{CH(C_6H_5)-CH_2-CH_2-CH{=}CH-CH_2}\right]_n$

calcium carbonate, clays/kaolin, mica, silica and silicates, talc, and wollastonite. Calcium carbonate is used to fill PVC, PE, PP, polyester, and EPDM, while talc is used in PP, PVC, PA, PE, and rubber. Wollastonite finds application in under-bonnet components for the automotive industry. Carbon, in forms varying from synthetic graphite to treated petroleum coke, also is employed. Carbon fibre has been used to reinforce plastics to produce extremely strong, light-weight structures with particular applications in sports equipment.

Other additives include colouring matter and flame retardants. Colour may be imparted to plastics using dyes or pigments. Those chosen frequently are of high specification because of their need to resist high temperatures, long exposure to daylight, and other harsh processing and environmental conditions.

Finally, plastics may contain small quantities of other compounds—including residual 'catalysts' and their decomposition products, thermal stabilizers, processing aids (like dispersion agents for the efficient manufacture of masterbatches, to improve cycle times in moulding, and release agents), anti-oxidants, and light stabilizers.

Laminated Polymer Assemblies

Complex assemblies of polymers also have been developed significantly during the twenty years since the first edition. Lamination and co-extrusion may be considered together since the result of each process is very similar—that is, a layered assembly of dissimilar plastics. Lamination involves the pressing together of separate films under pressure and/or heat, with or without an intervening adhesive coating. Any one of the individual films can be preprinted and the print is encapsulated. In co-extrusion alone, only the outer layers of the assembly can be printed, so that the print often is on the exposed wear layer.

Co-extruded films include combinations of LDPE/LLDPE, LDPE/MDPE/LDPE, and LDPE/LLDPE/LDPE (polyethylene of different densities or structural arrangements)—the last combination being used as a clear stretch-wrap film for food products. Plastic-coated boards are becoming a dominant substrate in the manufacture of cartons for the food industry. The cellulose-based board is coated with polyethylene which provides a barrier capable of performing a number of functions—as examples, protecting the contents, protecting the board from disintegration when being frozen and thawed, protecting the board against attack by the contents, containing odours, and providing a water-proof barrier.

As an example of another form, co-extruded blow mouldings have been introduced to the market as containers for liquid detergents. The process used is known as 'sandwich blow moulding' and provides a means by which recycled high density polyethylene can be incorporated again into a useful form by being employed between two layers of virgin material.

Extruded films for flexible packaging often are complex assemblies of laminates designed to meet what often are exacting specifications. Liquids may be packed safely in cartons made from board on which has been extruded an outer layer of low-density polyethylene and a series of inner layers comprising adhesive, poly(vinyl alcohol), adhesive and, again, low-density polyethylene. Foods, detergents, and pharmaceuticals may be packed and dispensed from tubes and pouches made from laminations assembled from layers of low-density white polythene, adhesive, poly(vinyl alcohol) or aluminium foil, adhesive, and finally clear low-density polyethylene. A film for the outer wrapping of many products comprises a layered structure of cast films of linear low-density polyethylene. Extrusion lamination techniques are used in the manufacture of deep-draw thermoformed containers for the cups of vending machines.

In summary, today a multiplicity of different types of plastic is required to be printed, many in compositions with a number of different additives and often in rather complex presentations (such as co-extrusions and laminates). This means in turn that there is need for a variety of approaches to the problems of printing at the speeds required and to obtain satisfactory results—not only in terms of the choice of technique but also the design of machinery capable of handling the pieces of work and of formulations of inks with the correct qualities of adhesion, drying, and colour.

Other Factors to be Considered

Other factors to be considered include the effect on disposal of the plastic as a consequence of its being printed. Packaging is an important source of waste for disposal and not only the nature of the plastic but also that of the printing ink used must be assessed with regard to air, soil, and water quality after disposal.

For a printing ink to print it must wet the surface being printed. If the surface is a complex mixture of components successful printing will depend on the species present—the bulk of the material having little effect on its printability.

For this reason much attention is being applied to surface science and to the interaction of coating materials with intended substrates. Some plastics—particularly polyethylene, polypropylene—and the silicone rubbers, cannot be covered with high-performance coatings without their surfaces being pre-treated.

Cleaning

Good printability of plastics depends upon the surfaces to be printed being clean—that is, free from any contaminants. Cleaning may be necessary to remove particulate matter—such as dust—or liquid contaminants—such as mould-release agents. Washing with solvent and ultrasonic degreasing are used in the electronics industry but rarely in packaging. In the printing of packaging film and containers, mechanical brushes coupled with elimination of electrostatic charges are used to remove surface dust and static electricity. Plasmas can be used also to clean contaminated surfaces by reaction with the contaminant and formation of volatile compounds that evaporate from the surface.

Roughening

Mechanical key of the print to the substrate can be increased by solvent treatment, sand-blasting, and chemical etching. Solvent treatment can clean and roughen the surface of the plastic but care is required to ensure that the ink is not printed on a detachable layer of swollen plastic. Sand-blasting can clean, roughen, and alter the chemical nature of the surface being abraded. Chemical etching cleans and probably roughens the surface as well as introducing new chemical species that allow ink to adhere to a greater degree. Etchants comprising chromic and sulphuric acids can be used on plastics such as polyethylene, polypropylene, acrylonitrile–butadiene–styrene (ABS), poly(phenylene oxide), poly(oxymethylene), polyether, and polystyrene. In the case of ABS the etchant dissolves away the butadiene; the amorphous areas of polyethylene and polypropylene can be removed selectively if chemical etching is preceded by solvent swelling. Metallic sodium etchants are used to replace the fluorine atoms in polytetrafluoroethylene with carboxyl or carbonyl groups. Nylon may be etched with a solution of phenol or iodine–potassium iodide. Polyethylene terephthalate may be etched with a 20% solution of sodium hydroxide, which probably hydrolyses the surface, increasing its polarity and reactivity while reducing the chain length—and as a consequence increasing its solubility and the attack of solvents in the printing ink. To improve coatability, the poly(oxymethylene) polymers may be treated by Du Pont's Satinizing process.

Flame Pre-treatment

Flame pre-treatment is the most common method of modifying the surface of mechanically strong articles made from polyethylene and polypropylene—for example polythene bottles and containers (see Chapter 13). A flame comprising air and gas contains excited species of oxygen and nitrogen which can oxidize the surfaces of polyolefins to give a new polar structure. The effect is to produce a modified surface which has a surface tension higher than the original polymer and higher than most printing inks. As an example, the polymer may have a surface tension of approximately 20 dyn cm^{-1}, the printing ink 35 dyn cm^{-1}, and the newly pre-treated surface of polyethylene a surface tension of 45 dyn cm^{-1}. Flame pre-treatment is reported to be used with plastics such as acetal and polyester as well as polyolefins.

Corona Discharge

The preferred method for the pre-treatment of webs containing polyethylene or polypropylene is corona discharge (see Chapter 13). The discharge produces a mixture of excited species including electrons, protons, excited atoms, and ions that are capable of breaking the carbon–carbon and carbon–hydrogen bonds found in many plastics. Rupture results in the formation of highly reactive free radicals which react with the oxygen and nitrogen species in the plasma to give polar surfaces containing hydroxyl, carbonyl, and carboxyl groups, and various nitrogen moieties. It is reported that corona discharge is effective for improving the adhesion of coatings to polyolefins, polyfluorocarbons, polyesters, poly(vinyl chloride), silicones, nylon, and polycarbonate.

Plasma Pre-treatment

A plasma has been described as the fourth state of matter after solids, liquids, and gases. Essentially it is a mixture of wholly or partially ionized gas or gases produced by subjecting the gas to a high input of energy from sources such as a flame, an electrical arc, glow discharges, lasers, or even shock waves. The gaseous state of the plasma consists of atoms, molecules, ions, metastables, excited states, and electrons, such that the concentrations of positive and negative species are roughly equal. Plasmas also are a rich source of ultra-violet light; in particular, high-energy vacuum UV. They can be 'doped' with elements such as mercury to produce highly intense sources of 254 and 365 nm UV radiation. In flame pre-treatment the flames and in corona discharge the discharges are plasma, but the term 'plasma pre-treatment', with respect to the surface modification of plastics, is confined to a cold plasma formed in a vacuum chamber by applying a suitable electrical voltage to the contained gas.

In a 'low-temperature plasma' the temperature of the chemical species within the plasma is lower than the temperature of the electrons in the gaseous mixture. The following species were shown to be generated in an oxygen plasma:

O^+, O^-, O, O_3, ionized ozone, metastabally excited O, and free electrons

Recombination of the above species led to the release of energy in the form of a blue glow and copious amounts of ultra-violet light. The UV photons had sufficient energy to break carbon–carbon and carbon–hydrogen bonds in the polymers:

$$R{-}C{-}C{-}R \xrightarrow{h\nu} 2R{-}C^{\cdot}$$

$$R{-}C{-}H \xrightarrow{h\nu} R{-}C^{\cdot} + H^{\cdot}$$

The carbon and hydrogen free radicals would react with the moieties contained in the plasma to produce polar surfaces.

Plastics for which the adhesion of coatings may be improved by plasma pre-treatment are reported to be: polyethylene, polypropylene, polycarbonates, polyacetals, polyaromatic esters, polyimides, polyamides, polyphenylene ether, polyacrylates, acetal homopolymer, and poly(ether imide).

Radiation Pre-treatment and Primers

Electromagnetic radiation interacts with matter by a number of mechanisms depending upon the wavelength of the radiation and the nature of the matter. Organic plastics comprise essentially a carbon–carbon backbone with pendant atoms—particularly hydrogen, and occasionally oxygen, nitrogen, chlorine, and sulphur. The bond strength of these linkages is of the order 50–100 kcal mol^{-1} and the energy of ultra-violet light at 400, 300, and 200 nm is approximately 70, 100, and 140 kcal mol^{-1}, respectively—so it can be seen that the typical mercury source emitting 254 and 365 nm radiation is capable of inducing chemical changes in a plastic. It has been reported that exposure of polyethylene film to ultra-violet light in fact does improve the adhesion of printing inks to such a substrate. The use of sensitizers or photo-initiators that may speed up the bond cleavage also has been reported (for example, the coating of polyethylene film with a solution of benzophenone followed by exposure to UV light). Another method suggests the use of photo-grafting of ink-receptive species on to the surface of a polymer. Using a number of polymer supports, including polyethylene and polypropylene, it was found possible to graft such monomers as styrene–MMA and polyfunctional acrylates if the photo-grafting were carried out in the presence of additives such as 0.2 M sulphuric acid, benzoin ethyl ether, lithium nitrate, and charge-transfer complexes such as the benzophenone/dimethylaminoethanol combination. This work showed also the possibility of grafting divinyl ethers and diepoxides on to the surfaces of polyolefins using a UV-initiated photo-cationic photo-grafting mechanism.

The electron beam may be used in the printing industry to cure coatings but the energy available by this means is too great for surface treatment and such exposure would affect the bulk properties of the polymer.

Pre-treatment may also involve the coating of the plastic with a primer, which may be applied to untreated or pre-treated material and dried by solvent evaporation or UV-curing.

In the preceding review of plastics suitable for printing I have tried to show the importance of knowing what it is that is being printed—for the ink can only adhere to whatever is at the surface of the article to be printed.

Physical Chemistry of Printing

The physical chemistry of printing involves the study of:

(i) the preparation of the ink
(ii) the method of printing
(iii) the substrate to be printed.

An ink is a coloured viscous liquid; it consists of a two-phase system—a solid phase dispersed in a liquid phase. The solid is coloured and commonly consists of a mixture of substances of different colours, each with its own distribution of particle size, shape, and surface topography. Often, the liquid phase is a mixture of synthetic polymers dissolved in a volatile solvent. The function of the polymers is to bind the particles of pigment together and to bond them to the substrate. The function of the solvent is to dissolve the solid resins to give a solution the viscosity of which allows efficient dispersion of the pigment—a paste the consistency of which is suitable for printing—and to evaporate at a rate suitable for economical production of a dry print. In addition, an ink will contain substances to help disperse the pigment, raise the speed of drying, help wet the surface, prevent surface defects, improve levelling, gloss, slip, resistance to marring and scratches, and stabilize the print with respect to thermal and photo-chemical degradation.

A printing ink has to carry sufficient colouring matter so that when spread as a very thin film it appears a colour of the correct shade and strength—yet in practice this highly concentrated dispersion has to flow over and wet many different types of surface before being deposited upon the substrate to be printed. The physical chemistry of printing involves surface tension and the wetting of a number of different materials under varying conditions of shear. These materials include various synthetic rubbers, metals, and photo-polymers, and the shear rate regimes range from idle to $10^6\ s^{-1}$.

The surface to be printed can be almost any shape or size, composed of almost any material, and be required to be printed at speeds of up to $15\ m\ s^{-1}$.

Evolution of Printing Processes

Juveniles and craft students often indulge in experiments with printing—such as potato printing and lino cuts—and these in turn may have had origins in the ancient craft of woodcuts. The earliest known woodcut made in Europe dates back to 1423. Essential features of woodcut printing are the cutting of a design into the surface of a block—in such a way that the surface is lowered and the design stands out in relief. Ink is rolled over the relief and the inked block pressed into contact with the sheet to be printed. Originally the press for transferring ink from the block to the paper was a modified wine press. The process of printing from wooden blocks is known as 'Xylography' and it was suggested that it had its beginnings in the first century *anno Domini.*

Communicating a message by woodcuts had the twin disadvantages of any damage to the surface being almost impossible to repair and that the storage of blocks took up much space. However, such problems were overcome progressively by (i) cutting separate wooden letters (1438); (ii) cutting separate metal letters (1450), and finally (iii) the casting of moveable type (1455). Credit for the invention of moveable type is commonly assigned to Johannes Gutenberg, who was working as a printer in Strasbourg in 1455. Relief printing, or—as it is more often known—'letterpress', has survived to this day and the process still is finding new markets.

Until recently, the essential elements in printing were: a printing plate coated with a printing ink, which was transferred to paper by or in a printing press. A recent process, ink-jet printing, invented by Richard G. Sweet in 1971, removes the need for a printing plate or a press.

Originally, printing was developed as a method of producing many copies of a written work. Later it became a method of producing many copies of drawings and works of art. During the second world war, 1939–45, the need to produce electrical equipment in large quantities led to the introduction of printed wires (as first proposed by Paul Eisler in 1936). The technique was so successful that a whole new branch of printing—the printed-circuit industry—came into being. The printing process used originally for this was screen printing—which was developed to the point where printed wires only 150 μm wide and 150 μm apart could be packed on a circuit board. The concept since has been extended into the sub-micrometre range by means of photo-micro-lithography, and research is being conducted into the production of circuits at the molecular level. Increasingly engineers are looking at printing techniques as the means by which they can fabricate other devices, including engines: one such engine is the glucose biosensor produced on plastic and fabricated entirely using printing techniques.

The processes may be classified broadly as the older contact printing techniques, and the newer non-contact processes, as illustrated in Figure 84.

The contact printing processes can be distinguished by comparing the image and the non-image areas of their respective printing plates (see Figure 85).

In letterpress printing a raised surface above a rigid plane is coated with ink forming the image area of the plate. Flexography is similar to letterpress in principle and differs only in the flexibility of its printing plate. In lithography the image and non-image areas are distinguished by their respective hydrophobic and hydrophilic natures. The lithographic plate is essentially planar, with oleophilic ink being accepted by the image areas and rejected by water lying on the surface of the non-image areas. Processes in which the ink lies in recessed

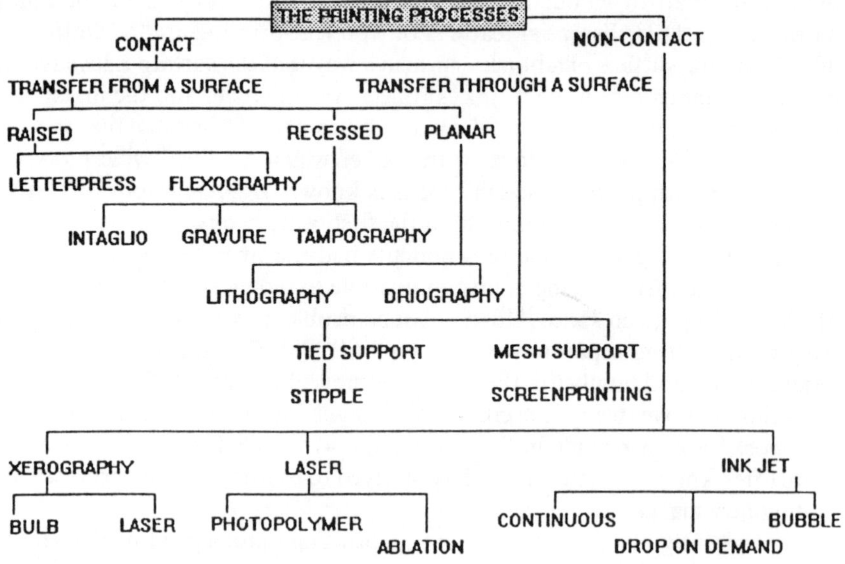

Figure 84 *A classification of processes for printing, contact and non-contact*

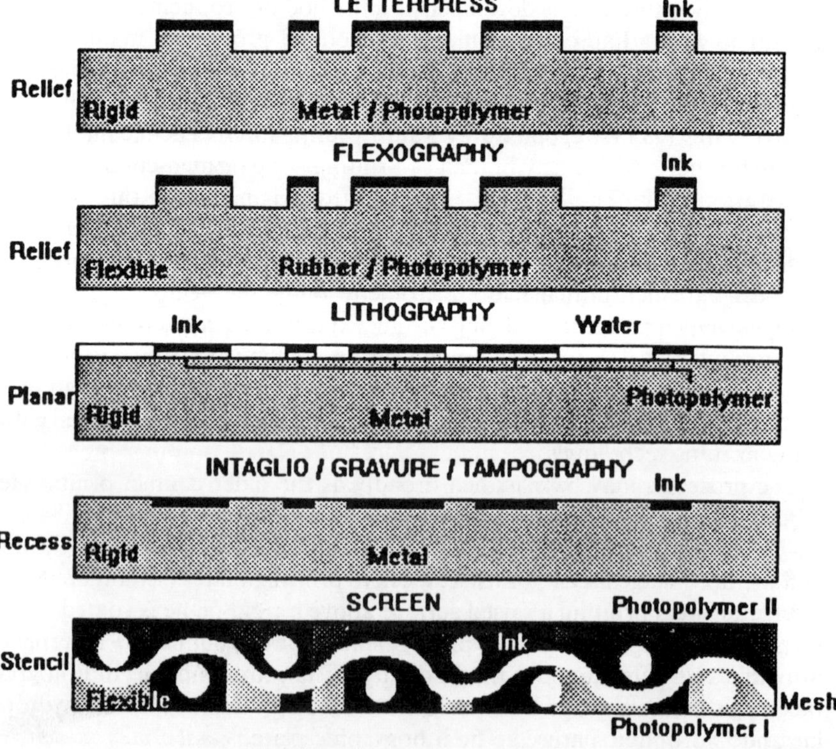

Figure 85 *Image and non-image areas in contact printing*

areas of the printing plate are: intaglio, gravure, and tampography. In screen-printing a frame is covered with a mesh on to which is fixed a stencil. The stencil preferentially occludes the open structure of the mesh, giving an image area which lets ink through and a non-image area in which the passage of ink is blocked.

All the contact printing processes except flexography were originally flat-bed techniques based on reciprocating mechanics. Flexography was developed as a high-speed printing process, and the other contact processes have all been developed into rotary machines. Screen-printing is the only process in which the ink passes through the printing plate; in all other contact printing ink is transferred from the surface of the plate.

The term 'non-impact' applies to those processes which do not rely on pressure to transfer ink from the plate to substrate. Of these, ink-jet is a truly non-contact form of non-impact printing.

Figure 86 summarizes the features of the processes, in terms of machinery available, types of plate, composition of inks, methods of drying, and applications, and Figures 87 to 95 show the basic elements of each process.

Figures 96 and 97 show printed effects obtainable on sheet, blow mouldings, and extruded products.

Pre-press

'Pre-press' is a term used for all the processes and techniques that are available to get information both graphic and pictorial from the originator to the press. The originator may in fact be a war correspondent transmitting a report from a field of battle as it is happening—with the printed message coming off the press within hours of its transmission. Such reports are possible because of the ability to send digitized messages *via* satellites orbiting in space to desk-top publishing systems which decode them into letter form on electronic displays. The message on the screen then is captured, edited, and assembled into pages with any pictorial information that may have been received. The pages thus assembled electronically are checked either on the VDU or as a proof copy from an electronic printer and when accepted by the editor are sent electronically 'direct to plate'. The printing plate is exposed by laser-scanning and the whole process from reporter to report is printed in a matter of hours. Such a system is possible only because of the advances in microelectronics and computer programming. However chemists and physicists also have contributed by providing techniques for rapid access to hard copy based on photophysics (lasers) and photochemistry (photo-imaging systems).

Pre-press with respect to the printing of plastics is associated with the electronic digitization of original artwork and its production directly on a VDU screen, using one of the very powerful graphics computer programmes that now are available. The graphics for decorated plastic packaging are being generated by computer and proofed by modern electronic printing techniques that give an accurate facsimile of how the final print will appear on the polythene carrier bag or detergent bottle.

(a)

Process	*Stock*	*Maximum print area*/mm	*Impression pressure*/kg cm^{-2}
Letterpress			
Flat bed	Sheet	510 by 380	15 to 70
Cylinder	Sheet	568 by 820	
Rotary	Continuous film	600 by 2040	
Flexography	Continuous film	3000 by 2000	
Intaglio			
Flat bed	Sheet	800 by 700	
Cylinder	Continuous film	800 by 2000	97
Gravure	Continuous film	6000 by 15 000	
Tampography			
Flat	Three-dimensional	300 by 200	
Cylinder	Three-dimensional		
Lithography	Sheet	940 by 1280	20
Driography	Continuous film	578 by 2000	
Screen printing			
Flat bed	Sheet/Three-dimensional	609 by 2080	
Cylinder	Three-dimensional	1100 by 1680	
Rotary	Continuous film	1850 by 1850	
Xerography	Sheet	841 by 1189	
Laser	Three-dimensional	297 by 420	None
Ink-jet			None
Continuous	Three-dimensional	13 mm wide	None
Drop on demand	Three-dimensional	25 mm wide	None
Bubble	Three-dimensional	3 mm wide	None

(b)

Process	*Shape on machine*	*Composition*	*Ink deposit*
Letterpress			
Flat bed	Flat	Metal or	Above plane
Cylinder	Cylinder	Photo-polymer	of plate
Rotary	Cylinder		
Flexography	Cylinder	Rubber/ Photo-polymer	Above plane of plate
Intaglio			
Flat bed	Flat	Metal	Below plane
Cylinder	Cylinder		of plate
Gravure	Cylinder	Metal	Below plane of plate
Tampography			
Flat	Flat	Metal or	Below plane
Cylinder	Cylinder	Photo-polymer	of plate
Lithography	Cylinder	Bimetallic or	In plane
Driography	Cylinder	Photo-polymer	of plate
Screen printing			
Flat bed	Flat	Photo-polymer	In plane
Cylinder	Flat	Photo-polymer	of plate
Rotary	Cylinder	Metal	
Xerography	Cylinder	Photo-conductive ZnS	In plane of plate
Laser	None	None	None
Ink-jet			
Continuous	None	None	None
Drop on demand	None	None	None
Bubble	None	None	None

Figure 86 *Characteristics of printing processes:* (a) *Types of machine*; (b) *Types and compositions of plates*

(c)

Process	*Binders*	*Solvent*	*Boiling range/* °C	*Viscosity/* Poise	*Ink thickness/* μm	*Print thickness/* μm
Letterpress						
Flat bed	Vegetable oils	Vegetable oils				
	Rosin esters	Mineral oils			3–6	2–4
Cylinder	Synthetic resins		270–330	10–500		
	Acrylate esters	Acrylate esters				
Rotary	Mineral oils	Mineral oils				
Flexography	Shellac	Water	80–100	0.5–5.0	3–4	1–2
	Synthetic polymers	Alcohols				
Intaglio						
Flat bed	Vegetable oils	Vegetable oils	270–330		20	20
Cylinder	Rosin esters	Mineral oils				
Gravure	Synthetic polymers	Esters, ketones,			7–24	4–12
		Glycol esters,	80–120	0.3–2.0		
		Aromatic				
		hydrocarbons				
Tampography						
Flat	Synthetic polymers	Esters, Ketones,			≤ 20	≤ 10
Cylinder		Glycol ethers,	120–180	10–20		
		Aromatic				
		hydrocarbons				
Lithography	Acrylate esters	Acrylated esters			2	1–2
	Vegetable oils	Vegetable oils		100–3000		
	Rosin esters	Mineral oils	270–330			
Driography	Siliconized polymers	—	—	—	2	1–2
Screen Printing						
Flat bed	Vegetable oil	Mineral oils				
	Cellulosics	Aromatic	120–250		8–30	4–15
Cylinder	Rosin esters	hydrocarbons		15–20		
	Synthetic polymers	Esters, Ketones				
Rotary	Acrylated esters	Glycol ethers,				
		Acrylated esters				
Xerography	Rosin esters	None	—	—		
	Synthetic polymers					
Laser	None	None	—	—		
Ink-jet						
Continuous	Vegetable oil	Water		1–80		
	Rosin ester	Alcohols	80–100	No. 4 Cup		
Drop on demand	Cellosics	Esters				
	Synthetic polymers	Ketones				
Bubble	Acrylated ester					

Figure 86 *Characteristics of printing processes:* (c) *Typical formulations and properties of inks*

(d)(i)

Process	
Letterpress	
Flat bed	Racking in air, stacking in piles
Cylinder	Racking in air, stacking in piles
Rotary	Gas fired hot air
Flexography	Hot air, IR, UV, Electrobeam, Radio-frequency (microwave)
Intaglio	
Flat-bed	Racking in air, stacking in piles
Cylinder	Racking in air, stacking in piles
Gravure	Hot air
Tampography	
Flat	Racking in air, IR
Cylinder	
Lithography	Racking in air, stacking piles, IR, UV, Electron beam
Driography	Racking in air, stacking in piles, IR, UV, Electron beam
Screen Printing	
Flat bed	Racking in air, stacking in piles, hot air, IR, UV
Cylinder	Radio-frequency (microwave)
Rotary	Hot air
Xerography	—
Laser	—
Ink-jet	
Continuous	Hot air, UV
Drop on demand	
Bubble	

(d)(ii)

Process	
Letterpress	
Flat bed	Absorption, precipitation, air oxidation polymerization, UV polymerization, evaporation, polycondensation
Cylinder	Physical barrier (spray powder)
Rotary	
Flexography	Absorption, evaporation, polycondensation, UV-initiated polymerization, molecular resonance heating (microwave)
Intaglio	
Flat-bed	Air oxidation polymerization
Cylinder	
Gravure	Absorption, evaporation
Tampography	
Flat	Evaporation, air oxidation polymerization, polycondensation
Cylinder	
Lithography	Absorption, evaporation, air oxidation polymerization, UV initiated polymerization Physical barrier; spray powder, emulsion coating
Driography	as for Lithography
Screen Printing	
Flat bed	Evaporation, absorption, air oxidation polymerization, UV-initiated polymerization, polycondensation
Cylinder	Molecular resonance heating (microwave)
Rotary	
Xerography	—
Laser	—
Ink-jet	
Continuous	Evaporation, UV initiated polymerization
Drop on demand	Evaporation, UV initiated polymerization
Bubble	Evaporation

Figure 86 *Characteristics of printing processes:* (d) *Methods of drying inks:* (i) *Drying methods;* (ii) *Drying processes*

(e)

Process	*Stock*
Letterpress	
Flat bed	Traditionally paper printing but now used only in specific specialized areas
Cylinder	
Rotary	Newspapers, yoghurt cups, labels
Flexography	Wrappings, carrier bags, newspapers
Intaglio	
Flat-bed	Fine art work printing
Cylinder	Bank note printing
Gravure	Magazines
Tampography	
Flat	Complex shapes such as toiletry bottles
Cylinder	Spark plugs
Lithography	Brochures, catalogues, provincial newspapers, labels
Driography	as for Lithography
Screen Printing	
Flat bed	Point of sale advertising, posters, printed circuit boards, T-shirts, labels
Cylinder	Point of sale advertising, posters
Rotary	Fabrics, wall coverings
Xerography	Office documentation
Laser	Office documentation
Ink-jet	
Continuous	Identification coding
Drop on demand	Identification coding
Bubble	Office documentation

Figure 86 *Characteristics of printing processes:* (e) *Applications*

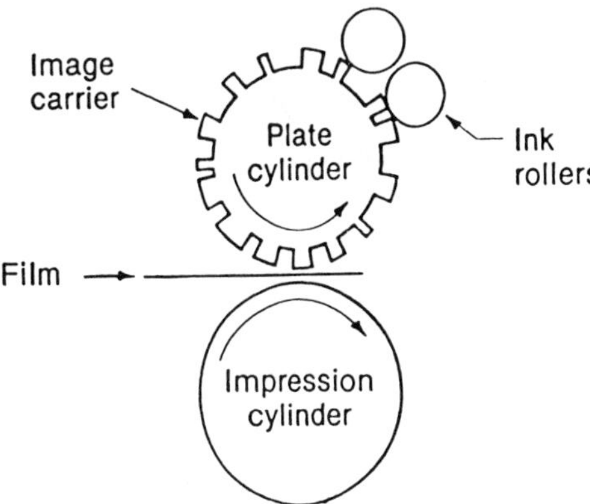

Figure 87 *Rotary letterpress*

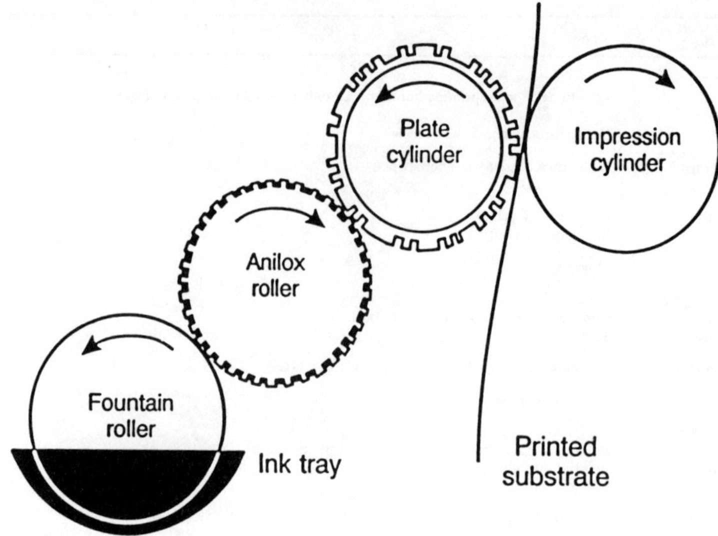

Figure 88 *Flexography*

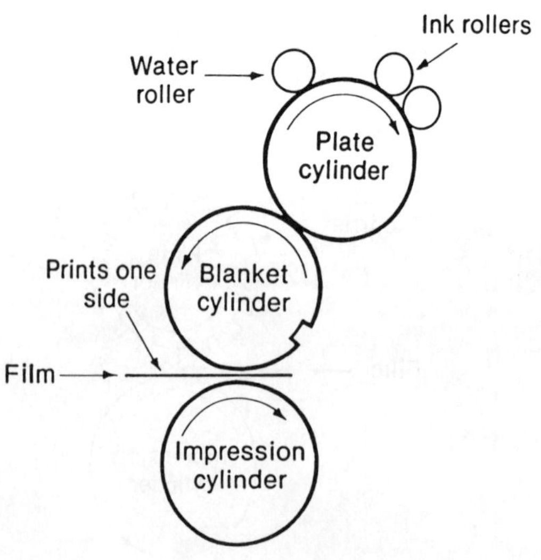

Figure 89 *Lithography*

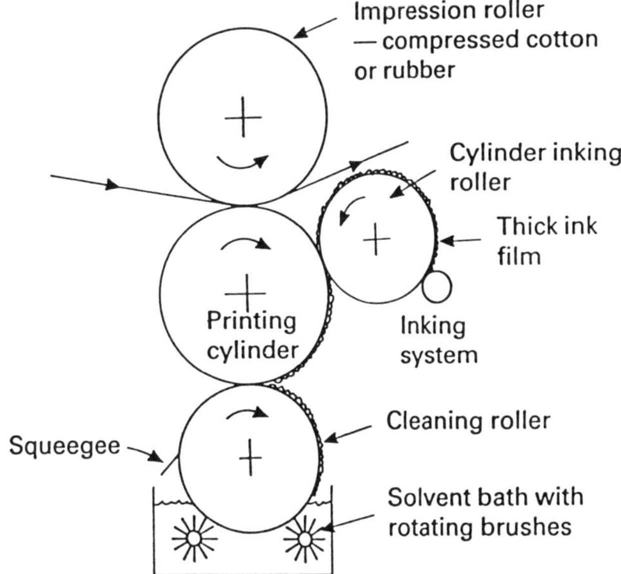

Figure 90 *Intaglio*

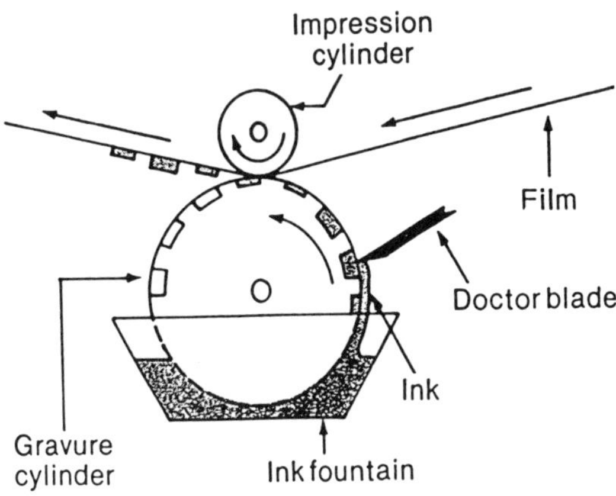

Figure 91 *Gravure*

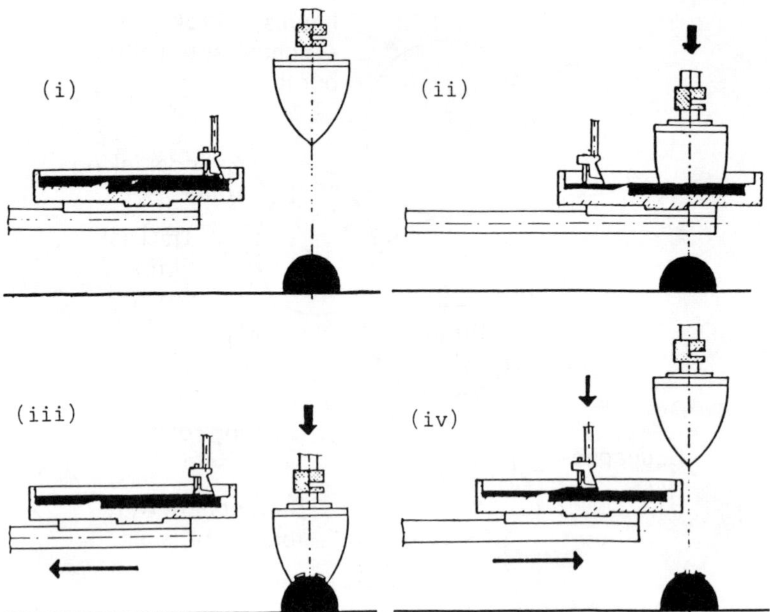

Figure 92 *Tampography:* (i) *The plate is coated with ink;* (ii) *Doctor blade clears the plate and the pad takes an impression of the engraving;* (iii) *Doctor blade covers the plate and pad is impressed on the object to be printed;* (iv) *Pad is released and doctor blade clears the plate*

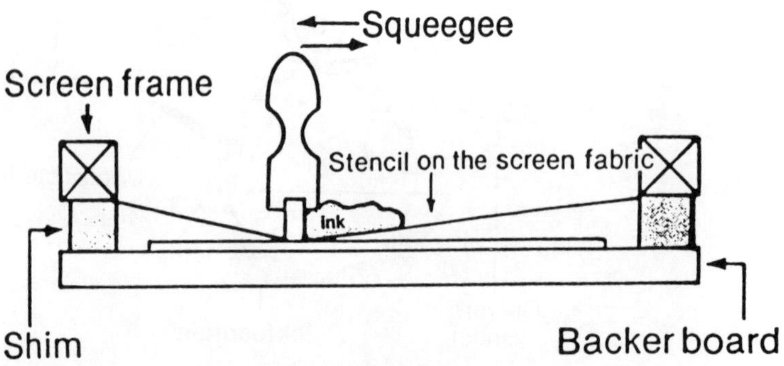

Figure 93 *Screen printing*

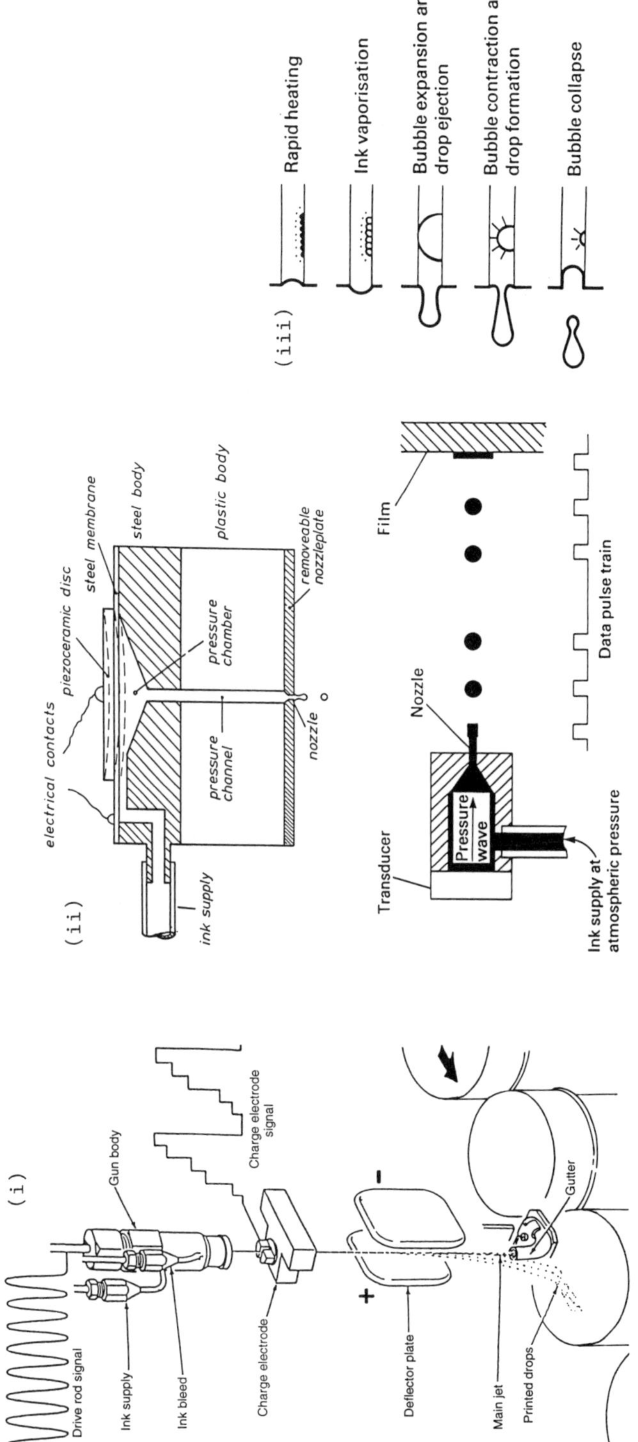

Figure 94 *Ink-jet printing:* (i) *Continuous;* (ii) *Impulse or drop on demand;* (iii) *Bubble jet*

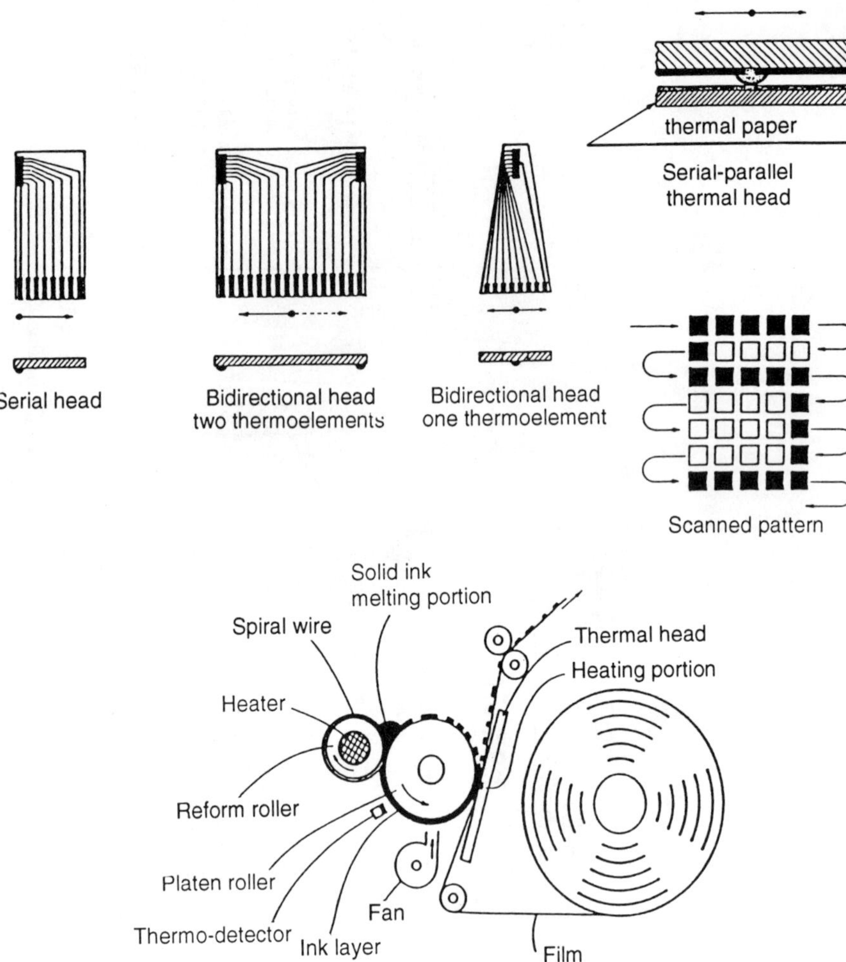

Figure 95 *Thermal transfer:* (i) *Typical print heads;* (ii) *Cross-section of machine*

Figure 96 *A presentation by* Scintillex *for the retail sale of toiletries and other fancy goods fabricated or moulded from plastics: the transparent pouches would protect and display the bright printed designs within (1960s)*
(Photo Plastics Historical Society)

Figure 97 *Containers blow-moulded or extruded for a variety of household goods and toiletries. The* Sqezy *(centre) was made by* Cascelloid, *Leicester, from flexible plastic tube with metal ends, and it is understood to have been the first developed in Britain. Most of the others were made wholly from plastics, particularly polyethylenes of various densities. The containers demonstrate the use of silk-screen printing in a variety of colours for work of this kind.*
(Photo Plastics Historical Society)

Microelectronics and computer programming have allowed the printing processes to be analysed, and sensed remotely so that many of the variables in the printing are monitored, adjusted, and controlled automatically from a console. Such fundamentals as flow of ink from the duct, colour strength, ink viscosity, temperature, and humidity all are monitored and checked against pre-determined limits, so that they can be adjusted to maintain consistency of print quality.

Letterpress Printing

Letterpress printing is the oldest of the mechanical printing processes and its most notable advance was the casting of movable type by Gutenberg in 1455. It has been developed from a slow reciprocating flatbed process to an extremely fast technique using rotary cylinders. Only thin films of ink can be deposited but they can be printed with great accuracy and evenness of deposit over a very wide area of print. The substrate can be in the form of a sheet, web, or cylinder

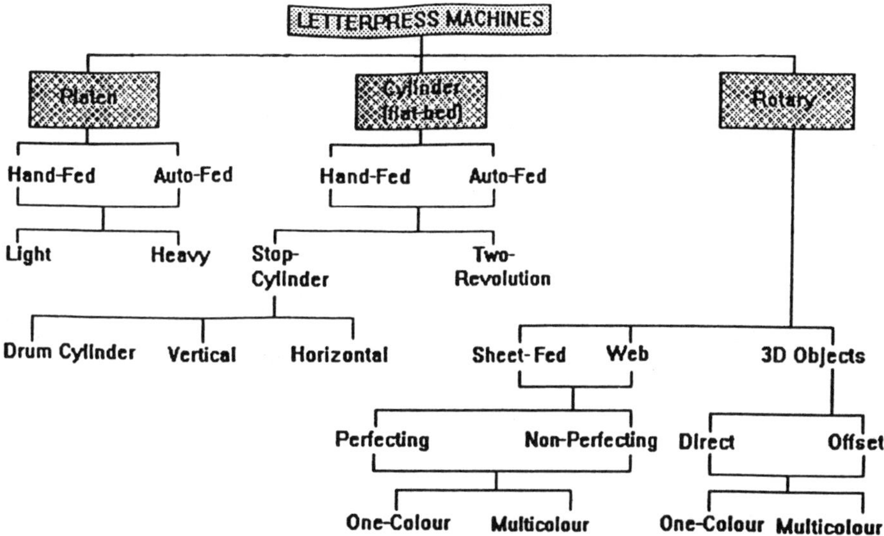

Figure 98 *The evolution of letterpress printing*

and may range in composition from paper and plastics to metals. Traditionally, to achieve an even print the ink is distributed over a very large train of rollers which expose a thin continuous film of ink to the atmosphere. Hence, the ink must be very stable with respect to evaporation and oxidation. Inks for the purpose are formulated from liquid resin precursors or from solid resins dissolved in solvents with very high boiling points and low rates of evaporation. For fast production speeds accelerated methods of drying the ink are employed. Letterpress inks can be formulated from a wide variety of resins and solvents and give satisfactory results on many different substrates.

Letterpress declined almost to the point of oblivion in the area of general or jobbing printing but still is widely used in the production of newspapers and for the printing at high speed of plastic containers such as yoghurt cups and margarine tubs.

(i) The Press

Figure 98 provides an outline of the evolution of letterpress.

Because of the mechanical limit set by the enormous pressure involved in transferring ink to paper, the platen technique can print only areas up to 510 by 380 mm. However, rotary machines with rollers 2 m long, diameters capable of printing four pages of newsprint per revolution, and running at 40 000 rph are producing some of today's newspapers.

All printing machines appear to be a myriad of cascading rollers of various diameters and compositions. The roller train is necessary for the even distribution of ink from the ink duct to the printing plate. Until the 18th century the method of applying ink to the plate was by a leather pad known as the 'ink ball'.

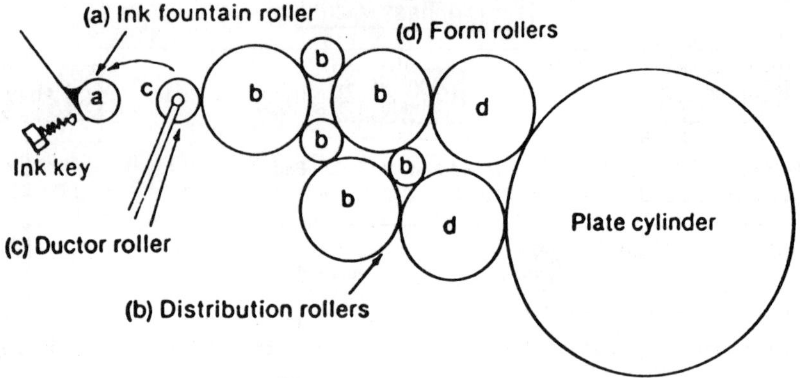

Figure 99 *A typical inking system*

The pad was replaced by a leather-covered roller attached to the press and so allowing the mechanical inking of the plate. However, the leather-covered roller had a stitched join that left an uneven streak of ink on the plate. This problem was solved by the introduction of wooden rollers covered with a composition of glue and syrup. Until the introduction of synthetic rubber rollers after 1945 many rollers were covered with a 'composition' of gelatin and glycerine (which was first patented by Usher in 1860). Natural rubber and gelled linseed oil also have been used as covering materials for printing rollers. Today, printing rollers covered with butyl, nitrile, or polyurethane rubbers are the most common. Coverings now are available that are compatible with the many solvents found in the wide range of inks used in modern rotary letterpress printing. When in the 'sixties UV-curing inks were introduced the rollers fitted for use with oil-based inks had to be replaced with a composition that resisted the acrylate monomers used in the newer inks. There seems still to be some controversy over claims that it is possible to use a blend of nitrile rubber and poly(vinyl chloride) to cover printing rollers for use with both oil-based inks and UV-curing inks based on acrylate.

The general configuration of rollers from the ink duct to the printing plate is common in those processes which use 'paste' inks, that is, letterpress, intaglio, and lithography. A typical arrangement is shown in Figure 99.

On most letterpress and lithographic machines the ink issues from a duct, the gap of which is set by a series of threaded keys. This method of ink flow to the press requires considerable skill and attention in order to minimize time in setting up and waste. A recent attempt to overcome this was the introduction of keyless metering. An Anilox roller, similar to that used in flexography, rotates in the ink duct and a doctor blade scrapes the surface free of ink. The roller has a ceramic composition with cells etched into its surface. When the cells are full of ink a rubber roller is brought into contact and ink is drawn from the cells on to the surface of the rubber roller. Only one more inking roller is required to ink up the plate. The Anilox roller is being fitted to a number of rotary newspaper presses and is being experimented with in lithographic printing. Metering of ink from the duct *via* such a roller may very well in future become the normal method by which a printing plate is inked.

Over the years letterpress printing has declined into niche markets, two of which are the printing of plastic self-adhesive labels on narrow-web rotary-offset letterpress machines that are fitted with UV drying and the common-impression rotary-offset letterpress and UV-curing on mandrel of plastic containers such as yoghurt tubs and toothpaste tubes. In uses such as these the most common self-adhesive plastic is PVC and the most usual for tubs and tubes are polystyrene, PVC, ABS, and polypropylene.

(ii) The Plates

Plate-making for letterpress has been revolutionized by the introduction of photo-polymer plates and computer-generated artwork. The modern photo-polymer letterpress plate is a block of plastic carrying an image raised above its plane that resembles in appearance an original Chinese woodcut. However, its production was not by hand cutting but by exposure through a photographic negative of a photo-polymer that in exposed areas polymerized to a compressible plastic and elsewhere remained solvent-soluble. Washing of the plastic with a suitable solvent dissolved the unexposed photo-polymer, leaving behind a relief image suitable for letterpress printing. Fonts of movable type exist now only as bytes of information in a computer programme, and are seen to be individual and 'movable' only on a VDU screen. Systems now are available which allow artwork to be scanned directly to the plate, thus removing the need for cameras and photography based on silver halide chemistry.

Photo-polymer letterpress plates were patented by Gates in 1945 but the most successful commercial systems were based on the work of Plambeck in 1956, which led to the introduction in 1957 of Du Pont's Dycril plate and nearly ten years later of the Nyloprint system from BASF. The Dycril and the Nyloprint systems are very similar to the exposure and development of a photographic plate but there is another popular method of producing such plates using a liquid photo-polymer system. Letterpress plates for newspapers based on liquid photo-polymers were introduced in 1971.

In the context of printing plastics, the Nyloprint photo-polymer plate appears to be the most popular in the areas of narrow-web rotary-letterpress label printing and rotary-offset letterpress printing of plastic tubs and tubes.

The main advantages of the photo-polymer system include safer operations that do not use hot metals, such as zinc or magnesium, or plating solutions for the deposition of copper. Processing involves the use of developers such as water, dilute sodium hydroxide, and dilute solutions of sodium carbonate. Because each plate is a copy of the original (not a duplicate from a mould that has been used many times) the print quality is consistent and more predictable. While some of the pressures for change were environmental a bonus is that the overall costs of producing photo-polymer plates are lower than those for the traditional metal plates.

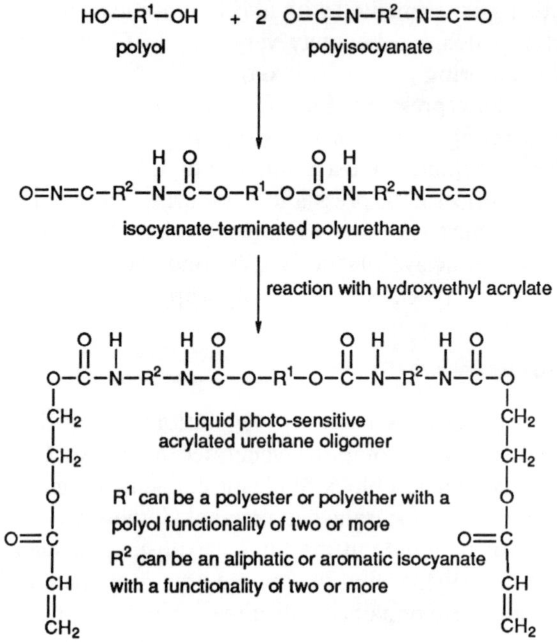

$$HO{-}R^1{-}OH \quad + \; 2\; O{=}C{=}N{-}R^2{-}N{=}C{=}O$$

polyol polyisocyanate

↓

$$O{=}N{=}C{-}R^2{-}N(H){-}C({=}O){-}O{-}R^1{-}O{-}C({=}O){-}N(H){-}R^2{-}N{=}C{=}O$$

isocyanate-terminated polyurethane

↓ reaction with hydroxyethyl acrylate

$$CH_2{=}CH{-}C({=}O){-}O{-}CH_2{-}CH_2{-}O{-}C({=}O){-}N(H){-}R^2{-}N(H){-}C({=}O){-}O{-}R^1{-}O{-}C({=}O){-}N(H){-}R^2{-}N(H){-}C({=}O){-}O{-}CH_2{-}CH_2{-}O{-}C({=}O){-}CH{=}CH_2$$

Liquid photo-sensitive acrylated urethane oligomer

R^1 can be a polyester or polyether with a polyol functionality of two or more

R^2 can be an aliphatic or aromatic isocyanate with a functionality of two or more

Figure 100 *Reaction of a liquid photo-polymer for letterpress printing*

The chemistry involved in both systems involves the photo-initiated, free-radical polymerization of polyfunctional ethylenically unsaturated oligomers and monomers.

Figure 100 illustrates the reaction of a liquid photo-polymer. Figure 101 summarizes the commercial systems.

Properties of a Letterpress Ink

The physical properties of a letterpress printing ink are linked with the printing process and the substrate to be printed. Because the letterpress machine has many rollers to distribute the ink as a thin film to the plate, the ink must contain either no solvents or only those with high boiling points and low rates of evaporation. Provided the solvents boil in the range 250 °C to 320 °C and do not attack the metal or rubber rollers they can be drawn from any chemical group. Solvents of this nature will dissolve a wide range of 'resins' and polymers, giving rise to binders capable of producing inks that will adhere to a wide range of substrates. The inventory of materials from which letterpress inks can be formulated is considerable and this has allowed the process to remain important in some very significant uses. The consistency of the letterpress ink must allow easy transfer down the roller train and its 'tack' value (its resistance to splitting) must be adjusted correctly so that it does not act as an adhesive between the plate and the substrate. The 'tack' of the ink and the 'pick value' of

Company	Trade Name	Binder	Monomer	Appearance	Wash-out
Asahi	APR	Unsaturated polyester of poly(ethylene oxide) and fumaric acid	Triethylene glycol diacrylate	Liquid	Borax solution
BASF	Nyloprint	Polyamide	Bifunctional acrylamide	Solid	Alcohol
Du Pont	Dycril	Cellulose	Pentaerythritol triacrylate	Solid	Dilute sodium hydroxide
Dynaflex	Dynaflex	Poly(vinyl alcohol)/ poly(vinyl acetate)		Solid	Water
Grace	Letterflex	Polyurethane with allyl groups	Mercaptan	Liquid	Air knife or water
Kansai Paint		Polyurethane	Acrylic acid ester	Solid	Dilute sodium hydroxide
Nippon Paint	Napp	Poly(vinyl alcohol)/ Poly(vinyl acetate)	Acrylic ester	Solid	Water
Polychrome	Flexomer	Polyurethane	Triethylopropylene triacrylate		Trichloroethylene
Teijin	Tevista	Unsaturated polyester		Liquid	Dilute sodium carbonate
Time	Tilon	Polyamide	Methylene bis-acrylamide	Solid	
Tokyo Okka	Toplon	Polyamide	Phenylene bis-acrylamide	Solid	Alcohol

Figure 101 *Commercial photo-polymer systems*

the substrate can be determined by means of arbitrary comparative tests. Letterpress inks have relatively high tack values—10 to 20, on the Mander–Kidd Scale. The consistency of the ink must be controlled further with respect to its flow. If the ink is too 'soft' and too 'long' it may be forced down the sides of the raised image, with the effect of increasing unnecessarily the image area. Such an ink may exude from between the plate and substrate under the high pressures involved and form a halo around the print. This is a characteristic of letterpress printing and can only be minimized, not eliminated.

Because of the wide latitude available in formulating letterpress inks, many methods of drying can be used.

Inks may be deposited with a thickness in the range of 3 to 6 μm. Dry prints, if they are made from non-volatile materials and are not adsorbed into the substrate, will have a similar film thickness.

As it travels from duct to substrate an ink will be subjected to a number of different shearing forces. A letterpress ink is a non-Newtonian liquid and will have a different viscosity at each of the rates of shear stress to which it is subjected. It is important for good printing that the ink has the correct viscosity at each of these rates. Recently, instruments have been introduced that characterize an ink in rheological terms over long periods of time, over a wide range of temperatures, and over a wide range of strains—particularly those at the very low end of the scale. Instrumentation also is being fitted to presses to monitor the rheological characteristics of the ink continuously while they are running. (Figure 102 presents deformation processes in diagrammatic form.)

Besides having to flow from ducts and to split efficiently between rollers, a letterpress ink also has to wet quickly all the surfaces with which it comes into contact. On the press these may be various metals, rubbers, and plastics, while the substrate can be any one of a wide variety of materials. For a liquid to wet a surface it must have a lower surface tension than the substrate. The surface tension of a liquid is related to the contact angle that the liquid makes with the substrate. The contact angle is defined as nil for complete wetting and greater than 90° for increasing non-wetting tendencies.

In printing, a film of ink is formed by wetting the surface with the compression force of the rollers. This force spreads the ink over the surface and into any capillaries that may be present. Spreading and penetration are controlled thermodynamically and kinetically. Measurement of the contact angle can be used to determine the thermodynamics of wetting. This angle can be used also to determine the contribution that polarity and dispersive forces of the liquid make to the wetting of the surface.

A letterpress printing ink is a dispersion of pigment particles in a polymer solution. Deviation from ideality is assured by the presence of polymers of high molecular weight dissolved in solvents of differing polarity so as to disperse particles of differing polarity, shape, and size.

There is a relationship between concentration of polymer and viscosity, and the viscosity of the solvent (more particularly, its polarity) also affects the final viscosity of the ink. In inks, the nature of the viscosity at high concentrations of polymer is important. Solvents may be considered to be viscous liquids and

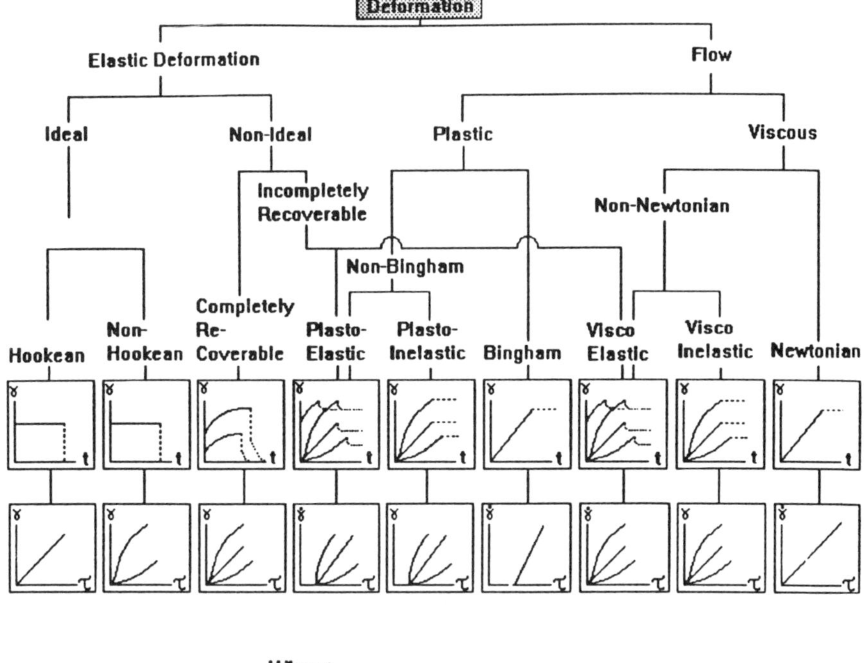

Where:

γ = shear strain

$\dot{\gamma}$ = shear strain rate

τ = shear stress

t = time

Continuous curves represent deformation under stress.

Dotted curves represent behaviour after the removal of stress.

The upper series of diagrams represents shear strain against time.

The lower series of diagrams represents shear strain rate against shear stress for an arbitrary time.

Figure 102 *Graphical representation of deformation processes*

polymers to be elastic solids. Solutions of polymer in solvent thus could be thought of as being visco-elastic—that is, such a solution would have some solidus character and some viscous character. Viscous deformations are irreversible whereas elastic deformations may be reversed: therefore, viscous deformation consumes energy whereas elastic deformation stores it.

The force and the period over which the force is applied determine the proportion of solidus and liquidus character. A letterpress ink is subjected to differing forces over differing periods of time and (since it is a mixture of viscous and elastic components) its solidus/viscous character will differ as it passes through the various stages of printing. At high concentrations, polymer solutions are treated as having interfering chains with flow being considered on the basis of their shape and ability to transfer momentum. There have been numerous studies but the result still is only an indication of the relationship between a particular polymer in a solvent at a given concentration and the viscosity of the resulting solution. As instances, it is possible to interpret the outcome in terms of the conformation of the polymer or the ratio in it of crystalline to amorphous phase. The rheological properties of the polymer solution binding the pigment to the substrate may be either beneficial or detrimental to printing qualities. The continuous phase can be used to adjust the yield value of the ink, so giving it the correct balance of structure and flow, but on the other hand it may make the ink dilatant to the extent that it will not transfer down the roller train.

The dispersion of pigments in non-Newtonian polymer solutions takes the resulting fluid even further away from the ideal—because the particles may be unequal in size, varied in shape, and active in a physio-chemical sense. On precipitation, pigments will flocculate in a range of particle sizes. A narrow band will tend to give a well-defined rheological dispersion, whereas, a broad band, with the ratio of sizes varying from batch to batch, will give inks the rheology of which also may vary from batch to batch. With regard to the effect of the shape of the particles, coatings containing pigments that are 'spherical' tend to be more Newtonian whilst 'rod'-like pigments tend to give coatings that are more non-Newtonian and structured.

Inks are made to specific shades, and mixtures of pigments are used to achieve these. Hence, the rheology of ink in one colour can vary from that of another. Carbon black provides a good example of the same pigment being used to produce inks of the same colour but with very different rheological characteristics. Pigments based on it can be made in various sizes, with different surface areas and various degrees of agglomeration and surface activity (physical and chemical). Therefore it is possible to make black inks with rheological characteristics that depend largely on the grade of carbon black dispersed in the binder (different configurations result in differing degrees of dispersion). Any change in the method of manufacture, particularly in the type of equipment used to disperse the pigment in the binder, will lead to deviation away from the rheological parameters set for the standard ink.

Pigment dispersion is begun with the removal of air from the pigment–air interface and its replacement by the binder, forming a pigment–binder interface. A further stage is the removal of adsorbed species (such as water) from its surface. Maximum dispersion is achieved when each primary pigment particle is coated with binder. The terms 'milling' and 'grinding' are used to describe the process of dispersion but they are not crushing processes and do not break up the primary particles. These tend to aggregate and the function of the dispersion machinery is to break up the aggregates and agglomerates. Even when dispersed completely into their primary particles, pigments have a tendency to clump together (a process known as 'flocculation'). To help the binder wet the pigment, to speed up wetting, and to prevent reflocculation, additives are used.

These can have a physio-chemical attraction for the surfaces of the particles and even react chemically to form permanent bonds.

The processes involved in wetting a pigment are similar to those in forming a surface coating. However, in most coating operations no large forces are required to apply the film, whereas in pigment dispersion a considerable amount of energy is consumed in separating and coating the individual particles. Often the energy for dispersion is applied through the shearing forces generated between the nip of rollers running at different speeds. Such an action tears the agglomerate apart to expose pigment surfaces to binders. A unit often used for the production of printing inks is the triple-roll mill, with the pigment binder pre-dispersion running between the nip of the three finely ground and polished rollers—each turning at a different speed. The passage of ink over these rollers is similar to the distribution of an ink down the roller train of a printing press. In the nip between the rollers of a triple-roll mill pressures arise similar to those at the point of print in a letterpress machine. However, in the triple-roll mill the forces are shearing the ink whereas at the point of print in a press the ink is being compressed and decompressed almost immediately—giving rise to extensive cavitation. If the cavitation is excessive, the ink can form fibrils which will fracture and release particles of ink. As the making and breaking of such fibrils takes place rapidly the atmosphere around a press soon can become contaminated with these particles—an aerosol condition known in the trade as 'ink fly'.

A knowledge of the rheology of the binder and of the ink formed by dispersing the pigment in the binder is important in controlling the production of the ink and its printing on the press; several papers on these topics have been published.

Inks for letterpress may be considered to have their origins in the early paints and writing inks. Paints are applied by brush, writing ink by a pen, and printing by a press but the basic ingredients for all three are the same—colorant, binder, and solvent. In general they may be distinguished by their differing rheological characteristics. During brushing, paints have viscosities of 250 to 500 mPa, writing inks are almost Newtonian and have viscosities of 5 to 30 mPa, while letterpress inks are non-Newtonian and range from 1000 to 50 000 mPa at shear rates up to $10^6\ s^{-1}$.

In appearance letterpress inks look as though they are made from a coloured mixture of butter and syrup. Historically they were based on vegetable or mineral oils but with the need to print plastic containers and films at high speed came a departure from oil-based formulations. Since the surfaces were not absorbent and unsaturated vegetable oils polymerized slowly it was necessary to introduce drying times measured in minutes rather than days. Formulations introduced were based on polymers which had good specific adhesion, say, to polystyrene and were dissolved in high-boiling solvents that could be evaporated quite quickly by exposure to infra-red radiation (as examples, acrylic resins dissolved in diethylene glycol monobutyl ether). Polypropylene tubes are printed with inks that dry by a combination of evaporation and polycondensation polymerization. Many of the plastic containers in this type of work are printed with inks that dry at speeds close to the design limits of the printing process. Such inks dry using a photo-initiated reaction brought about by exposure to ultra-violet light. The exposure times are measured in seconds; the chemistry often is based on free-radical polymerization of acrylate monomers and oligomers.

UV-curing of printing inks was investigated in the early 'forties but not studied and brought to commercial fruition until about twenty years later. Rule 66 in California required the elimination or sharp reduction of emissions of organic volatiles—including printing inks—and a second source of pressure at that time for an alternative method of drying the inks was a shortage in the USA of natural gas for use as fuel (there was insufficient for the flame-drying of newsprint, then the common method of drying newsprint).

Chemistries available now for the UV-curing of printing inks include:

(i) photo-initiated free-radical polymerizatrion of ethylenically unsaturated compounds
(ii) photo-initiated cationic polymerization of diepoxides and divinyl ethers
(iii) the thiolene system
(iv) photo-generation of acids to catalyse polyesterification and polyetherification condensation reactions.

Of these, the most commonly used in letterpress printing is photo-initiated free-radical polymerization of acrylates.

The technical success of the UV-drying inks may be attributed to the introduction of a number of types of acrylated oligomer and photo-initiator. The broad categories of acrylated oligomers include polyesters, epoxies, urethanes, ethers, and polyacrylates. Photo-initiators generally are derived from compounds that on exposure to UV light cleave into free radicals or abstract hydrogen to form free-radical species.

Figure 103 shows the sequence of curing a letterpress ink, with benzilketal (an initiator often used) and epoxydiacrylate as acrylate oligomer. Free-radical polymerization of an acrylate oligomer can be initiated by both the methyl and benzoyl radical; it has been found that the latter makes the greater contribution, as shown in Figure 104.

There are two sources of UV light in common use for letterpress inks, both based on the excitation of mercury in the presence of an inert gas contained within a quartz tube. One system depends on applying a high voltage across two tungsten electrodes that are doped with thorium; the other is an electrode-less system relying on the resonance of microwave radiation.

In a system of the latter type the light is generated by focusing radiation (2450 MHz) from a magnetron (2 by 1.5 kW) on a narrow vitreous silica tube containing a few droplets of mercury and filled with an inert gas. The microwave energy generated is focused through two wave guides, in one of which is a probe that takes up some energy to fire a low-pressure mercury lamp behind a reflector housing; radiation from the lamp passes through small holes in the reflector into an electrode-less lamp containing droplets of mercury. Radiation at 254 nm from the start-up lamp causes the emission from the droplets of mercury of photo electrons, which are accelerated in the focused field. This

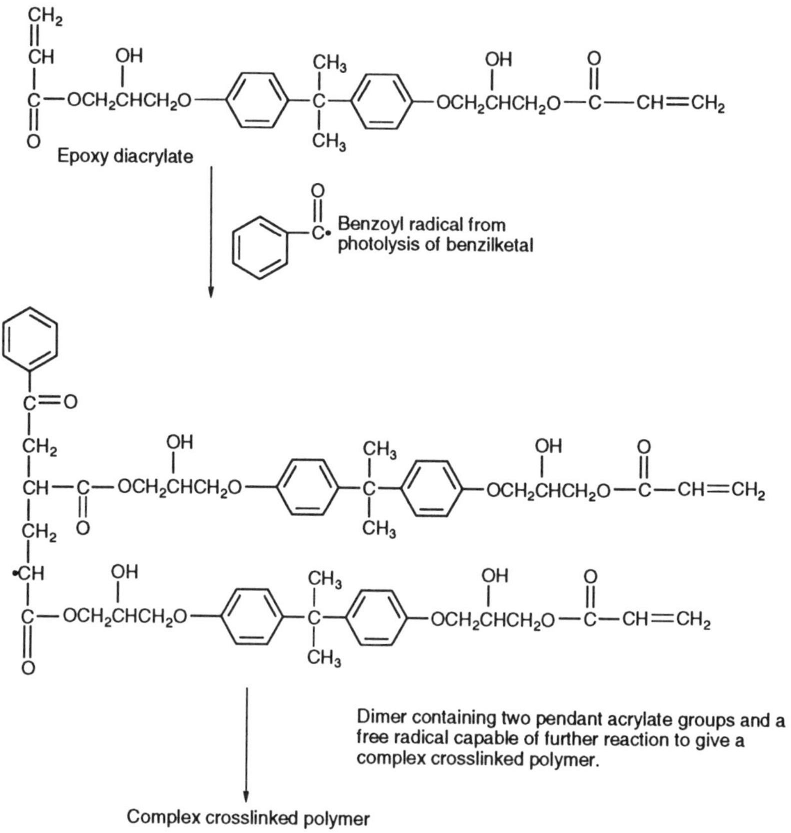

Epoxy diacrylate

Benzoyl radical from photolysis of benzilketal

Dimer containing two pendant acrylate groups and a free radical capable of further reaction to give a complex crosslinked polymer.

Complex crosslinked polymer

Figure 103 *Curing a letterpress ink using a photo-initiator*

has a warming effect which vaporizes the mercury. Excited mercury returns to the ground state with the emission of UV radiation.

Flexography

(i) The Press

Flexography is related to the letterpress process and, as the name implies, employs a flexible rather than rigid printing plate. It was developed originally in the 'twenties as a means of printing paper bags cheaply at very high speed. Mechanically it is the simplest of all the contact methods of printing and probably this is why it has become the fastest method of producing large volumes of print. Its simplicity and speed demand quick-drying inks based on very volatile organic solvents. Flexography uses inks of very low viscosity and gives very thin films of print. It deposits very small quantities of resin, which must bind efficiently relatively large amounts of pigment to the substrate.

There are two basic configurations for the roller systems used in flexography

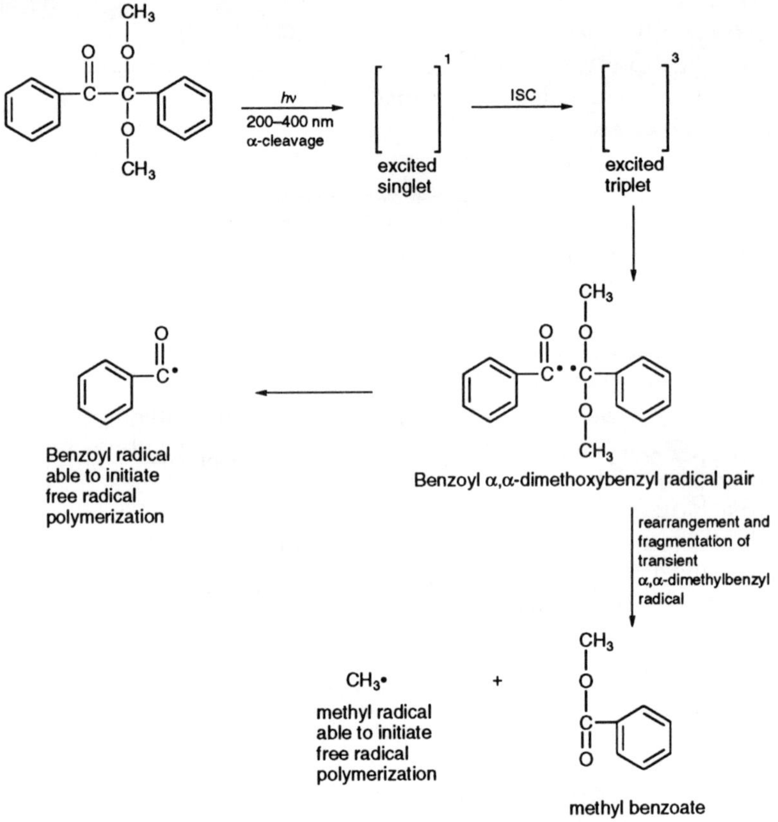

$h\nu$
200–400 nm
α-cleavage
excited singlet
ISC
excited triplet
Benzoyl radical able to initiate free radical polymerization
Benzoyl α,α-dimethoxybenzyl radical pair
rearrangement and fragmentation of transient α,α-dimethylbenzyl radical
$CH_3\bullet$ methyl radical able to initiate free radical polymerization
+
methyl benzoate

Figure 104 *Free radical polymerization using benzoyl initiator*

to meter the ink on to the roller that carries the flexographic plate. In the two-roller configuration the Anilox roller rotates in the ink well. The roller has on its surface an open box cell structure that, as it rotates, becomes filled with ink. Excessive ink is scraped away by a doctor blade—a steel knife set parallel to the roller with an angle of attack of about 30°. The method of metering ink is precise and consistent but it does suffer from excessive wear of the roller by the blade. However, as the cells are virtually rectangular their shape is maintained as the diameter of the roller is reduced gradually by wear.

In the three-roller configuration a rubber roller replaces the Anilox roller in the ink well. It runs in contact with the Anilox roller and as it rotates picks up a film of ink to fill the cells of the Anilox roller. In this instance the cells are in the form of inverted pyramids and if worn by a doctor blade would change in shape progressively. In this system they are run at different speeds and the surface of the Anilox roller is cleaned by the slip between it and the rubber one. The speed of the Anilox roller is varied against the constant speed of the rubber one, and this gives control over the amount of ink metered on to the printing roller. The resistance to wear has been enhanced by spraying the engraved steel roller with a ceramic layer.

Modern flexographic presses are being run at faster speeds to meet the demands for cheap good-quality printing of plastic film products such as bags, wrappings, labels and cartons. Webs of plastic film now are passing through multicolour flexo presses at speeds of 10 m s^{-1}. An inherent problem with a web press that produces labels of different sizes on a roller of constant diameter is that the diameter must be an exact multiple of the length of the label. However, this can be overcome by having quick-change units with rollers of different diameters (such roller systems allow the next job to be set up while a press is working).

The solvents used in the inks have low boiling points and high rates of evaporation so that presses can be run at their design speeds. The solvent content of a flexographic ink is high and the viscosity low, so that large amounts of volatile solvents can be used on a flexographic press. There may be problems associated with this, including difficulty in controlling and maintaining the visosity of the ink during a run. This has been overcome by employing a ‘closed loop’ system enclosing the pumping of the ink, monitoring viscosity, making adjustments as necessary, and delivering it to the press at the correct viscosity. Solvent recovery systems are required in order to comply with the requirements for the control of emissions under the Environmental Protection Act 1990.

(ii) The Plate

Plates for flexography are flexible wrap-round variants of the original flat-bed letterpress plates. At one time they were moulded from natural or synthetic rubbers, in a process similar to that for making photo-engravings for hot metal letterpress plates. However, photo-polymer plates now are the system of choice. There are two types: one in which the photo-polymer is in sheet form and the other in which it is a viscous liquid. In addition, for some time development has been directed towards the production of wrap-round sleeves of polymeric material covering completely the printing roller and which could be ablated by a computer-controlled laser to give a ‘direct to plate’ form of digitized art work for flexography.

Liquid photo-polymer flexographic plates are used commonly in the high-speed production of newspapers; sheet photo-polymer plates are used in printing plastic webs. Plates of the sheet type are made using materials and techniques similar to those described for letterpress photo-polymer plates.

Development continues: as an example—there is a move away from organic solvents to aqueous developers. The use of silver halide photographic film for imaging has been replaced by computer-controlled laser scanning.

(iii) The Ink

The choice of polymers and resins from which flexographic inks can be formulated is restricted by the need to dissolve them in combinations of alcohols, esters, and glycol ethers of low boiling point, or water, and to dry very quickly. These limitations restrict the choice of binder to methacrylics, nitrocellulose, polyamides, shellac, ethyl cellulose, cellulose acetate propionate, ketone aldehyde condensates, polyvinyl butyral and poly(vinyl chloride–vinyl acetate) copolymers. However, in order to dissolve some resins, small additions of strong rubber solvents such as aromatic and aliphatic hydrocarbons, and ketones, were necessary. Such strong solvents are added very cautiously, to avoid swelling or breakdown of the photo-polymer plate and of the rubber roller in the ink well of the three-roll system.

At present, the most important technical objective for the flexographic process is to reduce or eliminate the use of inks based on organic solvent. The research is directed mainly towards improving water-based inks but several reports on the use of UV-cured inks have been published. For applications in which no alternative to solvent-based inks is available at present the trend is towards eliminating hydrocarbon solvents.

Water-based flexographic inks use for the binder system a blend of polymer emulsion and polymer solution. The emulsion gives a continuous insoluble film in the presence of a coalescing solvent. It is necessary to employ a water-soluble polymer to allow the efficient and safe cleaning of presses after printing. The sensitivity to water of the printed film can be improved if the water-soluble resin is 'solubilized' with volatile compounds like ammonia or monoethanolamine. The presence of a coalescing solvent, a water-sensitive polymer, and the restricted number of polymers from which formulation is possible prevents complete replacement of solvent-based inks at present. Water-based inks have the important disadvantages of being slow to dry and of consuming more energy than the removal of organic solvents. Methods investigated for drying include forced hot air, infra-red, and microwave resonance heating.

The development of UV-cured flexographic inks has been retarded for want of suitable low-viscosity acrylated oligomers. Inks made solely on acrylated monomers are slow-curing and give brittle prints lacking in adhesion. Reducing the viscosity by adding solvents defeats the object of developing solvent-free inks. However, the curing of flexographic inks by means of photo-initiated cationic polymerization has been reported.

The essential feature of cationic polymerization is that the growing species is electrophilic and the monomer is nucleophilic. Cations may be electrophilic ions such as carbonium, sulphonium, and ferrocenium. So that propagation will take place to give useful polymers the corresponding counter-ions must be non-nucleophilic anions. In other words, competition for the electrophilic centre of the growing polymer chain must favour the more nucleophilic monomer. More than 500 compounds have been suggested for possible use as cationic monomers and they can be divided into two broad groups: alkenes, which contain electron-donating substrates which enhance the electron density of the carbon–carbon double bond, and cycloaliphatic compounds, with rings containing an electron-rich hetero-atom. An example of the alkene available commercially is triethylene glycol divinyl ether and an example of the cycloaliphatic heterocyclic is 3,4-epoxycyclohexylmethyl-3′,4′-epoxycyclohexane carboxylate.

Photo-initiators for cationic polymerization may be classified in four groups: the diazonium salts, the diaryliodonium salts, the triarylsulphonium salts, and the mixed ligand arene cyclopentadienyl Fe^{II} salts.

The usefulness for this purpose of triarylsulphonium hexafluoroantimonate comes from its considerable thermal stability, stability in the presence of highly reactive monomer, and highly efficient photolysis to yield reactive cations capable of initiating cationic polymerization. These properties arise from the unique chemical composition of the photo-initiator, the effectiveness of which can be shown to be a result of the presence of a very weak cation and a similarly weak anion.

Photo-initiated cationic polymerization combines two distinct chemical reactions. In the first, the cation for the initiator of polymerization is formed; in the second, the cation initiates polymerization of the cationic monomer. The first reaction is subject to the laws of photophysics and photochemistry, whilst the second is governed by the laws of thermochemistry.

Once the initiating cation has been formed the polymerization is controlled by temperature (technical difficulties may arise if the control is not adequate). A reaction scheme for cationic polymerization of an epoxide using triarylsulphonium hexafluoroantimonate as photo-initiator is shown in Figure 105. Termination of the polymerization often is adventitious, particularly with anions and other bases (theoretically the reaction can continue until the supply of monomer is exhausted).

Lithography

All high-speed high-quality printing processes share the characteristics of being very difficult, costly, and complex technologies that call for considerable inputs of skill, craft, and science. To the art of printing has been added science to the degree that a modern press can only operate with up-to-date knowledge from disciplines such as mechanical engineering, microelectronics, mathematics, physics, and chemistry.

Figure 106 takes lithography as an example and presents in diagram form the variables affecting success at the stages of pre-press, preparation of plate, choice of substrate, water quality, ink, printing proper, and variables attributable to the operator. Similar diagrams could be compiled for each of the other techniques.

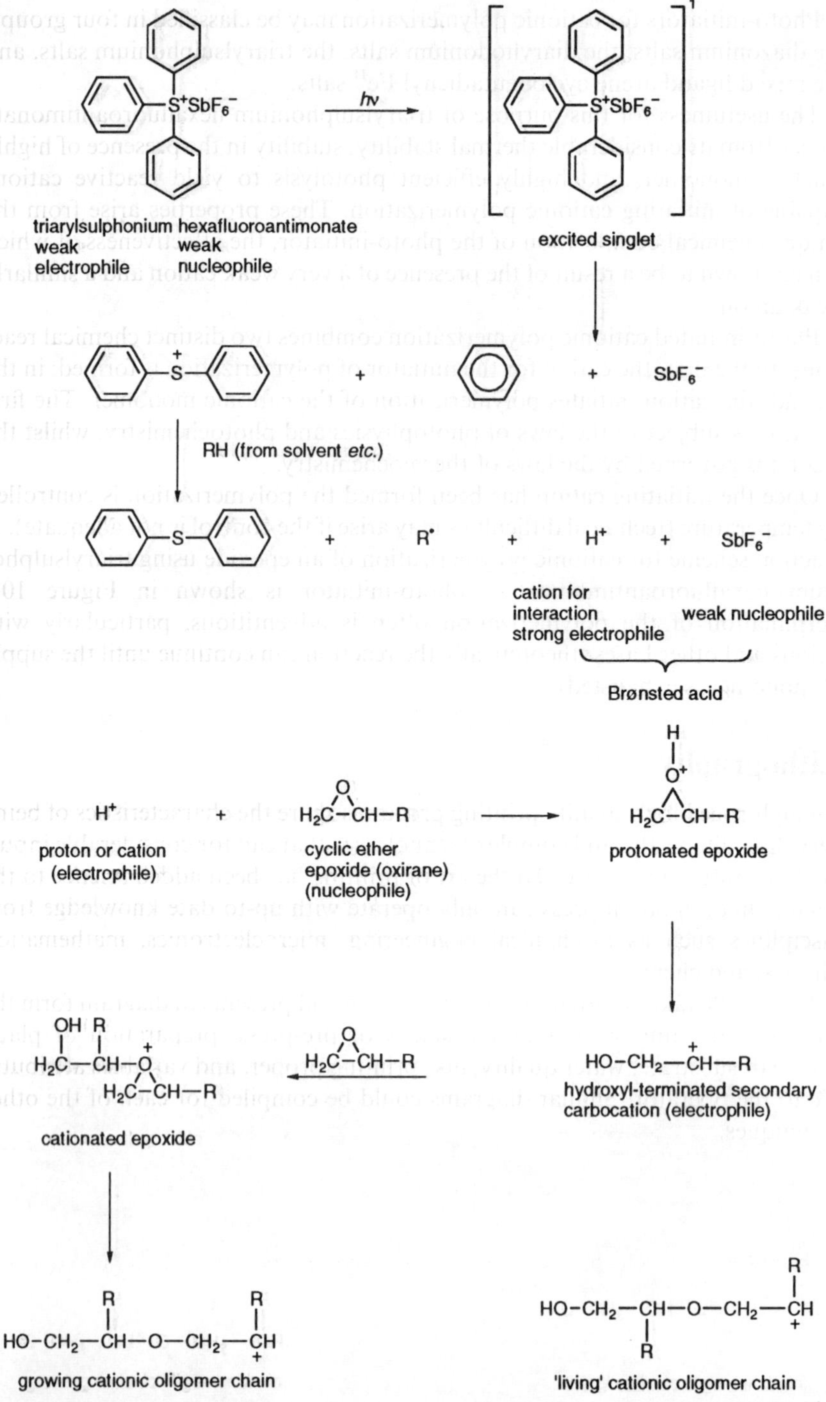

Figure 105 *A reaction scheme for cationic polymerization of an epoxide using triarylsulphonium hexafluoroantimonate as photo-initiator*

(i)

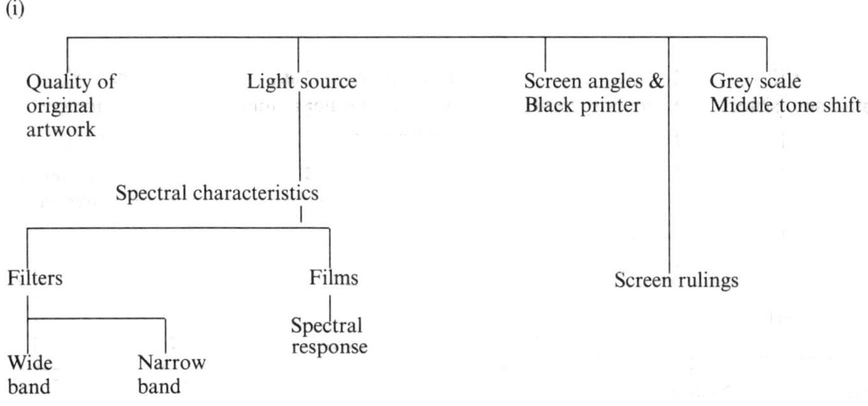

(ii)

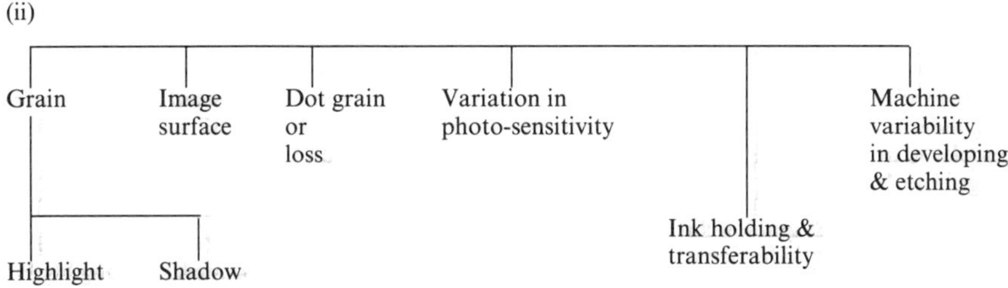

(iii)

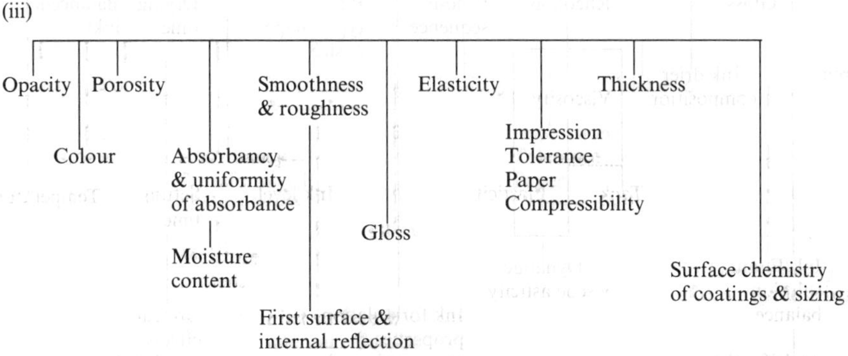

Figure 106 *Examples of variables in four-colour lithography:* (i) *At the pre-press stage;* (ii) *The plate;* (iii) *Substrate*

(iv)

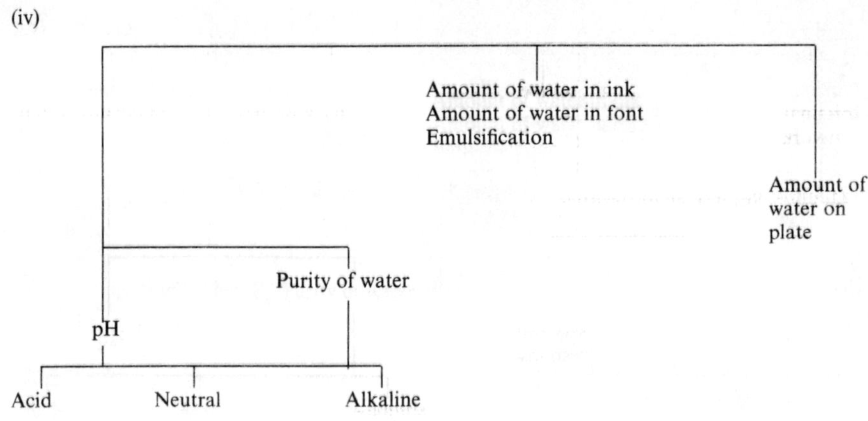

(v)

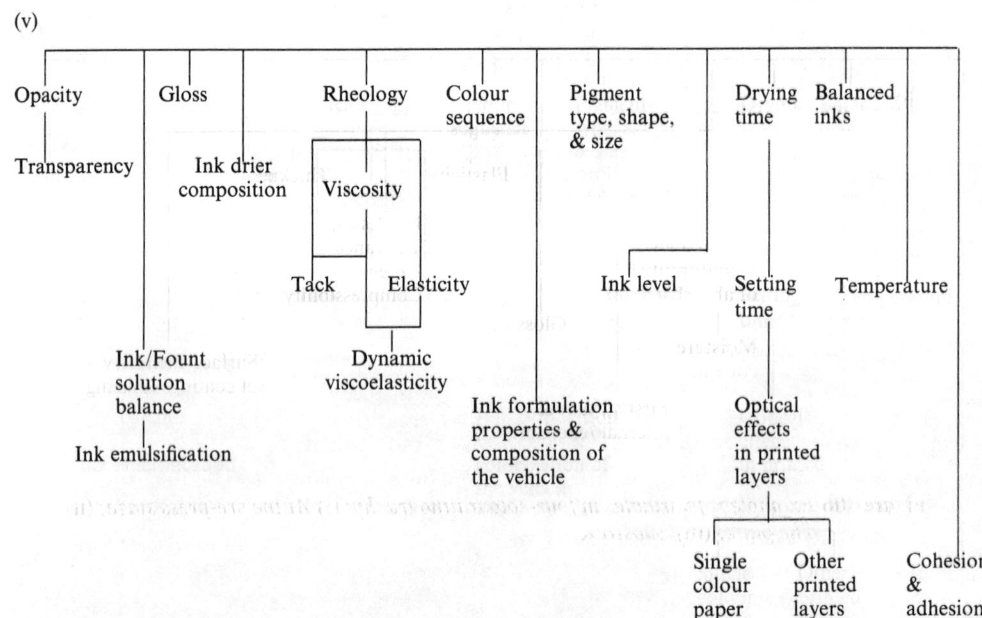

Figure 106 *Examples of variables in four-colour lithography:* (iv) *Water quality;* (v) *Ink*

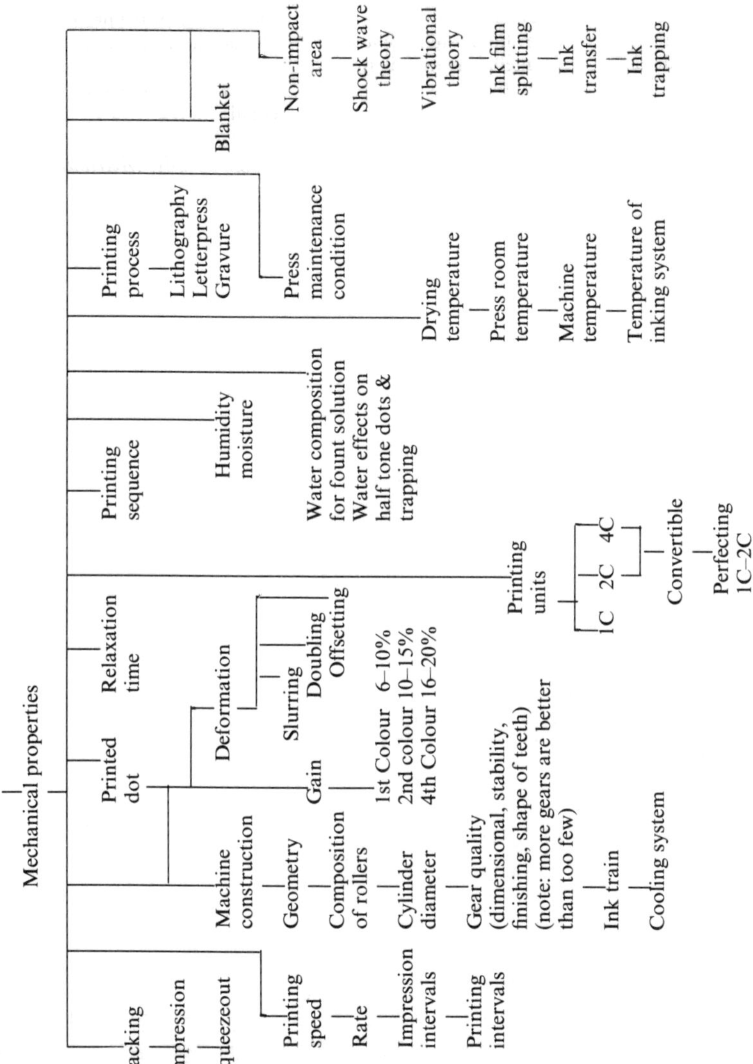

Figure 106 *Examples of variables in four-colour lithography:* (vi) *Factors in printing*

(vii)

Physical attributes	*Mental aptitudes*	*Trained*
Able to perceive lines, tones, hues, *etc.* To judge accurate register and alignment To make fastidious judgments	Mechanically minded Ability to understand and adapt to new methods and machines intelligent and highly motivated; a perfectionist	To notice variations in the plate, the ink and the running of the press Confident in handling equipment and materials Meticulous in attending to detail

Figure 106 *Examples of variables in four-colour lithography:* (vii) *Attributes of the operator*

CHAPTER 15

Vinyl Wallcoverings

W. G. NIVEN

Introduction

This chapter describes the use of poly(vinyl chloride) pastes in making an important range of consumer products.

Vinyl wallcoverings were introduced in the 'fifties but at that time were made by laminating calendered vinyl film or sheet to a woven textile backing. This gave tough, durable but relatively expensive coverings satisfactory for heavy-duty commercial or contract purposes but used only rarely for domestic decoration.

In the early 'sixties a vinyl wallcovering of lighter weight was developed, expressly for home use. This, known as the 'compact' or 'flat vinyl', comprised a paper base coated overall with PVC and decorated by printing and embossing. The new product featured attractive designs and bright colours, was easy to hang, durable, and convenient for cleaning—in other words, superior in many ways to traditional wallpapers.

Some of the early compact vinyls were produced by lamination of film but the preferred manufacturing method soon became direct coating of paper with PVC paste. Later, in the 'seventies, expandable pastes were brought into use, to give deeply textured wallcoverings with mechanical or chemical embossing of the layer of flexible expanded vinyl, while in the last decade rotary screen printing has been adopted and a new type of relief wallcovering introduced (called 'screen-blowns'), which became popular very quickly.

The manufacture of wallcoverings has become an important outlet for paste resin—some 80 000 tonnes in Western Europe in 1989, or 14.5% of the total demand for resin of this type.

Types of Vinyl Wallcovering

The main types of product may be classified as follows:

compact vinyls
'textured foam'—light weight
heavy weight: embossed mechanically
heavy weight: embossed chemically
'screen-blown' —high relief
low relief

Each is taken in turn:

(i) Compact Vinyls

These comprise a simple wallpaper base, of weight in the range 80 to 140 g m^{-2}, coated overall with PVC at weights from 80 to 150 g m^{-2} or more, decorated by printing and usually with a final hot embossing. Gravure, flexographic, and rotary screen printing are employed. Lacquer, to give a matt or pearlized appearance, can be applied over the print, and a wide variety of effects can be obtained. The embossment can be registered with the print so as to fit perfectly with it, and with light-reflecting surfaces or pearlescent lacquer the visual appeal can be remarkable.

(ii) 'Textured Foam'

With these products, a blowing agent is added to the PVC coating and at the temperatures of fusion this liberates a large volume of gas that expands the plastic to form a cellular structure. In essence, the result is a surface layer less dense but of greater thickness—the weights of coatings generally are lower (and hence less costly in terms of materials), and deeper embossings are possible. When wallcoverings of this type are made in heavier weights they may give a degree of sound and thermal insulation, although claims in this respect sometimes are exaggerated. The chemical embossing process used to create the effect of tiles and other special finishes is described later in this chapter.

(iii) Screen-Blown

A wide variety of decorative relief finishes can be produced by applying expandable PVC pastes by screen printing methods to selected areas of the paper substrate. (Screen printing also is described later.) These products range from high-relief decorations in white, designed for over-painting, to subtle multi-coloured low-relief designs.

The paper substrate usually is of weight in the range 110 to 140 g m^{-2}, in the form of two layers that can be peeled apart when the wallcovering is to be discarded after service.

Nowadays many of these products are offered in 'ready-pasted' form, which

TABLE XXII *Formulations for typical plastisols*

Constituent	*Parts by weight*
Dispersion (paste) resin	60 to 100
Filler resin	40 to nil
Plasticizers	45 to 100
Heat stabilizers	1 to 5
Inorganic filler	nil to 70
Pigment	1 to 10
Volatile diluent	nil to 10
Others	nil to 5

makes hanging easier. Their reverse side is coated with a starch or acrylic adhesive, which is activated before hanging by immersion in water for 15 to 45 seconds.

PVC Pastes

The pastes used are plastisols—finely divided polymer dispersed in plasticizer. They are mobile mixtures which may range in viscosity from liquids that can be poured readily to thick pastes—the viscosity depending essentially on the type of resin and the amount and type of plasticizer involved. Such mixtures can be applied to substrates by coating or by techniques akin to printing; after application they are converted into homogeneous flexible vinyls by heating to temperatures in the range 175 to 200°C, when the dispersed particles of resin dissolve in the plasticizer and are fused into a continuous structure.[1]

Formulation

As has been indicated, in practice there can be wide variation in the formulation of plastisols for particular purposes. Table XXII provides an indication of typical constituents and proportions but in commercial practice there may be differences in molecular weight of polymer, in distribution of particle size, in the presence and concentration of surface-agent, and so on. These differences affect performance during manufacture (as examples, in rheology and ease of fusion), and in service.[2,3] For making wallcoverings a resin of medium molecular weight usually is preferred (*K*-value 65 to 70); resins for expanded coatings often have high contents of emulsifier such as sodium lauryl sulphate—which is believed to explain at least partly their advantages for making cellular structures.

In some formulations filler or extender resin can be used—partly for economy and partly to reduce the viscosity (especially for harder products, with low plasticizer contents). Often, however, the particle sizes of such additives are of the order of 35 μm, much larger than are the particles of the paste resins (5 to 20 μm), and because of this they cannot be used in formulations for thin coatings—that is, of weights less than 200 g m^{-2}.

A considerable range of plasticizers is available but those used most commonly are phthalate esters such as dioctyl or dinonyl phthalate (DOP or DNP). If it should be necessary to reduce the temperature of fusion, or to increase the rate of fusion (as with some expandable plastisols) a fast-solvating plasticizer such as butyl benzyl phthalate may be employed. Besides primary plasticizers such as these, less efficient secondary plasticizers (like chlorinated paraffins) can be included with a view to reducing costs and improving flame retardance.

The quantity of plasticizer included in the formulation is determined primarily by the degrees of flexibility and hardness required in the finished wallcovering but the selection and amount can be varied a little with a view to obtaining paste with viscosity best suited to the processing. In this connection, it is a common practice to add a diluent or a solvent with lower boiling temperature—such as white spirit or a similar aliphatic hydrocarbon fraction, which is evaporated during processing before the fusion of the PVC.

The inclusion of heat stabilizers is essential to protect the system against thermal decomposition at elevated temperatures during processing. For this purpose, tin carboxylate esters or liquid calcium–zinc stabilizers are preferred. Thio-tin compounds are very effective as heat stabilizers but must be regarded with caution, bearing in mind that they can lead to unpleasant and unacceptable residual odours. Secondary stabilizers that can be used include epoxidized soya bean oil.

Fillers usually are natural chalks or dolomites, finely ground, and titanium dioxide the universal white pigment. Colouring is by selected inorganic and organic pigments, which are chosen because of good stability when exposed to light and heat, and the absence of 'bleed' in plasticizers. In general (particularly for environmental, health, and safety reasons) pigments based on heavy metals, like lead chromates, are avoided.

The 'others' mentioned in Table XXII can include a wide range of additives. Surface-active agents may be present to improve the wetting and dispersion of the polymer, so reducing the viscosity of the paste, and in some formulations viscosity is raised by means of thickening agents like finely divided silicas—particularly for heavier textures, to reduce sagging and distortion. Among the 'other' additives the most important is the chemical blowing agent and among a number of these the one used most widely for PVC is 1,1-azobisformamide, known generally as azodicarbonamide. This decomposes in the temperature range 195 to 230 °C—liberating 220 ml of gas per gram and leaving a colourless residue—and its temperature of decomposition is brought down to within the range of fusion of the vinyl by including in the formulation soaps of metals such as cadmium and zinc (which accelerate the reaction). Activators such as these may act also as heat stabilizers. The speed of blowing and the cell structure of the wallcovering can be controlled and determined by balancing the melt viscosity characteristics of the plastisol, the temperature of decomposition of the blowing system, and that of fusion of the vinyl.[4]

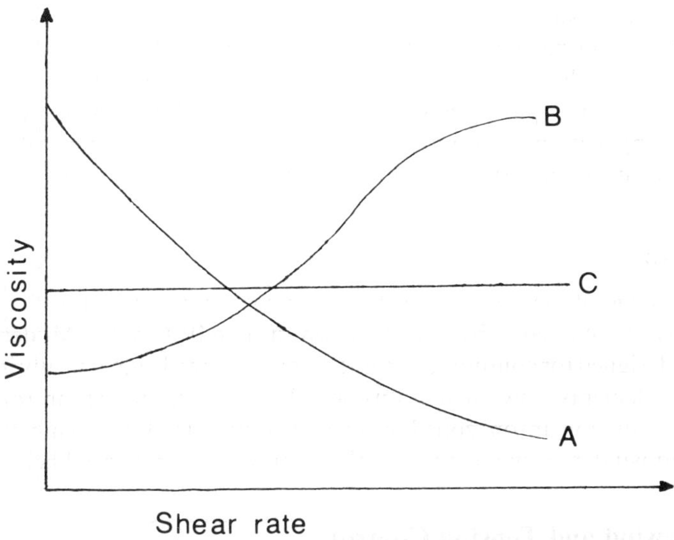

Figure 107 *Plastisol rheology:* (A) *Pseudoplastic;* (B) *Dilatent;* (C) *Newtonian*

Viscosity and Rheology

The apparent viscosity of PVC pastes varies with the shear rate applied, so that they rarely, if ever, exhibit truly Newtonian behaviour. The rheological properties of plastisols can range from pseudoplastic ('shear thinning') to dilatent ('shear thickening'), as illustrated in Figure 107.

There is a tendency also for the viscosity of a paste to increase as time goes on and as the plasticizer gradually solvates and swells the particles of polymer. Hence, in order to evaluate the rheology of a paste, the viscosity should be measured over a range of relevant shear rates at known intervals of time after mixing. Once the rheology of a given formulation has been established in this way the needs of production control usually can be satisfied by single-point measurement of viscosity by a simple rotational viscometer.

Overall rheology is determined by the selection of the dispersion resin but the viscosity of a paste is influenced by many factors—the types and amounts of the constituents in the formulation, the methods of mixing, and the temperature. Further brief comments on the effects of constituents would include:

(i) increased levels of plasticizer or the use of plasticizer with lower solvating effect will give pastes with lower viscosities and improve behaviour on aging; however, if during mixing the temperature exceeds 30 °C there will be an increase in solvation of the resin, with consequent rise in the viscosity and adverse effect on aging qualities

(ii) the reduction of viscosity by means of surfactants and diluents of low boiling temperature was mentioned earlier

(iii) increased loadings of filler will absorb and reduce partly the liquid phase, so increasing the viscosity of a paste.

For coating paper at high speed as a base for vinyl wallcoverings the plastisols should have essentially Newtonian flow and a low viscosity at low shear. On the other hand, pastes for rotary screen printing—particularly for high-relief effects—should have a marked pseudoplastic rheology. For both types of application (coating and printing) dilatent flow generally is to be avoided, as it can give rise to stringing, spitting and missed areas.

Coating

The first stage of manufacture of both compact and textured wallcoverings is direct coating of a paper base with an appropriate PVC paste. Modern coating lines are designed for continuous operation at high speed and typically include the following elements: unwind; tension control; coater; automatic control of weight of coating; suction apron; zoned oven; embossing calender; cooling drums; web guide and winder. Some features of these units are summarized below.

(i) Unwind and Tension Control

Paper base is supplied in roll form and together these units provide the coater with a continuous feed at the speed and tension required.

(ii) Coater

A great variety of techniques is available but as a basis for vinyl wallcoverings two only have become popular—reverse-roll coating and screen coating—both of which have the virtues of accuracy and high speed.

Screen coating is described later in this chapter. The system used most widely is reverse-roll coating, which is capable of applying uniform coating of weights ranging from 80 to several hundred g m^{-2}, within limits better than ± 5% and at speeds up to 150 m min^{-1}. The usual form of the equipment is three-rolls, nip-fed, as illustrated in Figure 108.

Essentially the machine comprises two precision-ground steel rollers and a rubber-covered backing roller. To refer to the Figure, the plastisol is held within edge-dams between the two steel rolls to form a metered layer of paste on the surface of the central applicator or transfer roll; the paper web passes over the backing roller in the opposite direction and the paste is deposited on it. The weight of the coating is determined by the metering gap and by the speed of the applicator roll in relation to that of the web. Normally the surface speed of the applicator is at least as fast as the web but preferably it should be faster (up to twice as fast), in order to wipe completely the metered layer of paste. Normally the metering roll at the top runs at very low speed; it is fitted with a doctor blade so that with each revolution a clean surface is presented at the feed nip.

In order to maintain and control the weight of coating it is necessary to ensure that the steel rollers are ground accurately and mounted in precision bearings in a rigid frame. Gradually in the course of use the steel rolls become

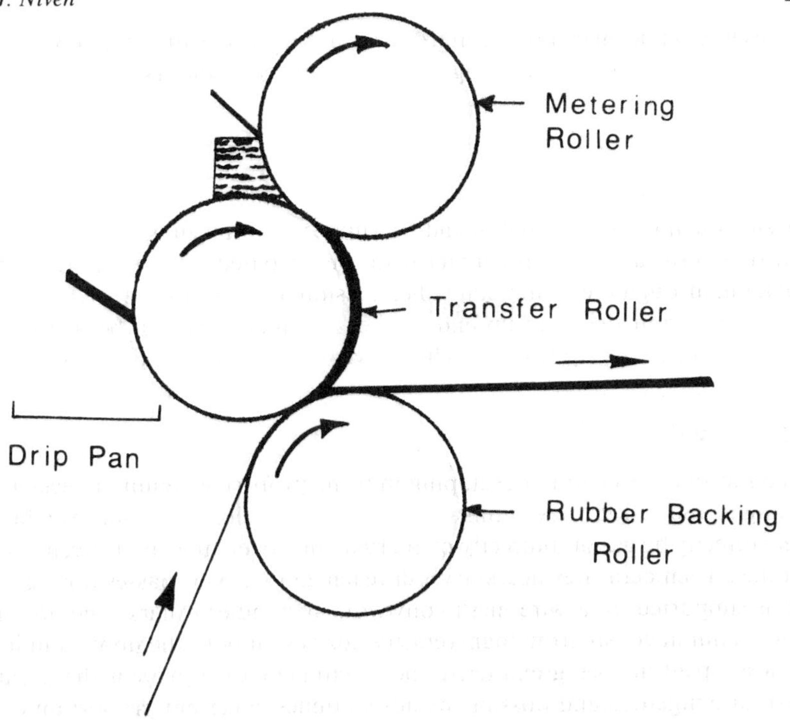

Figure 108 *Schematic representation of reverse-roll coating with three rolls*

scored and worn, and at intervals they have to be removed and re-ground so that the quality and uniformity of the coating remain consistent.

At low coating weights and high speeds considerable separation forces are generated between the steel rolls and deflection of the rolls can occur; the effect of this will be most pronounced in the middle, where the restraining influence of the bearings is least, and the coatings applied there will be thicker than at the edges. In modern machines this problem is countered by arranging the metering roll at a slight angle to the applicator (it can be seen when looking down on them that the axes of these rolls are not parallel); 'roll crossing' in this way gives clearance increasing outwards from the centre and hence an increased application of paste towards the edges of the web.

(iii) Automatic Control of Weight of Coating

The requirement for continuous production at speed implies that there must be convenient equipment for assessing weight and so ensuring uniformity. Almost always the instrument used for this work is the beta gauge, which records the attenuation of a beam of electrons from a weak radio-active source as it is passed through the coated web. The gauge can be calibrated with samples of known composition and weight, and the results are expressed in terms of 'mass per unit area'.

The electron beam can be arranged to scan the entire width of a moving web, providing data for coating weight in both the longitudinal and transverse directions.

(iv) Suction Apron

The suction apron comprises an endless rubber belt travelling round a vacuum box. It is perforated and in contact with the uncoated side of the paper web, providing through the coater controlled tension in the substrate (if the paper is pulled only from the wind-up end of the machine, there can be tearing and breaks). The suction apron also helps prevent curling at the edges of the web.

(v) Zoned Oven

Fusion of the coating and developing in it the properties required necessitates heating to 180 to 195°C, and this is done in a high-velocity hot-air oven heated either directly by gas or indirectly by thermal oil. Dwell time in the oven is short and the air temperatures necessary can reach 230°C. As it passes through, the web is supported on a wire mesh conveyor, or by idler rollers. The hot air is blown simultaneously from high-velocity nozzles on both the upper and lower surfaces, spent air being extracted and re-circulated (to prevent the accumulation of dangerous and possibly explosive fumes, a certain proportion of the air is exhausted and replaced by fresh).

So that the temperature of the web can be increased in stages the oven should be zoned in a minimum of three sections. Most pastes contain a volatile diluent and it is essential that the temperature in the first zone should be relatively low so that volatiles can be driven off before fusion occurs. Too rapid an application of heat leads to blistering of the coating as residual solvent escapes from the fused areas. In each of the zones careful control of the air temperature, velocity, and distribution is essential for uniform fusion across the full width and throughout the coating. Operating conditions for ovens for PVC plastisols were considered by A. C. Poppe.[5]

The textured (expanded) coatings usually are prepared in two stages: the spread paste is passed through the oven at 155 to 160°C and gelled—that is, it is changed from a liquid to a dry solid film, but at this stage lacks appreciable tensile strength. The gelled coating is coherent enough to be printed, by gravure, flexographic, or rotary screen methods, and after printing it is expanded by a second pass through the oven at air temperatures in the range 190 to 220°C. The rate of heating, the evolution of gas, and the degree of expansion are inter-related, so in this pass the control of distribution of heat and uniformity of temperature are even more important.

Air exhausted from the oven carries with it significant quantities of vapour (plasticizer, solvent, and products of oxidative decomposition); nowadays most ovens are fitted with some form of equipment to remove or reduce this—such as glass fibre candle filtration systems and electrostatic precipitators (which can remove up to 95% of plasticizer vapour but do not arrest white spirit and similar diluents). The most effective systems are integrated incinerators, from which heat can be recovered.

(vi) Embossing Calender

Usually, as the coated web emerges from the last zone of the oven it passes a bank of infra-red heaters (which maintain the temperature) and through a simple direct emboss nip fitted with a plain embossing roller. This treatment helps to smooth minor imperfections in the surface and gives a matt finish preparatory to printing. The roller is of steel with surface grit-blasted or (better) with a sprayed ceramic surface and water-cooled internally.

(vii) Cooling Drums

Before the coated base is wound up cooling is necessary. Some ovens include a cooling zone but more usually cooling is in ambient air as the material is passed round a number of water-cooled drums.

(viii) Web Guide and Winder

Together the guide and winder are used to wind the coated paper accurately into large reels ready for further processing.

Printing

The decoration of the wallcoverings usually involves appropriate printing techniques (the printing of plastics is reviewed in more detail in earlier chapters of this book). Early wallpapers were printed by means of stencils or wooden blocks which were carved to leave raised areas carrying the image or images. Later 'surface printing' was introduced—the first successful continuous printing process for wallpaper—which used rollers constructed on much the same principles as the wooden blocks. Processes of this kind are employed still, but usually on a limited scale for special products or to reproduce antique papers. Vinyl wallcoverings are printed by means of rotary gravure, flexography, or rotary screen processes.

(i) Gravure

Gravure makes it possible to reproduce the most delicate of tonal images and generally gives excellent results on the smooth and receptive surface of vinyl wallcovering base.

Traditionally the inks were based on organic solvents, comprising solutions of binder resins such as vinyl chloride polymers, vinyl chloride–acetate copolymers and acrylics in toluene/methyl ethyl ketone or ethanol/ethyl acetate—pigments, flattening agents, waxes, and other substances being added as required to enhance performance. More recently, because of the desire to reduce so far as possible the need for organic solvents, water-based ink systems have been developed and are offered commercially. Early difficulties with these (such as with adhesion of print and wetting) have been overcome.

(ii) Flexography

Many of the original surface printing machines for wallpapers were replaced by flexographic equipment, which is considered good for printing large areas of solid colour. However, modern developments in engraving by laser have made it possible for flexography to approach the quality of gravure in tonal work.

Generally inks for flexography are similar in composition to those for gravure but with an alcohol/ester solvent blend to reduce the extent of attack on rubber-covered rollers. Water-based inks also are being used successfully.

(iii) Rotary Screen Printing

The application of this form of printing, using PVC plastisols and water-based inks, is described later in this chapter.

Today, the printing of vinyl wallcoverings by gravure and flexographic methods should present few difficulties; the surface characteristics of the base material usually are of good standard but when using water-based inks attention may have to be paid to aspects such as surface tension and wettability.

The extensibility of the base is low, and it can be handled easily with a minimum of trouble with regard to register (although care is necessary to ensure maintenance of accurate register throughout a run). Modern presses are fitted with good automatic control of register. Consistent colour and good matching edge to edge also are essential. To maintain colour, the viscosity of the inks should be kept within prescribed limits throughout (again, automatic control is available). On modern presses high rates of output are possible—with automatic splicing of reels speeds of up to 250 m min^{-1}. Such presses also should be capable of rapid changes from one design to another, with limited losses in terms of time and material.

Mechanical Embossing

Since plasticized PVC is thermoplastic both compact and expanded wallcoverings can be given a surface texture by hot embossing. This usually is the last stage of production before trimming and reeling.

In a typical embossing unit the material first is passed over a steam-heated drum then under a bank of gas-fired or electric radiant heaters before entering a nip between a water-cooled metal roller engraved with the pattern to be embossed and a counter roller covered with resilient rubber. Normally the backing roller also is cooled, either internally or externally. Either roller may be driven, with the other free-running under the embossing pressure. The vinyl is softened by the initial heating and is moulded to the surface of the embossing roller before being chilled sufficiently to retain the embossment after it leaves the nip. The embossed material is passed over cooling drums before being reeled.

The speed of embossing is related to the thickness of the vinyl and to its thermal properties but rates from 40 to 80 m min^{-1} usually are achieved.

Compact wallcoverings are embossed in this manner and light expanded coatings (up to 120 g m^{-2}) can be expanded and embossed simultaneously. Heavier expanded coatings require a fixed-gap technique for embossing: with these the gelled vinyl base is printed then heated in an oven to bring the coating to the thickness required before it is passed through a nip with a fixed gap between the embossing and backing roller. The size of the gap depends on the depth and nature of the design to be produced but will be in the range 30% to 90% of the thickness of the expanded product. The two rollers are linked so that both are driven at the same surface speed. Expansion and embossing can be carried out in sequence in a continuous operation but this may not be easy to control and it may be preferable to expand and emboss in separate stages, so that the heating and cooling necessary for successful embossing can be controlled more precisely.

Faults in Embossing

In this form of embossing a variety of faults can arise. It is particularly important with wallcoverings to obtain visual uniformity, and especially that the embossment should be identical across the entire width. Even a slight variation—say, in the depth of the pattern—will be apparent when wallcovering is hung in large areas. Quite a common problem is the loss of embossing towards the edges of the web and this can be overcome only by taking great care to ensure uniformity of heating. Often, extra heaters are added to the machine to counter loss of heat from the edges as a web passes to the emboss nip.

Good and uniform reproduction of the pattern depends also upon adequate and uniform cooling of the embossing roller. In its simplest form this roller is no more than a hollow steel tube, with the cooling water entering at one end and leaving at the other; evidently such a configuration will give diminishing temperature across the roller and hence can result in variation in the standard of the work. Modified forms have been proposed and developed with spiral and twin-shell construction intended to increase the turbulence and efficacy of the coolant. The best results are obtained, particularly for running at high speed, with an embossing roller of large diameter and cooling water at low temperature.

Other problems in embossing can be caused by variations in the pressures applied, from an aged or worn backing cylinder, and from uneven coating of the feedstock. Provided its cause can be identified the steps necessary to correct any of these faults should be fairly clear.

Both compact and expanded wallcoverings can exhibit blistering at the embossing stage: with the former this is attributable to residues of solvents from printing inks and lacquers, which may evaporate explosively when exposed to the radiant heaters (in bad instances, solvent vapours may accumulate and ignite under the heaters, even setting fire to the web); with expanded materials there may be excess blowing agent, or it may have been dispersed unevenly and give rise to localized further blowing. Problems such as these often can be overcome by using instead of the radiant heaters a short high-efficiency hot-air oven (of length 6 to 8 m) as the primary source of heat. By this means, using air temperatures up to 220°C, compact and expanded vinyls can be embossed safely at speeds up to 100 m min^{-1}.

Embossing in Register

This important technique allows the production of compact wallcoverings where the embossing is superimposed in perfect register with a matching printed design. If the embossing includes light-reflecting surfaces and, say, pearlized lacquers, very effective and dramatic effects can be obtained.

In principle it would seem relatively easy to engrave an embossing cylinder and corresponding print cylinders so that embossment fits accurately with the printed images. In practice however some means of control is needed in order to maintain register during hot embossing and the basis for this is the small degree of elastic extensibility of the vinyl-coated paper base (of the order of 1.8% at break). In a machine fitted for embossing in register the printed base is heated and embossed under controlled tension between two positive nips. One nip, at the entry end of the machine, is driven, while the other is the emboss nip itself. Altering the tension on the web between the two alters its extension, and thus the length in which a design is repeated.

When an embossing roller and the corresponding print cylinders are being engraved as a set the diameter of the print cylinders is made deliberately smaller than that of the embossing roller (generally 0.7% less). This so-called 'disproportionation' is calculated so that the print repeat length will match the emboss repeat length when the extension of the printed vinyl web is 50%. While the work is in progress the relative positions on the web of the printed motif and the embossing roller are monitored constantly by an electronic control system: as the web enters the embossing nip an optical device notes the position of pitch marks printed at the edge in relation to the embossing roller as it revolves. The control system compares the two signals continually and if any divergence from perfect fit is noted there is an automatic slight adjustment (increase or decrease, as necessary) in the tension to bring print and embossing back into register.

Engraving Embossing Rollers

The manufacture of wallcoverings requires a wide variety of surface textures, ranging from simple matt finishes to complex light-reflecting effects with lenticular and multi-directional slash engraving.

For the most part the embossing rollers are made by milling, a process in which a small cylindrical hardened steel tool bearing a positive of the texture required is run under high pressure repeatedly around the roller to be engraved, until a complete negative is created on its surface. The tool must be applied

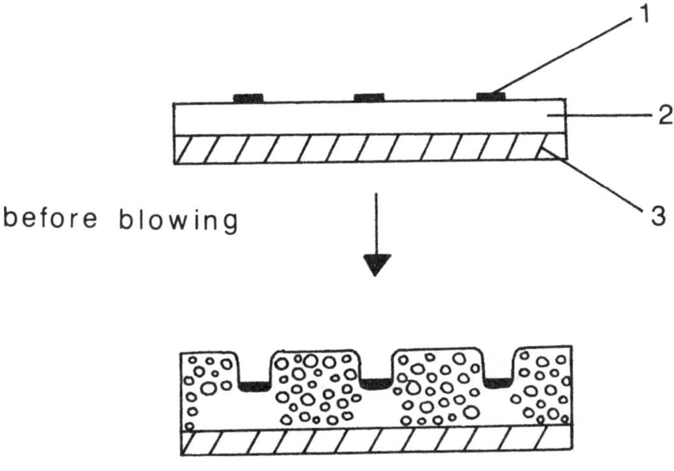

Figure 109 *The inhibition route to chemical emboss:* (1) *Ink containing inhibitor;* (2) *Gelled expandable PVC coating;* (3) *Substrate*

repeatedly, in stages, across the full width of the roller, and considerable skill is necessary to ensure that the joins are invisible. Often the engraved rollers are finished by hand, by acid etching, or shot-blasting. It then usually is chromium-plated in order to protect the surface against corrosion.

Very attractive finishes can be obtained from rollers formed by electro-deposition of nickel, including leather, textiles, and other natural and synthetic textures. This method allows the faithful reproduction of a finish across the full width of the cylinder and, since undercuts cannot be achieved by the milling process, gives a more realistic appearance in (as an example) woven designs.

Chemical Embossing

Strictly, this technique is not a method of embossing in the traditional sense of the word, but it is an important alternative method of creating finishes for textured expanded wallcoverings. Appropriate substrates are coated with an expandable plastisol containing a blowing system, and the coating is gelled and printed selectively with inks containing a reactant for the blowing system. The effect of the reactant is to modify the extent of expansion when heat is applied; it may activate the system, so that expansion is greater in the printed areas and they form the raised portions of the finish, or it may be an inhibitor—in which case the printed areas become 'valleys'. Since the reactant and the coloured inks are applied together, printing and finish are inherently in register. Figure 109 illustates this technique.

Reactions of the type described underlie several patented processes, among

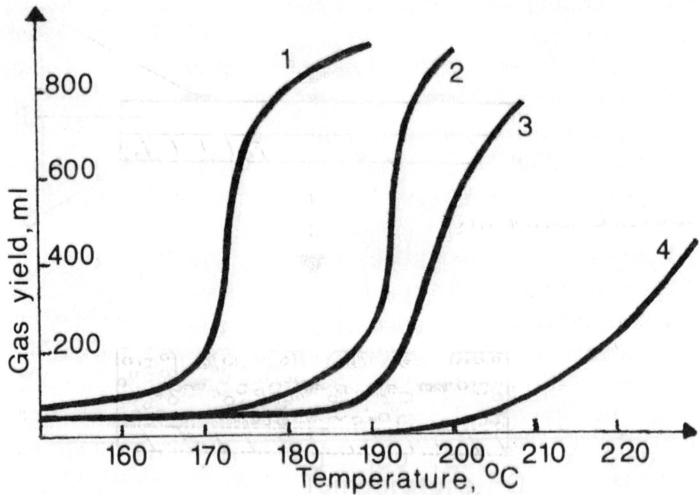

Figure 110 *The effect of reactants on temperature of decomposition and yield of gas from azodicarbonamide: reaction medium 400 g dioctyl phthalate, 3 g azodicarbonamide; reactants:* (1) *1 g zinc oxide;* (2) *1 g zinc oxide, 0.25 g trimellitic anhydride;* (3) *no reactant;* (4) *1.5 g thiourea*

which the Congoleum inhibition process is thought to be one of the most effective; the primary patents have expired and certainly it is used widely.[6]

Expansion by means of azodicarbonamide and associated metal salt activators was mentioned earlier: many substances have been suggested for use as inhibitors for such systems, their mode of action typically being to raise the temperature of decomposition of the blowing system and hence reduce the volume and speed of evolution of gas. Some inhibitors act by neutralizing metal salt activators so that they no longer have the effect of reducing the temperature of decomposition. Examples of this type include fumaric acid, trimellitic anhydride, and benztriazole. Others react with the azodicarbonamide and transform it into a different compound substantially resistant to the activation; thiourea is an example of this type. Inhibition in these different forms is exemplified in Figure 110.

(1) In the first stage, an expandable vinyl plastisol containing blowing system in a proportion from two to ten parts per hundred of resin is applied to a suitable substrate and gelled at 120 to 150°C. Zinc oxide is preferred as activator or 'kicker'; it is a good heat stabilizer for a vinyl composition, gives good expansion, and yet requires surprisingly small amounts of inhibitor to produce finishes of the standard required. At this stage, the degree of gelation is important; it must be good enough for the film to withstand printing without deformation, and without adhering to the print cylinders.

(2) The gelled film usually is printed by rotary gravure, although flexographic and screen methods also can be used. Where depressed or 'valley' areas are

required, inhibitor inks are applied. They consist of a solution or dispersion of binder resin, pigments, and from 5% to 20% of inhibitor, based on the weight of ink. There is a direct relationship between the quantity of inhibitor applied and the degree of contrast in the resulting finish. The normal depth of engraving on a gravure roll (from 12.5 to 50 μm) gives a means of controlling the contrast and producing different levels in the pattern at the surface; another way is by varying the concentration of inhibitor in the inks.

(3) Satisfactory inhibition requires full migration of the inhibitor into the vinyl layer: at the outset the rate of migration is quite rapid, so that with light-weight wallcoverings satisfactory results can be obtained by expansion immediately after printing; with heavier products the best results are obtained by allowing up to 24 hours between the printing and expansion stages. On the other hand, one must guard against excessive migration of the inhibitor, and particularly its movement from layer to layer in unexpanded material—which can lead to unwanted 'ghost' images in the product.

(4) A layer of transparent lacquer or PVC plastisol may be superimposed as desired to protect print and to control the gloss of the finish. A vinyl–acrylic lacquer may be applied in tandem with colour printing or, say, a clear plastisol coating by screen printing, usually in the last stage as the material passes to the oven for fusion and expansion.

(5) The oven temperatures used for expansion are in the range 190 to 220 °C; the period of time for which the material is heated depends on factors including the formulation and weight of the coating, and characteristics of the oven, but normally it is not critical. (If the dwell time is raised the inhibited layer may be expanded to a greater extent—so that enhanced contrast will not necessarily follow.)

Rotary Screen Printing

In a rotary screen system plastisols for compact and textured products are applied to suitable substrates continuously from seamless, cylindrical nickel screens. The plastisols are fed into the screens and pressed through perforations to the substrate by a squeegee of flexible steel or rubber; the substrate is supported by a counter-pressure roller, usually of chromium-plated steel. The principle is illustrated in Figure 111.

The technique is very versatile and adaptable. Plastisols may be applied overall or—by means of engraved screens—can be printed in selected areas, in weights that can be varied between 30 g m^{-2} and more than 500 g m^{-2}. Suitable substrates include paper, other non-woven materials, glass-fibre webs, and textiles. Water-based inks can be employed as coatings or in printed designs. For relief effects, expandable aqueous polyacrylate dispersions offer an alternative to expandable plastisol inks. They are dried and then expanded at 120 °C, a temperature much lower than is required for vinyls; with a multi-station machine, a wide variety of products can be made.

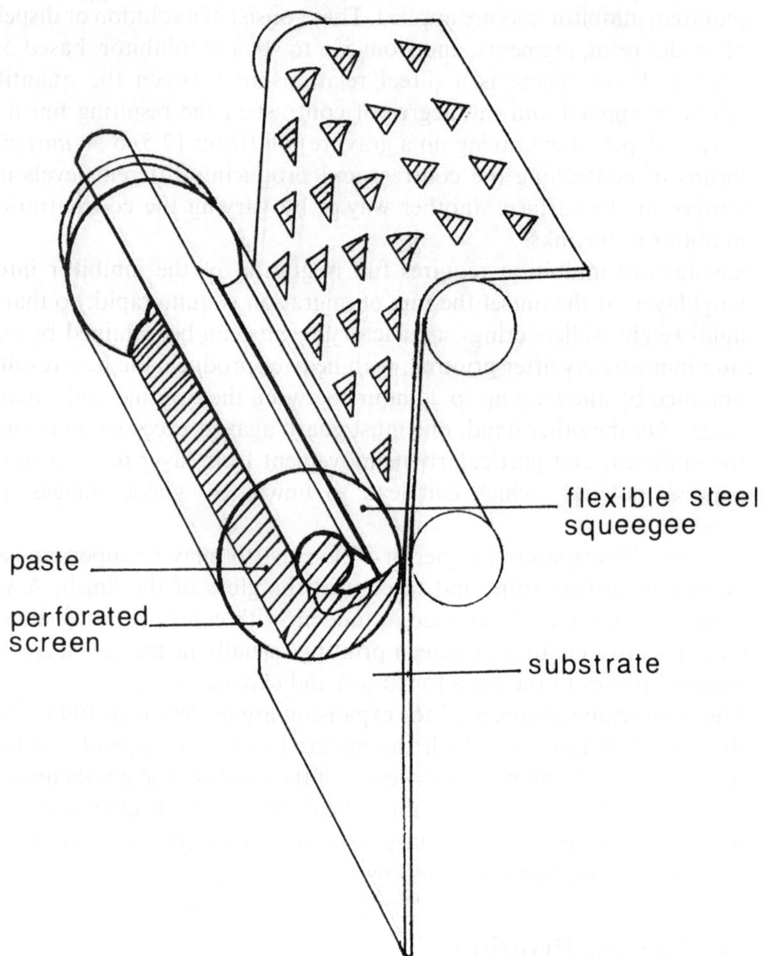

Figure 111 *Principle of rotary screen printing*

Typically, a screen printing line for wallcoverings would include the following elements: pre-conditioning unit; multi-station screen printer; suction apron; zoned oven; cooling zone or drums; web guide; splitting and reeling. Units of this nature were considered earlier in the section about coating but some special aspects of a line for screen printing are described below.

(i) Pre-Conditioning Unit

The paper feedstock includes moisture in the range 5% to 7% and it is dried by heating before being passed to the printing stations. Drying gives significant shrinkage in width and if this were to take place after the printing of the first colour lateral registration with other colours would not be possible. The pre-conditioning ensures that all the prints fit laterally.

(ii) Screen Printer

The line may include several printing stations in sequence, each one of which comprises essentially a print head, a vertical drier, and two cooling drums. Holders in the print head support the screen, hold it in tension, and drive it; the screen can be adjusted for lateral registration with others in the sequence.

A pneumatic pump feeds the printing paste and a sensor helps ensure that its level within the screen is kept fairly constant; application of the paste is by adjustment of the squeegee relative to the counter-pressure roller.

The web passes the screen vertically into the drier above; the drying or pre-gelling of the print is by means of hot air blown on both sides of the web, the air being heated by thermal oil or, more usually, by direct gas firing. The volume of air circulated and its temperature are controlled and can be varied. After drying or pre-gelling the web must be cooled so that it will not carry heat to the next screen, and this is done by passing it round two water-cooled cylinders.

The printing stations are connected mechanically: a common line shaft system drives the cooling cylinders, screens, and web so that once the screens have been brought into registration this can be maintained by the accuracy of the drive. (However, a facility to adjust registration is provided at each printing station.)

A line such as has been described is suitable for extended continuous operation but when changes of colour or design are required it is necessary to stop the work to make them: for changes of design, a set of screens is taken out and replaced by another; for colour, the existing screens are removed, cleaned, and put back. With a view to overcoming such delays, a 'non-stop design changing machine' now is available in which each printing station is provided with two screen print heads, arranged one above the other. While at each station one head is in use the second and its squeegee can be prepared. When the quantity required of the current run is complete, the second set of screens is moved into position, paste is pumped to them, and the squeegees are activated; simultaneously, the screens and squeegees of the first design are removed from the printing position. An automatic system registers the new colour or design and in this manner the change can be made without stopping the machine and with loss of material limited to about twenty or thirty metres. Figure 112 illustrates a six-station non-stop screen printing machine.

Machines with printing widths of 600, 900, 1200, and 1400 mm are standard. There are wider machines, which may be used to produce vinyl floorcoverings as well as wallcoverings.

In all types of machine the weight of plastisol and the qualities of coating and printing are determined by a correct balance of the following factors:

type of screen
type of squeegee blade, its adjustment and pressure
viscosity and flow behaviour of the paste
type of substrate
speed of machine.

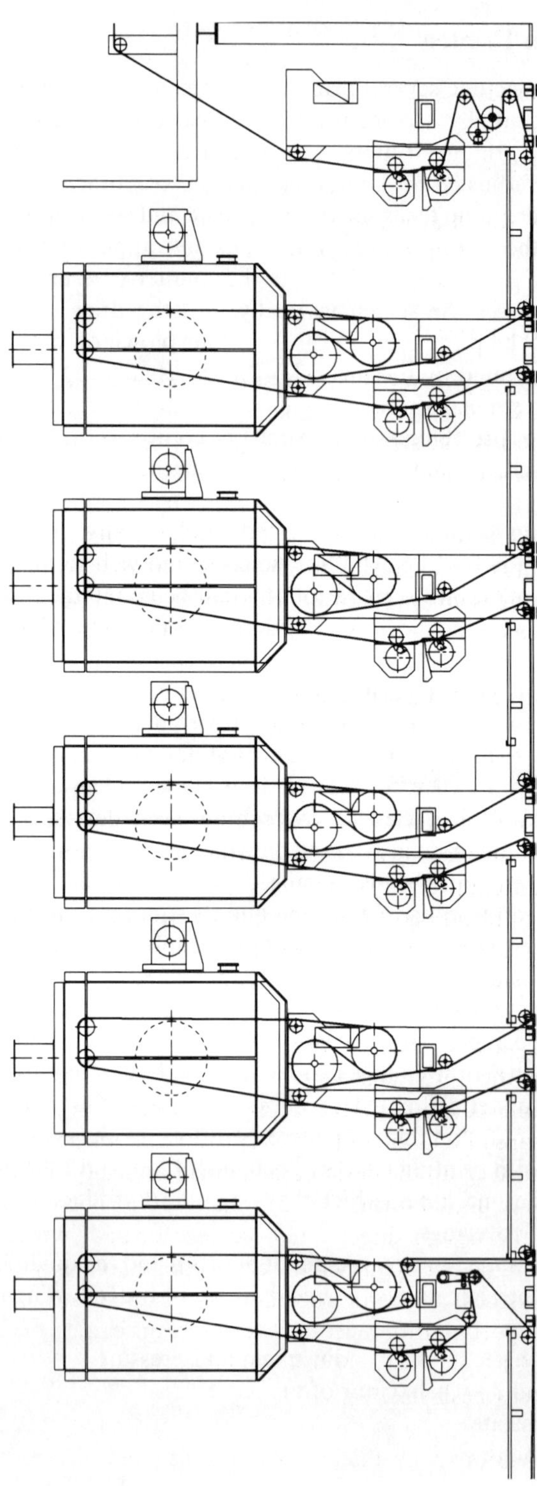

Figure 112 *A six-station, non-stop rotary screen printing machine*
(Courtesy Stork X-cel BV)

If desired, units for gravure or flexographic printing can be included in a multi-station screen printing line, so allowing the combination of different effects and finishes.

Rotary Screens

Two types of screen—lacquer and galvano—are in use, both produced by electro-deposition of nickel on an appropriate former.

(i) Lacquer Screens

In a screen of this type a regular pattern of uniform holes is created over the entire surface. Screens like this are used for coating.

To print a design, the screen has to be engraved. First it is coated overall with a light-sensitive lacquer, covering all the holes, and then a transparent photographic film carrying in black the design to be printed is wrapped carefully round the dried lacquered surface and the whole exposed to ultra-violet light. Exposure renders the lacquer insoluble in water whereas the unexposed parts under the black design are soluble and can be washed away, freeing the holes in the nickel screen in those areas. The next stage is baking to harden the lacquer remaining, and then end rings are fitted so that the screens can be subjected to tension and driven in the printing machine.

(ii) Galvano Screens

With galvano screens the design is produced at the time of manufacture, holes being left in the surface only in the areas desired. This approach permits the use of holes of different sizes, which can be used in the design to give local differences in application of plastisol and, when printing expandable products, relief differences from a single screen. This can be a useful feature, particularly when the number of printing stations on a machine is limited. On the other hand, the advantage over lacquer screens is outweighed to a degree by the higher cost of galvano, their limitation in mesh size, and (since they have lower wall thickness) their greater sensitivity to damage and deformation.

It has become possible in recent years to transfer print designs to lacquer screens by means of a laser controlled by computer: in this system the separate colour films making up the design are scanned optically and the resulting data stored on tape in digital form. The information then is used to direct the laser, which burns the design directly into screens coated overall with hardened lacquer. The 'engraving' is quick and very accurate, and, since the lacquer is not light-sensitive, stocks of pre-lacquered screens can be held in anticipation.

The most important differences between screens can be summarized as:

wall thickness
mesh number (mesh size)
open area.

Screens of various wall thicknesses may be used. The thicker screens are

TABLE XXIII *Features of rotary screens, with applications*

Feature				*Application*
Wall thickness/μm	*Mesh number*	*Open area/%*	*Hole diameter/μm*	
118	40Std	23	320	Coating; high-relief printing of textured products
200	40CH	40	400	High application; coating; printing textured products
100	60Std	14	170	Solid printing; printing textured products
130	60LR	14	170	Coating
85	80Std	12	115	Solid printing and expanded products; tonal prints
150	80SP	25	170	} Printing textured products; low to medium relief
150	100SP	20	120	}
105	105	15	105	Solid printing; low relief expanded products
100	125 Penta	15	80	Solid printing; fine detail
100	155 Penta	13	45	Solid printing; fine detail; tonal prints

more durable and stable for coating and printing, and for a given size of mesh give a larger application of plastisol.

The mesh number is the number of holes per linear inch and in this application useful numbers range from 40 to 155. (Table XXIII summarizes some important types and their applications.)

As a general rule, the higher the mesh number the finer the screen, the lower the weight of plastisol applied, and the better the quality of definition of print.

Some screens with mesh numbers in the range 11 to 40 are available with round openings but generally the opening is hexagonal in form, which gives the most favourable combination of permeability by the paste and firmness of screen.

The 'open area' is that percentage of the total surface that is open when viewed from inside the screen; in general, the higher the mesh number the lower the open area.

The range of rotary screens has been extended by the development of the 'Penta' screen, which has a higher open area for a given mesh number than a normal screen. This result is achieved by the use of openings in a different form: those in normal screen are of conical shape, with diameter diminishing towards the outer surface; with increasing mesh number the diameter of the opening becomes smaller, and with this the open area of the screen. In Penta screens the openings are of more cylindrical form, so that screens can be made with higher mesh numbers but still with good 'open areas'; with these, it is possible to achieve greater uniformity of print, sharper and more delicate contours, and half-tone effects. On occasion a residual image of the mesh can be seen on the edge of printed areas (the 'saw-tooth effect') but with Penta screens this fault is reduced to a minimum.

There are other screens with special combinations of mesh numbers and open area. Factors that influence the choice of screen for a particular purpose include:

the type of plastisol to be used
the weight of application required (and hence the degree of relief obtained in the expanded prints)
the nature and delicacy of the design.

Squeegees

The squeegee is one of the most important elements in successful screen work: it consists essentially of three components:

(i) a stainless steel pipe with openings and a clamp for the blade; (this serves both to feed plastisol to the screen and also to keep the blade in place)
(ii) a spreading plate, to ensure even distribution of plastisol across the screen
(iii) the steel blade proper—mounted accurately, evenly and firmly in the holder.

The weight of application of material and sharpness or otherwise of the print are linked with the thickness of the blade, its lip length, and the pressure applied to it. A short lip length will give comparatively low application of material and a sharp image; longer lip length will give high application weight but generally a less sharp image. As a guide, a blade 0.2 mm thick with lip length 28 to 32 mm will give high application of plastisol, at an extension of 25 to 27 mm an average application with good sharpness, and at 20 to 22 mm a low weight of application.

Thinner blades (*e.g.*, 0.15 mm) can be used successfully with water-based inks but show a tendency to 'float' when printing with plastisols.

Usually, for good printing, the tip of the squeegee blade should be arranged opposite the centre of the counter-pressure roller—where contact is made with the screen. Its vertical adjustment should be checked carefully on both sides of the screen to ensure an even alignment across the width. If coating with PVC, so as to raise the application weight, the tip can be placed above the centre of the counter-pressure roller.

The weight of application of a plastisol can be adjusted by varying the pressure applied. This has the effect of varying the angle between the blade and the screen and hence the fluid pressure in that wedge zone. As Figure 113 illustrates, a higher pressure (smaller angle) will result in a higher application weight.

Plastisols for Screen Printing

The general principles of formulation outlined earlier apply equally to compact and expandable plastisols for screen printing. Plastisols for compact products

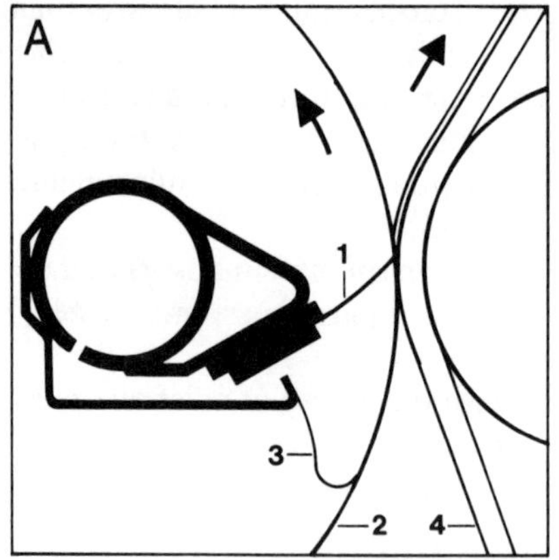

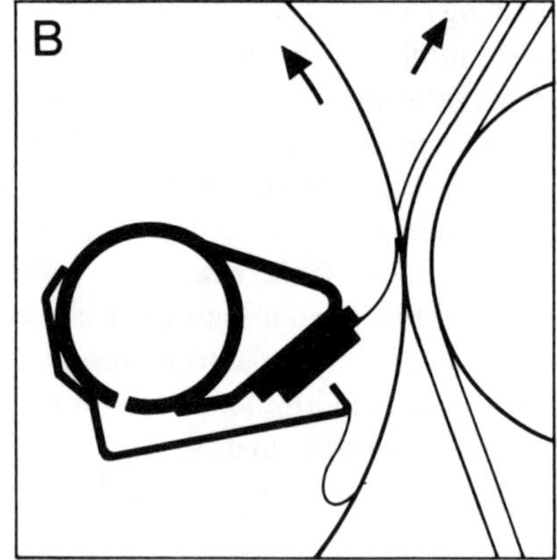

Figure 113 *The effect of squeegee pressure on the weight of paste applied:* (A) *Light pressure—low application;* (B) *Higher pressure—higher application;* (1) *Squeegee,* (2) *Screen,* (3) *Paste,* (4) *Substrate*

can be formulated as required for matt, high gloss, or metallic finishes; plastisols also can be given high or low ratios of expansion, according to the concentration of the blowing system included. Expanded prints may have smooth or textured surfaces: in the latter case, the surface of the expanded material is given micro-roughness by incorporating a small amount of a second blowing agent which has a temperature of decomposition lower than that of the primary azodicarbonamide system and/or of 1% to 2% isopropyl alcohol.

The rheology of the formulation is very important: printing pastes should be slightly pseudoplastic at low shear and at higher shear rates remain so or become more Newtonian. The degree of pseudoplasticity and the range of viscosity actually selected will depend upon the weight of application required and character of the print—heavier application weights demanding greater pseudoplasticity so that flow is minimized and the character and crispness of newly formed print are maintained.

Plastisols for coating may be less pseudoplastic and at low shear should have a low viscosity to give good flow and a smooth surface. Dilatency, which gives poor printing characteristics and 'scumming' of the plastisol on the outside of the screen, must be avoided. The main factor determining rheology of the paste is the resin selected, and this aspect therefore is most important.

Avoiding Faults

So that there will not be obvious and unsightly variations in effect when wallcoverings are hung, 'shading' and so forth, it is essential to ensure that print is uniform across the width of the web. This aspect becomes even more significant if the screen machines are printing 'double-width' material, to be split to form two finished webs, each 51 to 53 cm wide.

There are several machine checks that can be made to help ensure reproducible printing free from shading:

(i) for each type of screen and plastisol combination the most appropriate viscosity should be established and maintained rigorously throughout subsequent production (by checks of viscosity and by adjustment as necessary each time before printing is begun)

(ii) the feed of the plastisol should be uniform across the width of the screen and its level within the screen should be kept as constant as possible, by means of an electrical level probe controlling the feed pump

(iii) it is essential for both coating and printing that the counter-pressure roller be parallel to the screen; any deviation between the left and right side inevitably will lead to differences in application

(iv) normally also the height and pressure of the squeegee should be equal across the width of the screen (although sometimes slight differences in pressure from side to side may be employed to obviate other imbalances in the system)

(v) for products with textured prints overall (which are notoriously prone to shading) the best results usually can be obtained by printing with a screen of mesh number as high as possible consistent with the application weight and finished effect desired (Penta and special screens are useful here).

Plastisols containing titanium dioxide or other hard fillers have an appreciable abrasive effect on the steel squeegee blade. Normally this is of no consequence but when screens have only very small open ‘engraved’ areas and the residence time of the plastisol within the screen is lengthened, metal abraded from the squeegee can in the course of production discolour progressively the plastisol and the prints. Problems of this nature are particularly noticeable and unacceptable with white or pastel-coloured prints. They can be overcome by means of a bevel-edged polyester squeegee blade about 0.5 mm thick or by using a polyurethane rubber blade (the latter stuck by means of a two-phase curing adhesive to a 0.15 mm steel squeegee blade to give support).

Screen Coating

A rotary screen printer can be used to apply overall PVC pastes and other coatings. The system is capable of an accuracy in coating weight better than $\pm 5\%$ and has been run at speeds up to 140 m min^{-1}. With the appropriate screen, coating weights from 40 to over 500 g m^{-2} can be obtained. Generally, screens of mesh number 40 or 60 are the most useful—screens with thicker walls having greater durability and service life. When coating with PVC pastes the tip of the squeegee can be set above the centre of the counter-pressure roller, thus increasing application weight (see Figure 114). However, the blade must not be set too high, otherwise there may be wrinkling of the screen and damage to it. When operating with screens with a wall thickness greater than 0.2 mm the tip of the blade should not be placed more than 4 mm above the centre line; with thinner screens the over-centre position should not exceed 2 mm.

At the point of coating a series of dots of paste is extruded on the paper

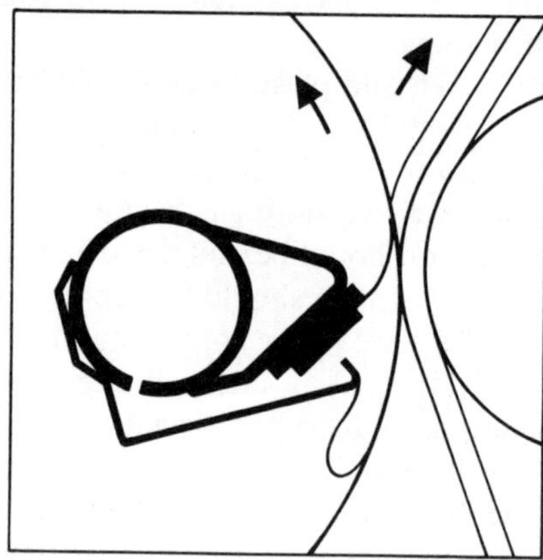

Figure 114 *Screen coating; increased delivery of paste when the blade of the squeegee is above the centre of the counter roller*

substrate. In order that the coating becomes continuous and smooth the paste must have good flow at low shear so the dots can link to form a coherent film. Smoothing to reduce residual roughness or effects such as 'orange-peel' may be done by means of a 'whisper' blade—a thin, flexible metal blade lying gently on the surface of the coating just after the web leaves the screen and passes over a support roller. The whisper blade must not act as a doctor blade and for this reason it should be thin (maximum thickness: 0.15 mm) and flexible—extending 5 cm or more from its holder.

Quality Standards[7,8]

Standards for vinyl and other wallcoverings in roll form are given in European Standards EN 233 to EN 235: 1989, published also as British Standard 1248 Parts 1 to 3: 1990.

These define and give methods of test for dimensions of rolls, washability, and colour fastness when exposed to light. Means of application, the matching of patterns, and the removal of wallcoverings also are discussed.

A further Standard, BS EN 259: 1992, gives performance characteristics for so-called 'heavy duty' vinyl wallcoverings—including an 'extra-scrubbable' washing test, and two higher levels of light-fastness, introducing also a test for resistance to impact. In the impact test, the sample of wallcovering is bonded to 12.5 mm plasterboard and struck at an angle of 30° with energy of 1 J by means of a precisely defined striker. When tested in this way it should exhibit no fracture of the surface which exposes the backing material or the substrate of the wallcovering.

References

1. H. A. Sarvetnick (ed.), 'Plastisols and Organosols', Van Nostrand Reinhold, New York, 1972.
2. R. H. Burgess (ed.), 'Manufacture and Processing of PVC', Applied Science Publishers, London, 1982.
3. L. Underdal, S. Lange, O. Palmgren and N. P. Thorshaug, 'Proceedings of the International Conference on PVC Processing', The Plastics and Rubber Institute, London, 1978.
4. J. H. Exelby, 'Proceedings of the 4th International Conference on PVC', The Plastics and Rubber Institute, London, 1990.
5. A. C. Poppe, *Vinyl Technol.*, 1987, **9**(4), 191.
6. U.S. Patent 3 293 094, Congoleum Nairn, Inc.
7. British Standards 1248 Parts 1 to 3: 1990.
8. European Standards EN 233 to 235: 1989, and BS EN 259: 1992.

Bibliography

The Bibliography is in two sections, as follows:

Section 1: References in alphabetical order by the surname of the author or editor first given

Section 2: References in alphabetical order by title, when the name of the author or editor is not stated or not known.

Section 1: References in alphabetical order by surname of author or editor

A. J. G. Allan, *J. Polym. Sci.*, 1959, **38**, 297.

E. A. Apps, 'Printing Ink Technology', Leonard Hill, London, 1958.

E. A. Apps, 'Inks for the Major Printing Processes' Leonard Hill, London, 1966.

G. Ardichvili, 'An attempt at a rational determination of the cambering of calender rolls', *Kautschuk*, 1938, **14**, 23.

P. C. Ashlee, 'The bogus butterfly collector: camphor and the early plastics industry', *Plastiquarian*, Spring 1991, **7**, 14–15.

J. T. Bergen and G. W. Scott Jr., 'Pressure distribution in the calendering of plastic materials', *J. Appl. Mechanics*, 1951, **18** (March), 101–106.

W. Berry, R. A. Rose, and C. R. Bruce, British Patent 723 631.

A. Bosini, *Materie plast.*, 1956, **22**, 9.

D. Brewis (ed.), 'Surface Analysis and Pre-treatment of Plastics & Metals', Applied Science Publishers, London, 1982.

J. Brown, 'Plastics calendering—types of machines and layouts', *Plast. Progr.*, 1951, 167–179.

W. J. Brown, 'Laminated Plastics', The Plastics Institute, London, 1961.

C. R. Bruce, US Patent 2 886 471.

K. F. Buchel, *Br. Plast.*, 1964, **37**(3), 142.

K. F. Buchel, *Adhaesion*, 1966, **10**, 506.

R. H. Burgess (ed.), 'Manufacturer and Processing of PVC', Applied Science Publishers, London, 1982.

A. Chambers, 'A Guide to Making Decorated Papers', Andre Deutsch, London, 1989.

L. G. Carpenter, 'Vacuum Technology', Adam Hilger, Bristol, 1983.

H. Collin, *Kunststofftechnik*, 1970, **9**(6), 217; **9**(7), 241.

J. G. Cook, 'Handbook of Polyolefin Fibres', Merrow Publishing, Watford, 1967.

J. K. Dennis and T. E. Such, 'Nickel and Chromium Plating', Butterworths, London, 1986.

G. E. Drabble, 'Applied Mechanics Made Simple', W. H. Allen, London, 1971.

M. Drucker and P. Finkelstein, 'Recipes for Surfaces', Cassell, London, 1990.

J. H. Exelby, 'Proceedings of the 4th International Conference on PVC', The Plastics and Rubber Institute, London, 1990.

L. Faulkner, 'How a compact disc is manufactured', *Plast. Rubber Int.*, 1992, **17**(1), 10–11.

A. H. Fawcett (ed.), 'High-Value Polymers', Special Publication No. 87, Royal Society of Chemistry, Cambridge, 1991.

R. W. Furness, 'The Practice of Plating Plastics', Robert Draper, Teddington, 1968.

R. Gacher and H. Muller (eds.), 'Plastics Additives Handbook', Hanser, Munich, 1987.

R. O. Gibson, 'The Discovery of Polythene', Royal Institute of Chemistry, London, Lecture Series 1964, No. 1.

K. J. Gooch, 'Designing better calenders', *Mod. Plast.*, 1957, **34** (March), 165–172.

I. Goodman, 'Synthetic Fibre-forming Polymers', Royal Institute of Chemistry, London, Lecture Series 1967, No. 3.

J. E. Gordon, 'The New Science of Strong Materials', Penguin, Harmondsworth, 1968.

H. J. Grow, US Patent 2 795 820.

N. Grassie, 'The Chemistry of High Polymer Degradation Processes', Butterworths, London, 1956.

N. S. Harris, 'Modern Vacuum Practise', McGraw-Hill, London, 1989.

G. Hatzmann, M. Herner, and G. Muller, *Kunststoffe*, 1975, **65**, 472.

W. F. Henderson, US Patent 2 502 841; British Patent 581 717.

K. Hodd, 'Weathering and degradation of plastics' (summary of symposium), *Anticorros. Mat. Proc.*, 1964, January/February, 19–23.

A. A. Hodgson, 'Fibrous Silicates', Royal Institute of Chemistry, London, Lecture Series 1965, No. 4.

L. Holland, W. Steckelmacher, and J. Yarwood, 'Vacuum Manual', E. & F. N. Spon, London, 1974.

J. W. Horton, R. v. Mazza, and H. Dym, *Proc. IEEE*, 1974, **52**, 1513.

P. V. Horton, US Patent 2 668 134.

B. I. Howe and J. E. Lambert, 'An analysis of the theory and operation of high speed steel roll calender stacks', *Pulp Paper Mag. Can.*, 1961, **62**(C), T139 (Convention issue).

D. W. Huke, 'Introduction to Natural and Synthetic Rubbers', Hutchinson, London, 1961.

C. W. Hurst and R. E. Schanzle, *Mod. Packag.*, 1966, **40**(2), 163.

A. Hutzenlaub, *Kunststoffe*, 1967, **57**, 163.

E. R. Inman, 'Organic Pigments', Royal Institute of Chemistry, London, Lecture Series 1967, No. 1.

W. R. Jones, 'Minerals in Industry', Penguin, Harmondsworth, 1963.

M. Kaufman, 'The First Century of Plastics', London, The Plastics and Rubber Institute, London, 1963.

W. H. Kreidl, US Patent 2 632 921; US Patent 2 704 382; US Patent 2 746 084.

M. F. Kritchever, US Patent 2 648 097; US Patent 2 683 894.

F. H. Lambert, 'Moulding of Plastics', George Newnes, London, 1948.

S. Leeds, *Tappi*, 1961, **44**(4), 244.

H. Makelt (R. Hardbottle, translator), 'Mechanical Presses', Edward Arnold, London, 1968.

S. K. Malhotra, 'Some studies on drilling of fibrous composites', *J. Mat. Proc. Technol.*, 1990, **24**, 291–300.

J. M. Margolis (ed.), 'Decorating Plastics', Hanser, Munich, 1986.

J. Max, *IRE Trans. Informat. Theory*, 1960, **IT-6**, 7.

T. F. McLaughlin, *Mod. Packag.*, 1960, **34**(1), 153.

T. F. McLaughlin, *SPE J.*, 1964, **10**, 20th Antec, Session iv, Paper 3.

J. P. McNamee, *Tappi*, 1965, **48**, 12, 673; 1967, **50**, 6, 308.

K. Mienes, 'Plastics in Europe', Temple Press Books, London. 1964.

J. Morgan, 'From milk to manicure sets: the casein process', *Plastiquarian*, Spring 1989, **2**, 12–13.

G. Muller and D. W. Baudrand, 'Plating ABS Plastics', Robert Draper, Teddington, 1970.

D. Nelson, 'Finding the stress hidden in parts', *Mach. Des.*, 1986, 20 November, 125–131.

J. F. O'Hanlon, 'A User's Guide to Vacuum Technology', John Wiley & Son, New York, 1980.

D. J. Pawson, A. P. Ameen, R. D. Short, P. Denison, and F. R. Jones, 'An investigation of the surface chemistry of poly(ether etherketone). I. The effect of oxygen plasma treatment on surface structure', *Surf. Interface Anal.*, **18**, 1 (January 1992), 13–22.

S. H. Pinner (ed.), 'Weathering and Degradation of Plastics', Columbine Press, Manchester, 1966.

S. H. Pinner (ed.), 'Modern Packaging Films', Butterworths, London, 1967.

S. H. Pinner and W. G. Simpson (eds.), 'Plastics: Surface and Finish', Butterworths, London, 1971.

A. C. Poppe, *Vinyl Technol.*, 1987, **9**, 4, 191.

H. D. Rees, 'The drilling characteristics of DMC materials', *Plast. Rubber Int.*, 1992, **17**(1), 6–9.

K. Rossman, *J. Polym. Sci.*, 1956, **19**, 141.

A. Roth, 'Vacuum Technology', North Holland Publishing, Amsterdam, 1982.

F. N. Rothaker, US Patent 2 864 755.

H. A. Sarvetnick (ed.), 'Plastisols and Organosols', Van Nostrand Reinhold, New York, 1972.

H. Schonhorn and R. H. Hansen, *J. Appl. Polym. Sci.*, 1968, **12**, 1231.

B. Sharp, 'Electronic speckle pattern interferometry (ESPI), an engineering approach', Newport Corporation, Fountain Valley, California, USA, n.d.

L. K. Sharples, Conference on Printing and Decorating Plastics, The Plastics Institute, Bristol, 1968.

E. Sheng, R. J. Heath, I. Sutherland, and D. M. Brewis, 'Surface modification of propylene by flame treatment—a study', *Plast. Rubber Int.*, 1991, **16**(4), 10–14.

P. B. Sherman, 'Additive influence on corona treatment', TAPPI Film Extrusion, 1991.

P. B. Sherman, D. Clarke, and J. Marriott, TAPPI Extrusion Coating Course, 1991.

D. Simpson and W. G. Simpson, 'An Introduction to Applications of Light Microscopy in Analysis', Royal Society of Chemistry, London, 1988.

W. G. Simpson, 'Technical properties and possible applications of some new plastics materials', *Manderstam Technical Digest (London)*, 1966, February, 19–23.

E. A. Smith, *SPE J.*, 1962, **18**(2), 157.

H. M. Stanley, 'The Petroleum-Chemicals Industry', Royal Institute of Chemistry, London, Lecture Series 1963, No. 4.

W. V. Titow, 'PVC Plastics', Elsevier Science Publishers, London, 1990.

Y. Toriyama, H. Akamoto, and M. Knazanchi, *IEE Trans.*, 1967, **Ei-2**.

L. Underdal, S. Lange, O. Palmgren, and N. P. Thorshaug, 'Proceedings of the International Conference on PVC Processing', The Plastics and Rubber Institute, London, 1978.

W. C. Wake, 'Adhesives', Royal Institute of Chemistry, London, Lecture Series 1966, No. 4.

J. A. Waller, 'Press Tools and Presswork', Portcullis Press, London, 1978.

A. Whelan and D. Dunning, 'The Dynisco Extrusion Processor's Handbook', published by the authors for Dynisco, Inc., 1st Edn., 1988.

A. Whelan and J. A. Brydson, 'The Kayeness Practical Rheology Handbook', published by the authors for Kayeness, Inc., 1st Edn., 1991.

A. Whelan and J. P. Goff, 'The Injection Moulding of Engineering Thermoplastics', Van Nostrand Reinhold, New York, 1991.

G. Weckler, *IEEE J. Solid-State Cirts.*, 1967, **SC-2**, 65.

H. E. Wechsberg and J. B. Webber, *Mod. Plast.*, 1959, **36**(11), 100.

R. Weiner (ed.), 'Electroplating of Plastics', Finishing Publications, Stevenage, 1977.

C. Williamson, 'The notebooks of Alexander Parkes', *Plastiquarian*, Spring 1989, **2**, 8.

C. Williamson, 'Shellac union cases', *Plastiquarian*, Summer 1989, **3**, 10.

L. E. Wolinski, US Patent 2 715 075; US Patent 2 715 076; US Patent 2 715 077.

J. A. Wordingham and P. Reboul, 'Concise Encyclopaedic Dictionary of Plastics', George Newnes, London, 1965.

H. C. Worsdall (ed.), 'The Paint Directory', Oil & Colour Chemists' Association, Wembley, 1991.

V. E. Yarsley, W. Flavell, P. S. Adamson, and N. C. Perkins, 'Cellulosic Plastics', Iliffe Books, London, 1964.

J. J. Ziccarelli, *Mod. Plast.*, 1962, **40**(3), 126.

Section 2: References in alphabetical order by title (name of author or editor not stated or not known)

American Society for Testing Materials:

- ASTM B 604: 1980 'Specification for decorative electroplated coatings of copper-nickel-chromium on plastics'.
- ASTM 02578–67 (re-approved 1972).

British Patents:

- 581 717; 723 631; 1 303 705.

British Standards:

- BS EN 438: 1991 'Decorative high-pressure laminates (HPL)—Sheets based on thermosetting resins. Part 1 Specifications; Part 2 Determination of properties'.
- BS 476 Part 7: 1987 'Fire tests on building materials and structures. Method for classification of the surface spread of flame of products'.
- BS 476 Parts 20 to 23: 1987 'Fire tests on building materials and structures. Methods for determination of the fire resistance of elements of construction'.
- BS 1248, Parts 1 to 3: 1990.
- BS 2782 'Methods of testing plastics'. Method 826A 'Determination of adhesion of print on plastics sheet'.

BS 4601: 1970 'Electroplated coatings of nickel plus chromium on plastics materials'.

BS 4965: 1991 'Decorative laminated plastics sheet veneered boards and panels'.

BS 6250 Part 3: 1991 'Domestic and contract furniture. Specification for performance requirements for cabinet furniture'.

BS 6853: 1987 'Fire precautions in the design and construction of railway passenger rolling stock'.

BS 7331: 1990 'Direct surfaced wood chipboard based on thermosetting resins'.

BS 7332: 1991 'Decorative continuous laminates (DCL) based on thermosetting resins'.

BS EN 259: 1992

Milton Keynes, dates as shown.

'Can It Be Made in Plastics or Rubber?', Plastics and Rubber Advisory Service, London, n.d.

'Code of Practice for Safety in the Use of Thermoforming Machines', British Plastics Federation, London, 1987.

'Directory of UK Companies Involved in the Recycling of Plastics', British Plastics Federation, London, 1991.

'Environmental Protection Act 1990', HMSO, London.

European and International Standards, European Standards (CEN, Brussels) and International Standards Organization, Geneva, various dates:

EN 233 to 235: 1989.

ISO 472: 1988, 'Plastics—Vocabulary'.

ISO 1043 Pt 1, 'Symbols—Basic polymers and their special characteristics'.

ISO 1043 Pt 2, 'Symbols—Fillers and reinforcing materials'.

ISO 1043 Pt 3, 'Symbols—Plasticizers',

ISO 4525: 1985, 'Metallic coatings—electroplated coatings of nickel plus chromium on plastics materials'.

ISO 8604: 1988, 'Plastics—Prepregs—definitions of terms and symbols for designations'.

ISO 11469, 'Plastics—Generic identification and marking of plastics products'.

'Guidance Notes' (Health and Safety Executive) HMSO, London, various dates, particularly:

EN 40/93 Occupational exposure limits 1993.

'Health and Safety at Work Act 1974', HMSO, London.

Leaflets issued by: British Industrial Plastics Limited; British Polymer Training Association; British Xylonite Company Limited; BX Plastics Limited; BXL Plastics Limited; Colloids Limited; Dawe Instruments Limited; Grant Spacey Limited; Guyson International Limited; Linx Printing Technology Limited; Lloyd Instruments Limited; Newport Corporation; R. Simon (Dryers) Limited; W. Canning Materials Limited.

'Modern Plastics Encyclopaedia', McGraw-Hill, New York, various dates.

'On-line automatic web control', *Eur. Plast. News*, 1991, **18**(8), 44–50.

'Report of the Departmental Committee on Celluloid', HMSO, London, 1913.

The Building Regulations 1991 Approved Document B, HMSO, London.

US Patents:

2 502 841; 2 632 921; 2 648 097; 2 668 134; 2 683 894; 2 704 382; 2 715 075; 2 715 076; 2 715 077; 2 746 084; 2 795 820; 2 864 755; 2 886 471; 3 293 094.

Index